Computational Biology

Volume 27

The *Computational Biology* series publishes the very latest, high-quality research devoted to specific issues in computer-assisted analysis of biological data. The main emphasis is on current scientific developments and innovative techniques in computational biology (bioinformatics), bringing to light methods from mathematics, statistics and computer science that directly address biological problems currently under investigation.

The series offers publications that present the state-of-the-art regarding the problems in question; show computational biology/bioinformatics methods at work; and finally discuss anticipated demands regarding developments in future methodology. Titles can range from focused monographs, to undergraduate and graduate textbooks, and professional text/reference works.

More information about this series at http://www.springer.com/series/5769

Fabricio Alves Barbosa da Silva
Nicolas Carels • Floriano Paes Silva Junior
Editors

Theoretical and Applied Aspects of Systems Biology

 Springer

Editors
Fabricio Alves Barbosa da Silva
Graduate Program in Computational Biology
Oswaldo Cruz Foundation
Rio de Janeiro, RJ, Brazil

Nicolas Carels
Graduate Program in Computational Biology
Oswaldo Cruz Foundation
Rio de Janeiro, RJ, Brazil

Floriano Paes Silva Junior
Graduate Program in Computational Biology
Oswaldo Cruz Foundation
Rio de Janeiro, RJ, Brazil

ISSN 1568-2684
Computational Biology
ISBN 978-3-030-09117-0   ISBN 978-3-319-74974-7   (eBook)
https://doi.org/10.1007/978-3-319-74974-7

This Springer imprint is published by the registered company Springer International Publishing AG, part
of Springer Nature.
The registered company address is: Gewerbestrasse 11, 6330 Cham, Switzerland

Preface

In the last decade, we have witnessed a transition from descriptive biology as approached through genomics and other omics to a systemic understanding of biological systems that was possible due to the impressive progresses in high-throughput technologies. The wave of data produced by these technologies is tremendous and offered an opportunity for big data as well as mathematical and computational modeling to take off. Now we testify exciting times where sciences integrate themselves for the benefit of solving specific purposes. Of course, medical sciences do not escape to this trend and it is a duty for Fiocruz to accompany these developments, to participate and to translate them in medical applications, and finally to transmit them to next generations. Indeed, the transmission of knowledge on cutting-edge developments in System Biology has been the purpose of the International Course of System Biology held in Rio de Janeiro in July 2017, which is now translated into the present book looking forward to becoming a useful resource to a much broader audience in this field.

This book is organized in two main sections. Although the whole book is made of contributions from researchers with a clear commitment to applied sciences, the first part brings a series of six chapters where the more fundamental aspects of computational and systems biology are addressed. The remaining seven chapters on the second part of the book deal with the application of such fundamentals on different biological problems.

We take the opportunity, here, to acknowledge the Brazilian Coordination for Improvement of Higher Level Personnel (CAPES), FIOCRUZ's Vice-Presidency of Research and Biological Collections (VPPCB), FIOCRUZ's Vice-Presidency of Production and Innovation in Health (VPPIS), the direction of the Oswaldo Cruz Institute (IOC), and the coordination of the Computational and Systems Biology program (PG-BCS) that made possible this event to occur by their financial and logistics support. We also would like to thank all other funding agencies that support research and education activities at our state, such as the Brazilian National Council for Scientific and Technological Development (CNPq) and the Rio de Janeiro State

Research Funding Agency (FAPERJ). Finally, we cannot emphasize enough how thankful we are for all authors contributing to the book for their dedication and generosity. A special thanks to Dr. Jonathan R. Karr for writing such a splendid foreword that much honored us and contributed to further elevate this book.

Oswaldo Cruz Foundation (FIOCRUZ)   Fabricio Alves Barbosa da Silva
Rio de Janeiro, Brazil   Nicolas Carels
                                                                  Floriano Paes Silva-Jr.

Foreword to *Theoretical and Applied Aspects of Systems Biology*

A central challenge of medicine is to understand how health and disease emerge from the myriad of molecular interactions among our genes and our environment. Over the past 30 years, researchers have invented a wide array of methods to measure our genes and our environment and numerous computational methods to analyze such molecular data. To date, these methods have been used to collect a wealth of data about a broad range of cellular processes, organisms, patients, and diseases. This has led to numerous insights into the functions of individual genes and the mechanisms of individual diseases, as well as diagnostic tests and treatments for several specific diseases. For example, cell-free fetal DNA sequencing can now be used to diagnose Down syndrome as early as 4 weeks after conception, breast cancers driven by mutations in HER2 are now frequently treated with anti-HER2 antibodies such as trastuzumab, and cytochrome P450 is now routinely genotyped to personalize the dosage of drugs metabolized by cytochrome P450 such as atorvastatin and tamoxifen.

However, despite this progress, we still do not have a comprehensive understanding of how our genes and our environment interact to determine our health. As a result, we still do not understand the genetics of most diseases, we still cannot tailor therapy to individual patients, and many patients continue to suffer from potentially treatable diseases. To better understand and treat disease, we must develop computational models of the molecular systems responsible for our health and disease. In particular, we must develop models that help researchers integrate complementary datasets, predict phenotype from genotype, design new drugs, and personalize medicine.

Overcoming the numerous technical challenges to achieve this vision will likely require the coordinated effort of an international community of modelers, experimentalists, software engineers, clinicians, and research sponsors. To prepare young scientists to address these challenges and to begin to form a community that can tackle them, Fabricio Alves Barbosa da Silva, Nicolas Carels, and Floriano Paes Silva Junior from the Oswaldo Cruz Foundation organized the International Course on Theoretical and Applied Aspects of Systems Biology in Rio de Janeiro, Brazil, in July 2017. The meeting featured ten lectures from scientists from Brazil, Canada,

Luxembourg, the Netherlands, Portugal, and the USA. The lectures taught students how to use multi-omics data and a wide variety of mathematical formalisms, including network modeling, Bayesian modeling, Petri Nets, flux balance analysis, ordinary differential equations, and hybrid modeling, to gain insights into cellular processes such as metabolism, transcriptional regulation, and signal transduction and advance infectious disease medicine, precision oncology, drug development, and synthetic biology. The meeting also featured a poster session for students to share their research.

This book edited by Fabricio Alves Barbosa da Silva, Nicolas Carels, and Floriano Paes Silva Junior summarizes the computational systems biology methods and applications that were presented at the course. The book includes eight chapters that summarize eight of the lectures, as well as five additional chapters contributed by other researchers who participated in the course on applications of systems biology to metagenomics and neuroscience.

Computational systems biology has great potential to transform biological science and medicine and numerous opportunities for young, ambitious scientists to make fundamental contributions to science and medicine. We invite you to use this book to learn about the state of the art and potential of systems biology, and we hope that you join our efforts to use systems biology to advance human health.

Icahn Institute for Genomics and Multiscale Biology   Jonathan R. Karr
New York, NY, USA
Icahn School of Medicine at Mount Sinai
New York, NY, USA

Contents

List of Editors and Contributors

Editors

Fabrício Alves Barbosa da Silva PROCC, Graduate Program in Computational and Systems Biology, Oswaldo Cruz Foundation (FIOCRUZ), Rio de Janeiro, Brazil

Nicolas Carels Laboratory of Biological System Modeling (LMSB), Centre of Technological Development in Health (CDTS), Graduate Program in Computational and Systems Biology, Oswaldo Cruz Foundation (FIOCRUZ), Rio de Janeiro, Brazil

Floriano Paes Silva-Jr LabBECFar/IOC, Graduate Program in Computational and Systems Biology, Graduate Program on Cellular and Molecular Biology, IOC-FIOCRUZ, Graduate Program on Pharmacogy and Medicinal Chemistry, ICB-UFRJ, Oswaldo Cruz Foundation (FIOCRUZ), Rio de Janeiro, Brazil

Contributors

Rodolpho Mattos Albano Departamento de Bioquímica, Rio de Janeiro State University, Rio de Janeiro, Brazil

Muhammad Ali Computational Biology Group, Luxembourg Centre for Systems Biomedicine (LCSB), University of Luxembourg, Luxembourg City, Luxembourg

Kele Belloze Escola de Informática & Computação, Centro Federal de Educação Tecnológica Celso Suckow da Fonseca (CEFET/RJ), Rio de Janeiro, Brazil

Rafaela Brum Departamento de Informática e Ciência da Computação, Instituto de Matemática e Estatística, Rio de Janeiro State University, Rio de Janeiro, Brazil

Rafael V. Carvalho Pontifical Catholic University of Goias, Goiânia, Brazil

Ana Paula D' Carvalho-Assef LAPIH/IOC, Oswaldo Cruz Foundation (FIOCRUZ), Rio de Janeiro, Brazil

Maria Clicia Stelling de Castro Universidade do Estado do Rio de Janeiro – UERJ, Rio de Janeiro, Brazil

Larissa Catharina Graduate Program in Computational and Systems Biology, Oswaldo Cruz Foundation (FIOCRUZ), Rio de Janeiro, BrazilSystem Biology to Access Target Relevance in the Research and Development of Molecular Inhibitors, Rio de Janeiro, Brazil

Jian Chen Department of Electrical and Computer Engineering, Drexel University, Philadelphia, PA, USA

Clarimar J. Coelho Pontifical Catholic University of Goias, Goiânia, Brazil

Alessandra Jordano Conforte Graduate Program in Computational and Systems Biology, Oswaldo Cruz Foundation (FIOCRUZ), Rio de Janeiro, Brazil

Celia Martins Cortez Universidade do Estado do Rio de Janeiro – UERJ, Rio de Janeiro, Brazil

Luis Alfredo Vidal de Carvalho Departamento de Medicina Preventiva, Faculdade de Medicina, UFRJ, Rio de Janeiro, Brazil

Laura Machado de Faria Oswaldo Cruz Foundation (FIOCRUZ), Rio de Janeiro, Brazil

Davi Augusto Caetano de Jesus Department of Statistics, Institute of Mathematics and Statistics, University of São Paulo, São Paulo, Brazil

Marcio Argollo de Menezes Instituto de Física, Universidade Federal Fluminense (UFF), Niterói, RJ, Brazil

Antonio del Sol Computational Biology Group, Luxembourg Centre for Systems Biomedicine (LCSB), University of Luxembourg, Luxembourg City, Luxembourg

Gregory Ditzler Department of Electrical and Computer Engineering, University of Arizona, Tuscon, AZ, USA

Ana Paula Barbosa do Nascimento Coordenação de Matemática Aplicada e Computacional, National Laboratory of Scientific Computing (LNCC), Petrópolis, Rio de Janeiro, Brazil

Marcelo Trindade dos Santos Coordenação de Matemática Aplicada e Computacional, National Laboratory of Scientific Computing (LNCC), Petrópolis, Rio de Janeiro, Brazil

Joshua Earl Department of Microbiology and Immunology, Drexel University College of Medicine, Philadelphia, PA, USA

Garth Ehrlich Department of Microbiology and Immunology, Drexel University College of Medicine, Philadelphia, PA, USA

Fernando Medeiros Filho Graduate Program in Computational and Systems Biology, Oswaldo Cruz Foundation (FIOCRUZ), Rio de Janeiro, Brazil

André Fujita Department of Computer Science, Institute of Mathematics and Statistics, University of São Paulo, São Paulo, Brazil

Thiago Giannini Graduate Program in Computational and Systems Biology, Oswaldo Cruz Foundation (FIOCRUZ), Rio de Janeiro, Brazil

Thiago C. Hirata Department of Clinical and Toxicological Analyses, School of Pharmaceutical Sciences, University of São Paulo, São Paulo, Brazil

Camila Kalil Universidade do Estado do Rio de Janeiro – UERJ, Rio de Janeiro, Brazil

Yemin Lan Department of Cell and Developmental Biology, Perelman School of Medicine, University of Pennsylvania, Philadelphia, PA, USA

Saeed Keshani Langroodi Department of Civil, Architectural, and Environmental Engineering, Drexel University, Philadelphia, PA, USA

Melissa Lever Department of Clinical and Toxicological Analyses, School of Pharmaceutical Sciences, University of São Paulo, São Paulo, Brazil

Milena Magalhães CDTS, Oswaldo Cruz Foundation (FIOCRUZ), Rio de Janeiro, Brazil

Thiago Merigueti Graduate Program in Computational and Systems Biology, Oswaldo Cruz Foundation (FIOCRUZ), Rio de Janeiro, Brazil

Helder I. Nakaya Department of Clinical and Toxicological Analyses, School of Pharmaceutical Sciences, University of São Paulo, São Paulo, Brazil

Alexandre Galvão Patriota Department of Statistics, Institute of Mathematics and Statistics, University of São Paulo, São Paulo, Brazil

Jacob R. Price Department of Civil, Architectural, and Environmental Engineering, Drexel University, Philadelphia, PA, USA

Erin R. Reichenberger U.S. Department of Agriculture, Agricultural Research Service, Eastern Regional Research Center, Wyndmoor, PA, USA

Edward A. Rietman Computer Science Department, University of Massachusetts Amherst, Amherst, MA, USA

Vanessa Rodrigues Universidade do Estado do Rio de Janeiro – UERJ, Rio de Janeiro, Brazil

Gail Rosen Department of Electrical and Computer Engineering, Drexel University, Philadelphia, PA, USA

Pedro S. T. Russo Department of Clinical and Toxicological Analyses, School of Pharmaceutical Sciences, University of São Paulo, São Paulo, Brazil

Domenico Sgariglia Programa de Engenharia de Sistemas e Computação, COPPE-UFRJ, Rio de Janeiro, Brazil

Dilson Silva Universidade do Estado do Rio de Janeiro – UERJ, Rio de Janeiro, Brazil

Tatiana Martins Tilli CDTS, Oswaldo Cruz Foundation (FIOCRUZ), Rio de Janeiro, Brazil

Jack Adam Tuszynski Department of Oncology, University of Alberta, Edmonton, AB, Canada

Department of Physics, University of Alberta, Edmonton, AB, Canada

Department of Mechanical and Aerospace Engineering, Politecnico di Torino, Turin, Italy

Fons J. Verbeek Leiden University, Leiden, The Netherlands

Maciel Calebe Vidal Department of Computer Science, Institute of Mathematics and Statistics, University of São Paulo, São Paulo, Brazil

Stephen Woloszynek Department of Electrical and Computer Engineering, Drexel University, Philadelphia, PA, USA

Zhengqiao Zhao Department of Electrical and Computer Engineering, Drexel University, Philadelphia, PA, USA

Part I
Fundamentals

Bio-modeling Using Petri Nets: A Computational Approach

Rafael V. Carvalho, Fons J. Verbeek, and Clarimar J. Coelho

Abstract Petri nets have been widely used to model and analyze biological system. The formalism comprises different types of paradigms, integrating qualitative and quantitative (i.e., stochastic, continuous, or hybrid) modeling and analysis techniques. In this chapter, we describe the Petri net formalism and a broad view of its structure and characteristics applied in the modeling process in systems biology. We present the different net classes of the formalism, its color extension, and model analysis. The objective is to provide a discussion on the Petri net formalism as basis for research in computational biology.

1 Modeling Systems Biology

Modeling of biological systems is evolving into the description, simulation and analysis of the behavior and interdependent relationship of biological phenomena. Since experimental research can be costly, time-consuming, or ethically infeasible, the result of theoretical and computational modeling is a great advantage and an important aspect of systems biology [1, 2].

In the core of systems biology relies the use of mathematical and computational methods to represent biological behavior, and modeling plays a crucial role in order to provide a system-level understanding of biological system. The aim is to represent life science at different levels of abstraction and description, i.e., molecules, cells, organs, or entire species, and their complex interactions, in order to understand the processes and emergent properties that happen with such systems. In this context, models can serve as:

R. V. Carvalho (✉) · C. J. Coelho
Pontifical Catholic University of Goias, Goiânia, Brazil

F. J. Verbeek
Leiden University, Leiden, The Netherlands
e-mail: f.j.verbeek@liacs.leidenuniv.nl

© Springer International Publishing AG, part of Springer Nature 2018
F. A. B. da Silva et al. (eds.), *Theoretical and Applied Aspects of Systems Biology*,
Computational Biology 27, https://doi.org/10.1007/978-3-319-74974-7_1

- Explanatory or pedagogical tools, using mathematical and computational methods to represent more explicitly the state of knowledge of biological behavior
- Analysis using models to test hypothesis about the system that best fit the data and, eventually, to predict outcomes of the model to different stimuli
- Instrumentation, designing alternative systems to circumvent experiments that are too costly, time-consuming, or ethically undesirable.

Models are typically conceptual, existing as an idea, a computer program, or a set of mathematical axioms which are valid in the considered model. For systems biology, the theory of concurrency is at the basis of most approaches that have been applied. Many different formal methods, languages, and modeling paradigms exist that depend on the information, the data, and the model properties that guide the modeling process. Mathematical models are traditionally used in biology to represent quantitative biological phenomena. Usually, they are equation-based models that formulate hypotheses about relations, variables, and magnitudes and how they change over time. Computational models formulate hypotheses about the biological mechanism process to understand the interactions between the system components. They are based on algorithmic process models that are executable and progress from state to state, not necessarily time dependent. They use computational formalisms/paradigms to specify an abstract execution that illustrates a biological phenomenon. Extended review about the distinction between computational and mathematical models is presented in [3, 4]. They debate about the dichotomies, applicability, and benefits of both approaches and how they are used to model biological phenomena.

Although the description of mathematical formalism is based on equations, in systems biology, the computer power is used to describe and analyze such models. Stochastic differential equations, ordinary differential equations, partial differential equations, and delay differential equations are examples of mathematical formalism that use process algebra, term rewriting systems, or different mathematical structure to model in systems biology. Moreover, there are mathematical formalisms that include methods inspired by biological phenomena, such as brane calculi [5], P systems [6], Biocham [7], and calculus of looping sequence [8].

Computational frameworks have provided means to describe and visualize biological networks: Cytoscape [9] is an open-source software platform for visualizing molecular interaction networks and biological pathways. It integrates these networks with annotations, gene expression profiles, and other state data. Another computational platform for visualization and manipulation of complex interaction networks is Osprey [10]. Moreover, there are computational initiatives to represent biological process like the systems biology markup language (SBML) [11], a free and open interchange format for representing computational models in systems biology [12]. Furthermore, there is the Biological Simulation Program for Intra- and Inter-Cellular Evaluation (Bio-SPICE) [13], an open-source framework and software toolset for systems biology that assists biological researchers in the analysis, modeling, and simulation of biological phenomena. It provides different tools across a distributed, heterogeneous community network.

The area of systems biology has benefited greatly from computational formalism based on concurrent systems to model, simulate, and visualize biological phenomena. Important computational formalisms have been successfully used in systems biology, i.e., calculus of communicating systems [14)], π calculus [15], process algebra for the modeling and the analysis of biochemical networks (Bio-PEPA) [16, 17], agent-based model [18, 19], Petri nets [20, 21], as well as variant of this method [22, 23]. The modeling formalism of Petri nets (PNs) has been used quite successfully in systems biology due its flexibility and strong emphasis on concurrency and local dependency. It comprises an abstract model of information flow, providing graphical representation and formal mathematical definition.

In the following, we will provide an overview about the Petri net formalism, its classes, definitions, and the key features of the formalism in systems biology.

2 Petri Nets

Petri nets were defined by Carl Adam Petri in his dissertation thesis in 1962 [24]. It was originally designed to represent discrete, concurrent processes of technical system. The aim was to define a mathematical formalism to represent and analyze causal systems with concurrent processes. The Petri net formalism combines an intuitive, unambiguous, qualitative bipartite graphical representation of arbitrary processes with a formal semantics. Petri nets were conceived as an abstract model of information flow. They are built from basically four different building blocks:

- *Places* represent passive nodes that refer to conditions, local states, or resources.
- *Transitions* represent active nodes that describe local state shifts, events, and activities in the system.
- *Tokens* are variable elements that represent current information on a condition or local state.
- *Directed arcs* are connectors that specify relationships between local state and local action by depicting the relation between transitions and places. It is along the arcs which tokens can traverse the model.

The static structure of the Petri net is described by these elements and how they are connected. The places (graphically represented by circles) are connected by arcs (graphically represented by arrows) with transition (graphically represented by a rectangle). The static structure is composed by transitions connected by a number of input places by directed arcs. The static structure also provides output places to which it is connected by a direct arc from transition. Direct connections between two places or two transitions are not allowed. The tokens (graphically represented by dots) can be distributed in the places in order to define a state of Petri net, referred to as a marking. The state space of a Petri net is the set of all possible markings in the model.

The dynamic properties of the system are governed by firing rule. It relates the transitions that can occur when enabled and then moves tokens around the places in a Petri net. A transition is considered enabled when its input places are sufficiently

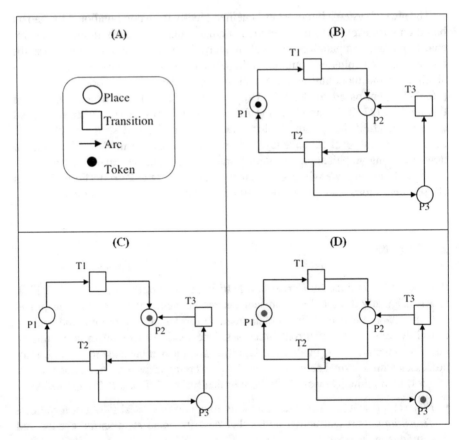

Fig. 1 An example of a Petri net model and its components. (**a**) Petri net consists of places, transitions, arcs, and tokens. (**b**) Simple example of a Petri net model where the place *P1* contain one token. (**c**) The transition *T1* is enabled by the input place *P1* and can fire, consuming the token from *P1* and producing tokens on its output place *P2*. (**d**) Transition *T2* is now enabled and can fire consuming the token from *P2* and producing tokens on its output places *P1* and *P3*

marked, and it fires by consuming and producing tokens. When they occur, tokens are consumed from input places and produced (fired) to output places. A transition can fire only if all the input places have at least one token present. The four main components of a general Petri net are shown in Fig. 1a. Figure 1b depicted a Petri net example, composed by three transitions (T1, T2, T3), three places (P1, P2, P3), and an initial mark (token in P1). Following the dynamic of the system, the transition T1 fires because its input place P1 is sufficiently marked by a token, producing one token on its output place P2; this event is depicted in Fig. 1c. Considering another step in the execution of this model, transition T2 is enabled and fires, consuming the token from its input place P2 and producing tokens on its output places P1 and P2. This event is depicted in Fig. 1d. In the example of Fig. 1, the fire sequence can run whenever one transition in the model is enabled (there is a token on its input place that satisfies the condition to the transition to fire). The fire sequence will stop only if there is no transition enabled to fire (end state).

Petri nets offer a number of attractive advantages to investigate biological system; we can state that the formalism:

- Has an intuitive graphical representation and directly executable model
- Has a strong mathematical foundation, providing a variety of analysis techniques
- Addresses structural and behavioral properties and also their relationship
- Integrates qualitative and quantitative methods as well as analysis techniques and simulation/animation
- Covers discrete and continuous, deterministic and stochastic, sequential and concurrent methods including hybrid techniques for quantitative and qualitative model and analysis
- Has a range of tools to support the implementation, simulation, and analysis of the models

The intuitive and graphical characteristics of Petri net became popular among computational systems biologists to describe biochemical reaction systems, where tokens are interpreted as single molecules of the species involved [25–27]. Koch et al. presented an extended study on the application of Petri nets in systems biology [28]. Moreover, the Petri net formalism provides a natural framework that integrates qualitative (given by the static structural topology) and quantitative (given by the time evolution of the token distribution) aspects tightly integrating different methods for model, simulation, and analysis as shown in [29].

Reddy et al. [21] and Hofestädt [30] were the pioneers in applying Petri net to biochemical systems. Reddy et al. presented a model of fructose metabolism using Petri net to represent the metabolic pathways, analyzing structural properties of the model. Hofestädt described the metabolic process based on expressed genes. He provided models from biosynthesis, protein biosynthesis, and cell communication processes considering the isoleucine biosynthesis in *E. coli*. There are numerous successful applications illustrating the versatility of Petri nets and their use for metabolic networks [31–36] and gene regulatory networks [37, 38]. Approaches that are based on signaling network [39–43], the human spliceosomal subunit U1 [44], and the mycobacterium infection dynamic process [45] have been modeled as Petri nets. Moreover, the Petri net formalism can provide an integration of models that represents biological behavior at different levels (i.e., molecular, cellular, organism, and process level) in a multi-scale hierarchical structure as described in [46, 47].

The basic standard class of the formalism consists of place/transition Petri nets or qualitative Petri nets (QPNs). These nets are discrete and have no association with time or probability. Possible behaviors of the system are analyzed in terms of causalities and dependencies, without any quantification. A Petri net model can be enhanced with special read and inhibitor arcs as a means of modeling activation or inhibition of activities, respectively. In addition, features can be added to allow one to connect sub-models in a hierarchical structure.

There are several ways to add values to a net for quantitative modeling. In continuous Petri nets (CPN), the discrete values of the net are replaced by continuous (real) values to represent concentrations over time [48]. In stochastic Petri nets (SPN), an exponentially distributed firing rate (waiting time) – typically state dependent and specified by a rate function – is associated with each transition [49]. Hybrid Petri nets (HPN) [50] allow one to combine continuous and stochastic features of the process to be modeled. Moreover, there are the colored Petri nets, proposed by Jensen [51, 52]. It is an extension to the Petri net formalism in which information is added in the form of "colors" (data types) assigned to tokens, allowing further operation and structure abstraction. The functional programming language Standard Meta Language (Standard ML) is used to manipulate and test data, providing a flexible way to create compact and parameterizable models. In a colored Petri nets, to regulate the occurrence of transitions, there are arc expressions that specify which tokens can flow over the arcs and guards that are in fact Boolean expressions used to decide which transitions instances exist.

The Petri net and its colored extension permit to organize the formalism in a set of modules as a family of related Petri net classes, sharing structure but being specialized by their kinetic information, dividing it in colored and uncolored: QPN–QPN^C (colored), SPN–SPN^C (colored), CPN–CPN^C (colored), HPN–HPN^C (colored). The conversion of these classes can be realized by a folding process. This process groups similar model components in one colored model by defining color set and the set of arc expression. The unfolding process dismembers a colored Petri net in one or more similar nets without colors. Moving between the colored and uncolored level changes the style of representation, but not necessary the net structure, though there may be loss of information in some direction. Heiner et al. [50] produced a classification of the different nets, and Fig. 2 depicts their structure paradigm of the Petri nets formalism.

In the following, we will provide a formal definition of the underlying Petri net classes and its colored extension.

2.1   *Qualitative Petri Net (QPN)*

A Petri net is represented by a directed, finite, bipartite graph, typically without isolated nodes. The three main components are places, transitions, and arcs. Places represent the discrete resources of the model, i.e., biochemical species. Transitions represent the events (activities) in the system i.e., biochemical reactions. Arcs carry stoichiometric information, called weight or multiplicity. Basically, places and transitions alternate on a path connected by consecutive arcs. Tokens on places represent the (discrete) number of elements or condition which may be understood as quantities of species (or level of concentration), number of molecules, or the simple presence of a gene. A particular arrangement of tokens over all places specifies the current markings of the system describing global states. In QPN, the firing of a transition is atomic and does not consume time. Therefore, they provide a purely qualitative modeling in systems biology.

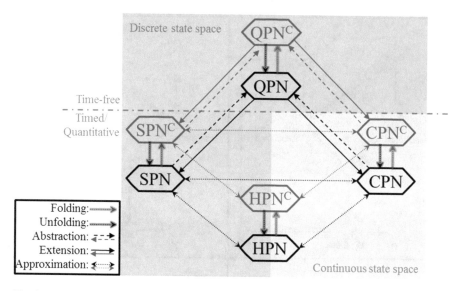

Fig. 2 Paradigm structure of the Petri nets formalism (According to [53])

Adapted from [10] a qualitative Petri net is defined as a tuple $N = (P, T, A, I)$ where:

- P is a finite nonempty set of places.
- T is a finite nonempty set of transitions such that

$$P \cap T = \varnothing \tag{1}$$

- A is a finite set of arcs, weighted by nonnegative integer values such that

$$A \subseteq (P \times T) \cup (T \times P) \to \mathbb{N} \tag{2}$$

- I is an initialization function (the initial marking) such that

$$I : P \to \mathbb{N} \tag{3}$$

An example that illustrates an enzymatic reaction modeled using QPN is depicted in Fig. 3a. The model represents a reaction: $S + E \leftrightarrow ES \to P + E$, where a substrate (S) reacts (associate) with an enzyme (E), producing an enzyme-substrate complex (ES), illustrated in Fig. 3b. This complex can dissociate into substrate and enzyme, depicted in Fig. 3c, or synthesize into a product (P) and enzyme (E), depicted in Fig. 3d.

The qualitative Petri nets have been extensively studied in Heiner et al. [22, 48] to model biological systems, and their detailed analysis is exercised step by step. In Blätke et al. [54], IL-6 signaling in the JAK/STAT signal transduction pathway

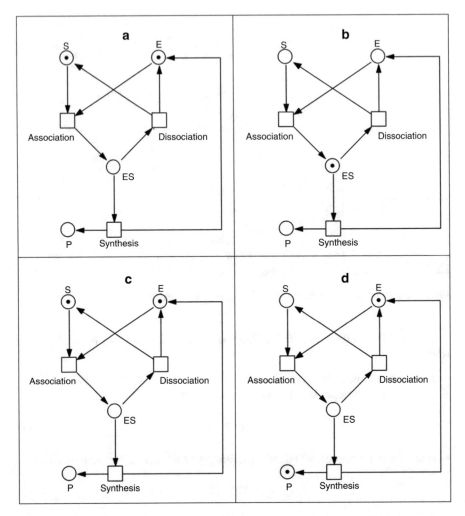

Fig. 3 Qualitative Petri net model of an enzymatic reaction. (**a**) The QPN model representation with an enzyme and a substrate. (**b**) The enzyme and substrate associate producing an enzyme-substrate complex. (**c**) The complex can disassociate into enzyme and substrate; or (**d**) the complex can synthesize into a product and an enzyme

serves as case study to illustrate a modular protein-centered modeling approach. Carvalho et al. [46] presents a hierarchical model using QPN to represent how the bacteria explore regulatory pathways, to evade host immune responses and enhance the infection inside the immune cell.

To be precise on the definition of a quantitative Petri nets (CPN and SPN), it is necessary to define:

- The preset of a node $x \in P \cup T$ is defined as

$$^{\circ}x := \{yy \in P \cup T | f(y, x) \neq 0\} \quad (4)$$

- The postset of a node $x \in P \cup T$ is defined as

$$x^{\circ} := \{yy \in P \cup T | f(x, y) \neq 0\} \quad (5)$$

2.2 Continuous Petri Net (CPN)

The distinguishing feature of continuous Petri nets is that the marking of a place is a real (positive) number and no longer an integer. The mark is called token value and can be considered as concentration, where one token value is assigned for each place. The instantaneous firing of a transition takes place like a continuous flow, and the firing event is determined by continuous deterministic rate functions, which are assigned to each transition.

In CPN, a transition is enabled if the token value of all pre-places is positive and greater than zero. Arbitrary firing rates can be defined by mathematical functions, e.g., mass-action kinetic and Michaelis-Menten kinetics to model a deterministic reaction. A reversible reaction can be modeled by using the negative firing rate. The semantic of continuous Petri net is given by the corresponding set of ordinary differential equations (ODEs), describing the continuous change over time on the token value of a given place. Therefore, the pre-transition flow results in a continuous increase, and the post-transition flow results in a continuous decrease. A CPN is the structured description of an ODE system.

Adapted from [55] a continuous Petri net is defined as a tuple $N = (P_{\text{cont}}, T_{\text{cont}}, A_{\text{cont}}, V, I_{\text{cont}})$, where:

- P_{cont} is a finite nonempty set of continuous places.
- T_{cont} is a finite nonempty set of continuous transitions which fire continuously over time.
- A_{cont} is a finite set of arcs, weighted by nonnegative real values:

$$A \subseteq (P_{\text{cont}} \times T_{\text{cont}}) \cup (T_{\text{cont}} \times P_{\text{cont}}) \rightarrow \mathbb{R}^{+} \quad (6)$$

- V is a function, which assigns a firing rate function V_t to each transition T_{cont}, such that

$$V : T_{\text{cont}} \rightarrow H_c \quad (7)$$

which assigns a continuous rate function h_t to each transition t, whereby

$$H_c = \bigcup_{t \in T_{\text{cont}}} \left\{ h_t | h_t : \mathbb{N}^{|^{\circ}t|} \rightarrow \mathbb{R}^{+} \right\} \quad (8)$$

is the set of all continuous firing rate functions and

$$V(t) = h_t \forall t \in T_{\text{cont}} \tag{9}$$

- I_{cont} is an initialization function (the initial marking) such that

$$I_{\text{cont}} : P \to \mathbb{R}^+ \tag{10}$$

The classical problem to be modeled using CPN is the Lotka-Volterra model (predator-prey model) [56, 57]. A continuous Petri net model that represents this problem is illustrated in Fig. 4a. This example describes the oscillation in the levels of the predator and prey population cycle defined by the reaction equation presented

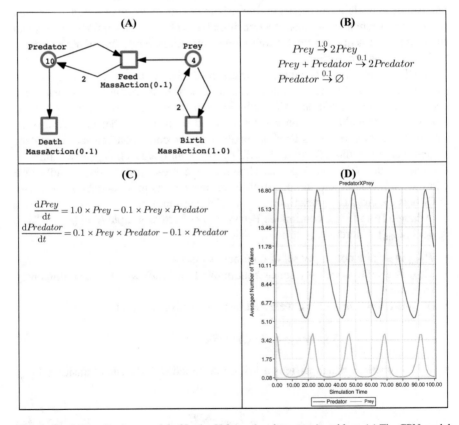

Fig. 4 Continuous Petri net model of Lotka-Volterra (predator-prey) problem. (**a**) The CPN model representation. (**b**) Reaction equations that represent the oscillation in the levels of the predator and prey population according to a specific rate for each function. (**c**) The set of ODEs equations describing the continuous change over time. (**d**) The continuous simulation of the reactions using mass-action kinetic function

in Fig. 4b. The ODEs used to solve this problem are defined in Fig. 4c. To simulate the model, we consider the initial parameters of 10 tokens of predator, 4 tokens of prey, a predator death rate of 0.1 tokens, a prey birth rate of 1.0 token, and a predator feed rate of 0.1 tokens. For the simulation, it is used a mass-action kinetic function [56] to solve the ODEs. The resulted oscillation of the population cycle is depicted in Fig. 4d, where a high number of predators lead to a low number of prey and vice versa. The Lotka-Volterra model is a famous problem, which describes the time-dependent oscillating behavior of biological and ecological systems [58].

2.3 Stochastic Petri Net (SPN)

Initially studied in [59–62] and summarized by Marsan [63], Stochastic Petri nets can be considered as a timed Petri net in which the timings have stochastic values, where a firing delay (rate functions) is associated with each transition. It specifies the amount of time that must elapse before the transition can fire. This firing delay is a random variable following an exponential probability distribution. The semantics of a SPN with exponentially distributed firing delays for all transitions are described by a continuous time Markov chain (CTMC). Their firing transition follows the standard firing rule of QPNs and does not consume time. The stochastic Petri nets can also be enhanced with modifier arcs, which allow pre-places to modify the firing rate of a transition without influence on its enabled state. There are also special transitions: (1) time is the set of deterministic transitions, whose firing delay is specified by an integer constant; (2) immediate transitions which have zero delay and are always of high priority; and (3) scheduled transitions, a special case of deterministic transition which is specified at an absolute point in the simulation time at which it might occur (it will always depend if it is enabled).

Adapted from [64], we define a stochastic Petri net as tuple $N = (P_{\text{disc}}, T_U, A_{\text{disc}}, V, I_{\text{stoch}})$, where:

- P_{disc} is a finite nonempty set of discrete places.
- T_U is the union of disjunctive transition sets

$$T_U = T_{\text{stoch}} \bigcup T_{\text{im}} \bigcup T_{\text{timed}} \bigcup T_{\text{scheduled}} \tag{11}$$

where:

1. T_{stoch} is the set of stochastic transitions with exponentially distributed waiting time.
2. T_{im} is the set of immediate transitions with waiting time zero.
3. T_{timed} is the set of deterministic transitions, which fire with a deterministic time delay.
4. $T_{\text{scheduled}}$ is the set of scheduled transitions, which fire at predefined firing time points.

- A_{disc} is a finite set of arcs, weighted by nonnegative integer values [cf. 2].
- V is a function such that

$$T_{stoch} \rightarrow H_s \tag{12}$$

which assigns a stochastic hazard function h_t to each transition t, whereby

$$H_s = \bigcup_{t \in T_{stoch}} \left\{ h_t | h_t : \mathbb{N}^{|^\circ t|} \rightarrow \mathbb{R}^+ \right\} \tag{13}$$

is the set of all stochastic hazard functions and

$$V(t) = h_t \forall t \in T_{stoch} \tag{14}$$

- I_{stoch} is an initialization function (the initial marking) such that

$$I_{stoch} : P \rightarrow \mathbb{R}^+ \tag{15}$$

The stochastic hazard function h_t defines the marking-dependent transition rate $\omega_t(m)$ for the transition t, i.e., $h_t = \omega_t(m)$. The domain of h_t is restricted to the set of input places of t, denoted by $^\circ t$ with $^\circ t = \{p \in P_{disc} | A_{disc}(p, t) \neq 0\}$, to enforce a close relation between network structure and hazard functions. Therefore, $\omega_t(m)$ actually depends on a sub-marking only.

To illustrate a stochastic Petri net model, we can use the same example of the enzymatic reaction presented in Fig. 3. However, for the SPN, it is necessary to associate rate functions with each transition. For this example, we assume that all transitions in the model have the same rate function (equal to 0.1). Figure 5a presents the SPN model and the reaction equations defined for the enzymatic reaction. To show the stochastic time-dependent dynamic behavior of the model, we simulate the model using the Gillespie stochastic simulation algorithm [65]. Figure 5b depicts the averaged results of 10,000 simulation runs for this example.

2.4 Hybrid Petri Net (HPN)

The hybrid Petri nets combine both discrete and continuous components in one model. It permits to represent, e.g., a biological switch in which continuous elements are turned on/off by discrete elements [66]. The HPN captures the randomness and fluctuation of the discrete stochastic model and allows at the same time a reasonable computation time. This goal is achieved by simulating fast reactions deterministically using ODE solvers, while simulating slow reaction stochastically.

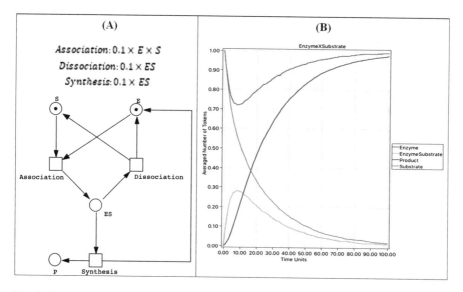

Fig. 5 Stochastic Petri net model of the enzymatic reactions. (**a**) The SPN model representation and the set of equations where the rate functions (probability distribution) of a reaction to occur are 0.1. (**b**) The stochastic simulation of the reactions using mass-action kinetic function for a 10,000 simulation runs

Hybrid Petri nets have been successfully applied in systems biology [67–70], and different extensions of HPN were proposed for different purposes [71]. Herajy et al. [72] introduced a more complete extension of hybrid Petri nets (generalized hybrid Petri nets) combining features of CPNs and SPNs, including three types of deterministic transitions.

Adapted from [72], hybrid Petri nets can be defined as a tuple $N = (P_h, T_h, A_h, V, I_h)$ where:

- P_h is a finite nonempty set of (discrete and continuous) places, whereby

$$P_h = P_{\text{cont}} \bigcup P_{\text{disc}} \qquad (16)$$

- T_h is the union of two disjunctive transition sets, continuous and stochastic [cf. 11], whereby

$$T_h = T_{\text{cont}} \bigcup T_U \qquad (17)$$

- A_h is a finite set of directed arcs

$$A_h = A_{\text{disc}} \bigcup A_{\text{cont}} \bigcup A_{\text{inhibit}} \bigcup A_{\text{read}} \bigcup A_{\text{equal}} \qquad (18)$$

where:

1. A_{inhibit} defines the set of inhibition arcs, such as $(P \times T) \rightarrow \mathbb{R}^+$ if $P \in P_{\text{cont}}$ or $(P \times T) \rightarrow \mathbb{N}^+$ if $P \in P_{\text{disc}}$.
2. A_{read} defines the set of read arcs, such as $(P \times T) \rightarrow \mathbb{R}^+$ if $P \in P_{\text{cont}}$ or $(P \times T) \rightarrow \mathbb{N}^+$ if $P \in P_{\text{disc}}$.
3. A_{equal} defines the set of equal arcs, such as $(P_{\text{disc}} \times T) \rightarrow \mathbb{N}^+$.

- V is a set of functions $V = \{f, g, d\}$ where:

1. $f : T_{\text{cont}} \rightarrow H_c$ is a function which assigns a rate function to each continuous transition [cf. 8].
2. $g : T_{\text{stoch}} \rightarrow H_s$ is a function which assigns a rate function to each stochastic transition [cf. 13].
3. $d : T_{\text{timed}} \bigcup T_{\text{scheduled}} \rightarrow \mathbb{R}^+$ is a function which assigns a constant time to each deterministic and scheduled transition representing the waiting time.

- I_h is the initial marking for both the continuous and discrete places, whereby

$$I_h = I_{\text{cont}} \bigcup I_{\text{stoch}} \tag{19}$$

The different HPN elements are connected with each other, such that they obey certain rules, e.g., it is not possible to connect a discrete place with a continuous transition using standard arcs. The semantics of continuous transitions are represented by a set of ordinary differential equations that require the existence of real values in the input and output places. Hence this is not allowed to take place for discrete places. Since one of the HPN objectives is to bring discrete and continuous parts together, there are some arcs, which allow the connections between discrete places and continuous transitions. Read, inhibitor, and equal arcs are some examples of these arcs.

To illustrate a hybrid Petri net model, Fig. 6 depicts a HPN of T7 phage, a simple biological network adapted from [72]. Table 1 lists the set of reactions that describe the interaction network. The HPN representation of the model is depicted in Fig. 6. The net is partitioned into discrete and continuous parts based on the reaction kinetic in Table 1. The reactions $R5$ and $R6$ are considered to be fast compared to the other four reactions. Therefore, they have been represented by two continuous transitions.

2.5   Colored Petri Net

Colored Petri net is a Petri net modeling concept, which extends quantitative and qualitative Petri nets by combining the capabilities of programing languages to describe data types and operations. It adds the concept of "color" to distinguish tokens and arc expressions that specify which token can flow over the arcs. Moreover, Boolean expressions (guards) can be defined in the transitions defining additional constrains to enable it.

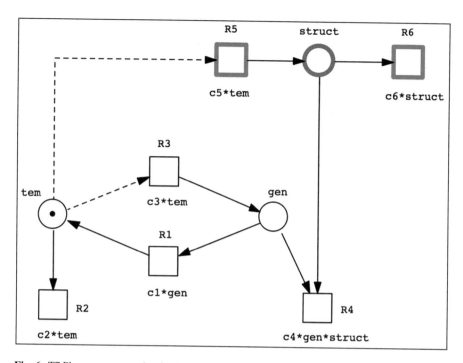

Fig. 6 T7 Phage – an example of a simple biological network modeled using HPN (Adapted from [72])

Table 1 List of the set of reactions that describe the interaction network of T7 phage biological network (Adapted from [72])

Name	Reaction	Propensity	Rate
R1	gen → tem	$c_1 \times gen$	$c_1 = 0.0025$
R2	tem → ∅	$c_2 \times tem$	$c_2 = 0.25$
R3	tem → tem + gen	$c_3 \times tem$	$c_3 = 1.0$
R4	gen + struct → "v"	$c_4 \times gen \times struct$	$c_4 = 7.5 \times 10^{E-6}$
R5	tem → tem + struct	$c_1 \times tem$	$c_5 = 1000$
R6	struct → ∅	$c_1 \times struct$	$c_6 = 1.99$

Adapted from [73], we use Type (Vars) to denote the set of types {Type(v)| $v \in$ Vars} of a typed set Vars. To denote the Boolean type, we use the set B consisting of the elements {false, true}.

A *multi-set m* over a nonempty set S is a *function m* : $S \rightarrow \mathbb{N}$. An element $s \in S$ is said to belong to the multi-set m if $m(s) \neq 0$, and then we write $s \in m$. The integer $m(s)$ is the number of appearances of the element s in m.

We represent a multi-set m over S by the formal sum:

$$\sum_{s \in S} m(s)'s \tag{20}$$

By S_{MS} we denote the set of all multi-sets over S.

A colored qualitative Petri net (QPN^C) is a tuple $(\Sigma, P, T, A, C, G, E, I)$, where:

- Σ is a finite nonempty set of types, called color sets.
- P is a finite nonempty set of places.
- T is a finite nonempty set of transitions.
- A is a finite set of arcs.
- C is a color function; it is defined from P to Σ.
- G is a guard function; it is defined from T to Boolean expressions such that

$$\forall t \in T : \left[\text{Type}\,(G(t)) = B \wedge \text{Type}\,(\text{Var}\,(G(t))) \subseteq \Sigma\right] \qquad (21)$$

- E is an arc expression function; it is defined from A to expressions such that

$$\forall a \in A : \left[\text{Type}\,(E(a)) = C\,(p(a)) \wedge \text{Type}\,(\text{Var}\,(E(a))) \subseteq \Sigma\right] \qquad (22)$$

where $p(a)$ is the place component of a.

- I is an initialization function (the initial marking); it is defined from P to multi-sets of colors such that

$$\forall p \in P : \left[\text{Type}\,(I(p)) \subseteq C(p)\right] \qquad (23)$$

In general, a marking associates with each place P a multi-set over $C(p)$, that is, a marking assigns to each place a multi-set of "colored tokens."

In the formal definition of the colored extension of the previous Petri net classes (CPN, SPN, HPN), we replace the set of transitions T and add the set of functions according to each class, such as:

- CPN^C : CPN $\cup$ QPN^C = $(\Sigma, P_{cont}, T_{cont}, A_{cont}, V, C, G, E, I_{cont})$
- SPN^C : SPN $\cup$ QPN^C = $(\Sigma, P_{disc}, T_U, A_{disc}, V, C, G, E, I_{stoch})$
- HPN^C : SPN $\cup$ QPN^C = $(\Sigma, P_h, T_h, A_h, V, C, G, E, I_h)$

In the colored Petri net, a transition is enabled (allowed to fire) if it has no input place or if each of its input places is sufficiently marked by tokens: i.e., the arc expressions evaluate to a multi-set of token colors that should be available in the corresponding preceding place. In addition, the guard of the transition – if present – should evaluate to be true for the given binding. When a transition fires, a multi-set of colored tokens is consumed (taken) from each of the preceding places, according to the evaluation of the expression on the arc. A multi-set of colored tokens is produced (added), in correspondence with the arc expression, to each successor place. The overall state space of the Petri net is determined by the firing sequences consisting of iterated occurrences of transitions [74].

A typical example of a colored Petri net and its components is depicted in Fig. 7. Here both the graphical representation and the values that make of the QPN^C are given. Figure 7a shows the declaration of the data values assigned to the net defining

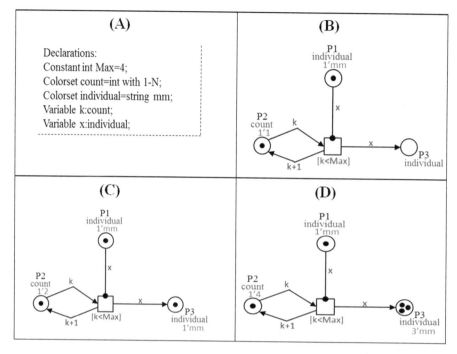

Fig. 7 Colored Petri net example. (**a**) Declarations of the data types and variables. (**b**) Colored Petri net components. (**c**) State of the net after firing the transition. (**d**) End state of the net; the transition is not able to fire since the condition of the guard is not satisfied

two color sets: *count* with integer values and *individual* with a string *mm*; a constant *Max* with an integer value; the two variables, k and x, used in arc expressions; and the guard of the transition. The color sets are assigned to places, and the tokens on each place will have a color from a color set assigned to the place. In Fig. 7b the net is defined with place *P1* (with color set *individual*) containing one token with the color *mm*, place *P2* (with color set *count*) containing one token with the integer value "1," and place *P3* (with color set *individual*) without token. The transition *T1* is connected to the place *P1* by a read arc that works as a test arc: if there is a token in *P1*, as described by the binding of x and that satisfies the firing rule, then the transition can occur, producing a token on the output places, but not consuming the token read in *P1*. Transition *T1* is connected to place *P2* by two arcs indicating that it will consume and produce tokens according to the arc expressions. Figure 7c shows the new marking obtained when *T1* has read *P1* and consumed the token in *P2*, adding a new token in *P2* and P3 according to the firing rule. Repeating this process, the firing sequence will stop in the end state as shown in Fig. 7d. In this case, the guard condition turns false since the value of the token in *P2* does not fulfill the condition.

3 Model Analysis

A reliable model is responsible in reproducing a biological process bringing evidences in reproducing a specific behavior and confidence on its predictive capability. Once a model has been constructed, it should serve as a clear description of the system and can be unequivocally communicated. Therefore, an analysis of the model is required in order to verify and validate its structure and behavior.

For the analysis of a model, it is necessary to refer to the model objective, the information, and data used in the modeling process. In the process of analysis, we can test the model: does the model represent the hypothesis at hand? Is the data concisely represented? Does the model reproduce the problem? Moreover, the analysis of the model must consider the characteristics of the biological system that have been used to define the model properties. Therefore, we can classify the models according to some properties:

- *Reversibility*: Checks if the initial condition of a model can be reached again. Therefore, the model has the capability of self-reinitialization, e.g., a chemical reaction can be classified as irreversible if nearly all of its reactants are used to form a product and there are no reactants to restart the reaction.
- *Periodicity*: Checks if it will be possible to reach a state in the model where a process is periodically active. In nature, we know calcium oscillation as an important periodic process that controls a wide variety of cellular mechanisms and is often organized into intracellular and intercellular calcium waves.
- *Stability*: Biological systems have to be stable in performing a certain function in a specific changeable environment.
- *Robustness*: Checks if the model can flexibly respond to environmental perturbations or other changes. It is implemented in biological systems by redundancy, feedback loops, and structural stability.
- *Activity*: Checks if part of the model can be active or inactive, depending on other activated/nonactivated components of the model. Proteins are an example where it can be activated or deactivated by phosphorylation or dephosphorylation process, respectively.
- *Determinism*: Deterministic models produce the same output for the same starting condition; there is no randomness in the development of subsequent state. Deterministic models can be distinguished between discrete and continuous models.
- *Stochasticity*: In stochastic models, the development of subsequent processes is additionally determined by a random element. This random element acknowledges that a subsequent state in the model is resulted of unknown causes (a phenomena that appear to vary in a random manner).
- *Discreteness*: Consider the countable amount (e.g., proteins, genes, etc.), where processes take place according to discrete rules. In contrast, continuous models do not consider countable entities; it works with concentration and reaction rates that depend on substance concentration.

Structural analysis of a model refers to the analysis of the structural properties, which only depend on the net topology. The structural properties of a Petri net model are directly dependent on the arrangement of places, transitions, and arcs. They are independent of the initial marking; they hold for any initial marking or are concerned with the existence of certain firing sequences from some initial marking. The structural analysis can be considered as an initial consistency check to prove that the model adheres to the assumptions and the modeling decision. The most important structural properties of a Petri net model are described in [22, 75].

One important (experimental) strategy to verify and validate the model is using simulation. The analysis of the results of a simulation can demonstrate the proper behavior of the model, and also it can contribute to possibly further refine the model, so the results approximate reality as close as possible. Although model simulations will never replace laboratory experiments, a model allows one to probe system behavior in ways that would not be possible in the lab. Simulations can be carried out quickly (often in seconds) and incur no real cost. Model behavior can be explored in conditions that are otherwise difficult to achieve in laboratory settings. Every aspect of model behavior can be observed at all time points. Furthermore, model analysis yields insights into why a system behaves the way it does, thus providing links between network structure modeled and observed in real experiments.

4 Tools

There are many tools available on the Internet as well as web platforms that can be used for the implementation of Petri net models [17, 32, 76–84]. Most of the tools are confined to specific classes and/or not support extensions, portability, or analysis. However, few tools can provide an extensible experience with Petri net methods [85–87]. We assessed that Snoopy software [53] is the most complete tool available to design and evaluate models in Petri nets. It supports a set of related important Petri net classes, i.e., QPN, SPN, CPN, and HBPN, and their colored extensions. Snoopy provides analysis techniques, e.g., animation and simulation, and also exports properties between classes and to external analysis tools, i.e., Charlie [88] and Marcie [89]. Recently, the Snoopy Steering and Simulation Server tool (S4) [34, 90] was released as an extension of the Snoopy simulation to perform stochastic simulations in multi-core servers. This extension provides a better performance for simulations of big models as well as to change stochastic properties of the model while it is running. The Snoopy is platform independent and freely available for all relevant platforms, i.e., Linux, Windows, and Mac OS platforms.

5 Conclusion and Discussion

Experimental and computational approaches can be combined to systematically investigate biological systems. The integration of modeling and experimentation is essential in the creation of a computational system that can simulate events and predict outcomes of biological behavior. Since experimental research can be costly, time-consuming, or ethically infeasible, the results of in silico experiments are a great advantage and an important aspect of systems biology.

In order to understand the structural and dynamic properties of complex biological systems, models are needed that can describe the entities involved and their interaction. In this chapter, we introduced computational methods that can be used to model a biological system, emphasizing the Petri net formalism. This method combines an intuitive graphical representation with a strong mathematical foundation, integrating qualitative and quantitative aspects in a range of tools that support implementation, simulation, and analysis. We also presented the structured family of Petri net classes, given a brief introduction and formal definition of each class. Our objective is to give a general idea of how to use Petri nets for modeling and analyze biological system with different Petri net modeling paradigms.

Research in bio-modeling and Petri nets reflects an approach to systems biology research, in which the integration of modeling and visualization aspects is essential. The characteristics discussed in this chapter suggest that Petri net is an important method to model, simulate, and analyze biological behavior. We provided plenty of pointers to related literature where the interested reader may find the inspiration to apply this formalism in systems biology. The collaboration between biology and computer science researchers both with an experimental and a formal background is crucial to develop systems that can contribute to our understanding of biological phenomena.

Acknowledgments This work was supported by Fundação de Amparo a Pesquisa do Estado de Goiás (FAPEG) and CNPq.

References

1. Kitano H. Systems biology: a brief overview. Science. 2002;295:1662–4.
2. Aderem A. Systems biology: its practice and challenges. Cell. 2005;121:511–3.
3. Fisher J, Henzinger TA. Executable cell biology. Nat Biotechnol. 2007;25:1239–49.
4. Hunt CA, Ropella GEP, Park S, Engelberg J. Dichotomies between computational and mathematical models. Nat Biotechnol. 2008;26:737–9.
5. Cardelli L. Brane calculi-interactions of biological membranes. Proc C. 2005; 3082:257–78.
6. Pun G. A guide to membrane computing. Theor Comput Sci. 2002;287:73–100.
7. Calzone L, Fages FF, Soliman S. BIOCHAM: an environment for modeling biological systems and formalizing experimental knowledge. Bioinformatics. 2006;22:1805–7.
8. Barbuti R, Caravagna G, Maggiolo-Schettini A, Milazzo P, Pardini G. The calculus of looping sequences. In: Proceedings of the Formal methods for the design of computer, communication, and software systems 8th international conference on Formal methods for computational systems biology. Berlin: Springer-Verlag; 2008. p. 387–423.

9. Shannon P, Markiel A, Ozier O, Baliga NS, Wang JT, Ramage D, et al. Cytoscape: a software environment for integrated models of biomolecular interaction networks. Genome Res. 2003;13:2498–504.
10. Breitkreutz B-J, Stark C, Tyers M. Osprey: a network visualization system. Genome Biol. 2002;3:1–6.
11. Hucka M, Finney A, Sauro HM, Bolouri H, Doyle JC, Kitano H, et al. The systems biology markup language (SBML): a medium for representation and exchange of biochemical network models. Bioinformatics. 2003;19:524–31.
12. Hucka M. Systems biology markup language (SBML). In: Encyclopedia of systems biology. New York: Springer New York; 2013. p. 2057–63.
13. Garvey TD, Lincoln P, Pedersen CJ, Martin D, Johnson M. BioSPICE: access to the most current computational tools for biologists. Omi A J Integr Biol. 2003;7:411–20.
14. Danos V, Krivine J. Formal molecular biology done in CCS-R. Electron Notes Theor Comput Sci. Elsevier. 2007;180:31–49.
15. Regev A, Silverman W, Shapiro E. Representation and simulation of biochemical processes using the pi-calculus process algebra. Pac Symp Biocomput. 2001;6:459–70.
16. Akman O, Ciocchetta F, Degasperi A, Guerriero M. Modelling biological clocks with bio-PEPA: stochasticity and robustness for the Neurospora crassa circadian network. In: Degano P, Gorrieri R, editors. Computational methods in systems biology SE – 4. Berlin: Springer; 2009. p. 52–67.
17. Ciocchetta F, Hillston J. Bio-PEPA: a framework for the modelling and analysis of biological systems. Theor Comput Sci. Elsevier Science Publishers Ltd. 2009;410:3065–84.
18. González PP, Cárdenas M, Camacho D, Franyuti A, Rosas O, Lagúnez-Otero J. Cellulat: an agent-based intracellular signalling model. Biosystems. 2003;68:171–85.
19. Segovia-Juarez JL, Ganguli S, Kirschner D. Identifying control mechanisms of granuloma formation during M. tuberculosis infection using an agent-based model. J Theor Biol. 2004;231:357–76.
20. Murata T. Petri nets: properties, analysis and applications. Proc IEEE. 1989;77:541–80.
21. Reddy VN, Mavrovouniotis ML, Liebman MN. Petri net representations in metabolic pathways. In: Hunter L, Searls DB, Jude W. Shavlik, editors. Proceedings of the 1st international conference on intelligent systems for molecular biology. Menlo Park: AAAI Press; 1993. p. 328–36.
22. Heiner M, Gilbert D, Donaldson R. Petri nets for systems and synthetic biology. In: Bernardo M, Degano P, Zavattaro G, editors. Form methods computational systems biology, vol. 5016. Berlin/Heidelberg: Springer; 2008. p. 215–64.
23. Jensen K. Coloured Petri nets : basic concepts, analysis methods and practical use. Berlin: Springer; 1997.
24. Petri CA. Kommunikation mit Automaten. Vol. Doktor. Fakultät Math Phys. 1962. p. 128.
25. Cordero F, Horváth A, Manini D, Napione L, De Pierro M, Pavan S, et al. Simplification of a complex signal transduction model using invariants and flow equivalent servers. Theor Comput Sci. 2011;412:6036–57.
26. Koch I, Junker BH, Heiner M. Application of Petri net theory for modelling and validation of the sucrose breakdown pathway in the potato tuber. Bioinformatics. 2005;21:1219–26.
27. Blätke MA, Heiner M, Marwan W. Predicting phenotype from genotype through automatically composed Petri nets. In: Gilbert D, Heiner M, editors. Computational methods in systems biology. Lecture notes in computer science, vol. 7605. Berlin/Heidelberg: Springer; 2012. p. 87–106.
28. Koch I, Reisig W, Schreiber F. In: Koch I, Reisig W, Schreiber F, editors. Modeling in systems biology: the Petri net approach. London: Springer-Verlag; 2011. p. 25.
29. Peleg M, Rubin D, Altman RB. Using Petri net tools to study properties and dynamics of biological systems. J Am Med Inf Assoc. 2005;12:181–99.
30. Hofestädt R. A Petri net application to model metabolic processes. Syst Anal Model Simul. 1994;16:113–22.

31. Albergante L, Timmis J, Beattie L, Kaye PM. A Petri net model of granulomatous inflammation: implications for IL-10 mediated control of Leishmania donovani infection. PLoS Comput Biol Public Libr Sci. 2013;9:e1003334.
32. Gilbert D, Heiner M, Lehrack S. A unifying framework for modelling and analysing biochemical pathways using Petri nets. In: Calder M, Gilmore S, editors. Computational methods in systems biology. Berlin/Heidelberg: Springer; 2007. p. 200–16.
33. Heiner M, Koch I, Will J. Model validation of biological pathways using Petri nets—demonstrated for apoptosis. Biosystems. 2004;75:15–28.
34. Herajy M, Heiner M. Petri net-based collaborative simulation and steering of biochemical reaction networks. Fundam Informaticae. 2014;129:49–67.
35. Wingender E. In: Wingender E, editor. Biological Petri nets. Amsterdam: IOS Press; 2011.
36. Doi A, Fujita S, Matsuno H, Nagasaki M, Miyano S. Constructing biological pathway models with hybrid functional Petri nets. In Silico Biol. 2004;4:271–91.
37. Chaouiya C, Remy E, Ruet P, Thieffry D. Qualitative modelling of genetic networks: from logical regulatory graphs to standard Petri nets. Appl Theory Petri Nets. 2004;2004:137–56.
38. Chaouiya C. Petri net modelling of biological networks. Brief Bioinf. 2007;8:210–9.
39. Matsuno H, Tanaka Y, Aoshima H, Doi A, Matsui M, Miyano S. Biopathways representation and simulation on hybrid functional Petri net. Stud Health Technol Inform. 2011;162:77–91.
40. Breitling R, Gilbert D, Heiner M, Orton R. A structured approach for the engineering of biochemical network models, illustrated for signalling pathways. Brief Bioinform. 2008;9: 404–21.
41. Li C, Ge QW, Nakata M, Matsuno H, Miyano S. Modelling and simulation of signal transductions in an apoptosis pathway by using timed Petri nets. J Biosci. 2007;32:113–27.
42. Hardy S, Robillard PN. Petri net-based method for the analysis of the dynamics of signal propagation in signaling pathways. Bioinformatics. 2008;24:209–17.
43. Sackmann A, Heiner M, Koch I. Application of Petri net based analysis techniques to signal transduction pathways. BMC Bioinf. 2006;7:482.
44. Kielbassa J, Bortfeldt R, Schuster S, Koch I. Modeling of the U1 snRNP assembly pathway in alternative splicing in human cells using Petri nets. Comput Biol Chem. 2009;33:46–61.
45. Carvalho RV, Kleijn J, Meijer AH, Verbeek FJ. Modeling innate immune response to early mycobacterium infection. Comput Math Methods Med. 2012;2012:790482.
46. Carvalho RV, Kleijn J, Verbeek FJ. A multi-scale extensive Petri net model of the bacterial – macrophage interaction. In: Heiner M, editor. 5th International workshop on biological processes & Petri nets. Tunis: CEUR Workshop Proceedings; 2014. p. 15–29.
47. Carvalho R, van den Heuvel J, Kleijn J, Verbeek F. Coupling of Petri net models of the mycobacterial infection process and innate immune response. Comput Multidiscip Digit Publ Inst. 2015;3:150–76.
48. Heiner M, Gilbert D. How might Petri nets enhance your systems biology toolkit. In: Kristensen LM, Petrucci L, editors. Applications and theory of Petri nets. PETRI NETS 2011. Lecture notes in computer science, vol. 6709. Berlin/Heidelberg: Springer; 2011. p. 17–37.
49. Mura I, Csikász-Nagy A. Stochastic Petri net extension of a yeast cell cycle model. J Theor Biol. 2008;254:850–60.
50. David R, Alla H. On hybrid Petri nets. Discrete event dynamic systems: theory and applications, vol. 11. Boston: Kluwer Academic Publishers; 2001. p. 9–40.
51. Jensen K. Coloured Petri nets and the invariant-method. Theor Comput Sci. Elsevier. 1981;14:317–36.
52. Jensen K. Coloured Petri nets. Brauer W, Reisig W, Rozenberg G. Theor Comput Sci. Springer Berlin Heidelberg; 2009; 254:248–299.
53. Heiner M, Herajy M, Liu F, Rohr C, Schwarick M. Snoopy – a unifying Petri net tool. Haddad Serge, Pomello Lucia. Lect Notes Comput Sci. Hamburg: Springer; 2012; 7347:398–407.
54. Blätke MA, Dittrich A, Rohr C, Heiner M, Schaper F, Marwan W. JAK/STAT signalling – an executable model assembled from molecule-centred modules demonstrating a module-oriented database concept for systems and synthetic biology. Mol BioSyst. 2013;9:1290.

55. Gilbert D, Heiner M. From Petri nets to differential equations – an integrative approach for biochemical network analysis. In: Donatelli S, Thiagarajan PS, editors. Petri nets and other models of concurrency – ICATPN 2006, Lecture Notes Computer Science, vol. 4024. Berlin: Springer; 2006. p. 181–200.

56. Lotka AJ. Undamped oscillations derived from the law of mass action. J Am Chem Soc. 1920;42:1595–9.

57. Hitchcock S. Extinction probabilities in predator-prey models. J Appl Probab. 1986;23(1): 1–13.

58. Hofestädt R, Thelen S. Quantitative modeling of biochemical networks. In Silico Biol. 1998;1:39–53.

59. Symons FJW. Introduction to numerical Petri nets, a general graphical model of concurrent processing systems. Aust Telecommun Res. 1980;14:28–33.

60. Symons FJW. The description and definition of queueing systems by numerical Petri nets. Aust Telecommun Res. 1980;13:20–31.

61. Natkin S. Les Reseaux de Petri Stochastique et leur Application a l'Evaluation des Systèmes Informatiques. Tesis doctorales. CNAM, Paris; 1980.

62. Molloy MK. On the integration of delay and throughput measures in distributed processing models. Ph.D. Dissertation. University of California, Los Angeles. 1981.

63. Marsan MA. Stochastic Petri nets: an elementary introduction. In: Rozenberg G, editor. Advances in Petri Nets 1989. APN 1988. Lecture Notes in Computer Science, vol. 424. Berlin/Heidelberg: Springer; 1990. p. 1–29.

64. Heiner M, Lehrack S, Gilbert D, Marwan W. Extended stochastic Petri nets for model-based design of wetlab experiments. In: Priami C, Back RJ, Petre I, editors. Transactions on computational systems biology XI. Lecture notes in computer science, vol. 5750. Berlin/Heidelberg: Springer; 2009. p. 138–63.

65. Gillespie DT. Exact stochastic simulation of coupled chemical reactions. J Phys Chem Am Chem Soc. 1977;81:2340–61.

66. Alla H, David R. Continuous and hybrid Petri nets. J Circ Syst Comput. 1998;8:159–88.

67. Matsuno H, Nagasaki M, Miyano S. Hybrid Petri net based modeling for biological pathway simulation. Nat Comput. 2011;10:1099–120.

68. Kiehl TR, Mattheyses RM, Simmons MK. Hybrid simulation of cellular behavior. Bioinformatics. 2004;20:316–22.

69. Marchetti L, Priami C, Thanh VH. HRSSA – efficient hybrid stochastic simulation for spatially homogeneous biochemical reaction networks. J Comput Phys. 2016;317:301–17.

70. Herajy M, Heiner M. Accelerated simulation of hybrid biological models with quasi-disjoint deterministic and stochastic subnets. In: Cinquemani E, Donzé A, editors. Hybrid systems biology. HSB 2016. Lecture notes in computer science, vol. 9957. Cham: Springer; 2016. p. 20–38.

71. David R, Alia H. Discrete, continuous, and hybrid petri nets. Discrete, continuous, and hybrid Petri nets. 2005. p. 1–524.

72. Herajy M, Heiner M. Hybrid representation and simulation of stiff biochemical networks. Nonlinear Anal Hybrid Syst. 2012;6:942–59.

73. Liu F, Heiner M. Colored Petri nets to model and simulate biological systems. In: Donatelli S, Kleijn J, Machado RJ, Fernandes JM, editors. Recent advances in Petri Nets and concurrency. CEUR workshop proceedings, vol. 827. 2012. p. 71–85.

74. Desel J, Juhás G. "What Is a Petri Net?" Informal answers for the informed reader. In: Ehrig H, Padberg J, Juhás G, Rozenberg G, editors. Unifying Petri nets. Lecture notes in computer science, vol. 2128. Berlin/Heidelberg: Springer; 2001. p. 1–25.

75. Liu F, Heiner M. Petri nets for modeling and analyzing biochemical reaction networks. In: Chen M, Hofestädt R, editors. Approaches in integrative bioinformatics. Berlin: Springer; 2014. p. 245–72.

76. Weber M, Kindler E. The Petri net kernel. In: Ehrig H, Reisig W, Rozenberg G, Weber H, editors. Petri net technology for communication-based systems SE. Berlin: Springer; 2003. p. 109–23.

77. Grahlmann B, Best E. PEP – more than a Petri net tool. In: Margaria T, Steffen B, editors. Tools and algorithms for the construction and analysis of Systems. TACAS 1996. Lecture notes in computer science, vol. 1055. Berlin/Heidelberg: Springer; 1996. p. 397–401.
78. Kindler E. The ePNK: an extensible Petri net tool for PNML. In: Kristensen LM, Petrucci L, editors. Applications and theory of Petri nets. PETRI NETS 2011. Lecture notes in computer science, vol. 6709. Berlin/Heidelberg: Springer; 2011. p. 318–27.
79. Verbeek E, van der Aalst WMP. Woflan 2.0 A Petri-net-based workflow diagnosis tool. In: Nielsen M, Simpson D, editors. Application and theory of Petri nets 2000. ICATPN 2000. Lecture notes in computer science, vol. 1825. Berlin/Heidelberg: Springer; 2000. p. 475–84.
80. Davidrajuh R. Developing a new Petri net tool for simulation of discrete event systems. In: Proceedings – 2nd Asia international conference on modelling and simulation, AMS 2008. Malaysia: Kuala Lumpur; 2008. p. 861–866.
81. Gao J, Li L, Wu X, Wei DQ. BioNetSim: a Petri net-based modeling tool for simulations of biochemical processes. Protein Cell. 2012;3:225–9.
82. Franck P. Quickly prototyping petri nets tools with {SNAKES}. In: Proceedings of the 1st international conference on Simulation tools and techniques for communications, networks and systems & workshops (Simutools '08). ICST, Brussels, Belgium; 2008. p. 1–17.
83. Zimmermann A, Freiheit J, German R, Hommel G. Petri net modelling and performability evaluation with TimeNET 3.0. In: Haverkort B, Bohnenkamp H, Smith C, editors. Computer performance evaluation modelling techniques and tools SE – 14. Berlin: Springer; 2000. p. 188–202.
84. Berthomieu B, Ribet P-O, Vernadat F. The tool TINA – construction of abstract state spaces for Petri nets and time Petri nets. Int J Prod Res. Taylor & Francis Group. 2004;42:2741–56.
85. Jensen K, Kristensen LM, Wells L. Coloured Petri nets and CPN tools for modelling and validation of concurrent systems. Int J Softw Tools Technol Transf. 2007;9:213–54.
86. Chiola G, Franceschinis G, Gaeta R, Ribaudo M. GreatSPN 1.7: graphical editor and analyzer for timed and stochastic Petri nets. Perform Eval. 1995;24:47–68.
87. Nagasaki M, Saito A, Jeong E, Li C, Kojima K, Ikeda E, et al. Cell illustrator 4.0: a computational platform for systems biology. In Silico Biol. 2010;10:5–26.
88. Heiner M, Schwarick M, Wegener J. Charlie – an extensible Petri net analysis tool. In: Devillers RR, Valmari A, editors. Petri nets 2015. Brussels: Springer; 2015.
89. Heiner M, Rohr C, Schwarick M. MARCIE – model checking and reachability analysis done efficiently. In: Petri nets 2013. Berlin: Springer; 2013. p. 389–99.
90. Herajy M. Computational steering of multi-scale biochemical networks. Cottbus: Brandenburgischen Technische Universität; 2013.

Modeling Gene Transcriptional Regulation: A Primer

Marcelo Trindade dos Santos, Ana Paula Barbosa do Nascimento, Fernando Medeiros Filho, and Fabricio Alves Barbosa da Silva

Abstract The main goal of Systems Biology nowadays, from a broad perspective, is to explain how a living organism performs its basic activities of growth, maintenance, and reproduction. To attain this objective, investigation on a living phenomenon spans at least three levels of interactions: metabolic, transcriptional regulation, and signaling. A common aspect within these levels is biological phenomena control. In this text, we present an introduction to transcriptional regulation and its mathematical and computational modeling. From ubiquitous carbon source uptake to antibiotic resistance mechanisms exhibited by some bacteria, description of biological phenomena can always be associated with a certain control level, which is, directly or not, associated with transcriptional regulation. Our contribution here is to make explicit what are the consequences of making a transition from verbal (and visual) descriptive biological language to predictive domains of mathematical and computational modeling, showing what are the limitations and advantages this transition can imply.

1 Introduction

Living organisms are capable of performing basic activities associated with its growth, maintenance, and reproduction. This means that they are capable of sensing their surrounding environment, perceiving its nutritional composition and physical characteristics, and transducing these perceptions, or signals, to modify their own internal organization and behavior. In bacteria, for instance, environment nutritional composition directly affects their metabolic capabilities. Cells perceive nutritional

M. Trindade dos Santos (✉)
National Laboratory for Scientific Computation, LNCC/MCTIC, Petrópolis, RJ, Brazil
e-mail: msantos@lncc.br

A. P. B. do Nascimento · F. Medeiros Filho · F. A. B. da Silva
Fundação Oswaldo Cruz - FIOCRUZ, RJ, Brazil
e-mail: fabricio.silva@fiocruz.br

© Springer International Publishing AG, part of Springer Nature 2018
F. A. B. da Silva et al. (eds.), *Theoretical and Applied Aspects of Systems Biology*, Computational Biology 27, https://doi.org/10.1007/978-3-319-74974-7_2

profiles of environment, and these perceptions induce specific enzymatic expression profiles capable to process the nutrients available. Gene transcriptional regulation mediates this interrelation between signaling and phenotypes modification. Modeling these phenomena is the subject of this work. Regulation of transcription is also present in a variety of basic cellular processes such as cell motility, DNA replication, and bacterial multidrug resistance to antibiotics [1].

Gene transcriptional regulation is a molecular mechanism used by cells to express their distinct possible phenotypes. It was first described by Jacob and Monod [2] as an explanation for the diauxic shift in *Escherichia coli* growth curves. Growth curves with two phases, according to Jacob and Monod, were a consequence of transcriptional regulation of the *lac* operon, induced by the presence and/or absence of glucose and lactose, glucose being the preferable carbon source. This seminal work was published only 8 years after proposition of the spatial distribution of DNA molecule [3], and these works paved the road for the development of Modern Biology.

In last two decades, we have seen Systems Biology rising [4], and nowadays it is a completely established research field. Mathematical and computational modeling has acquired a fundamental importance, emphasizing the importance of hypothesis-driven research process. Mathematical and computational models, in principle, are predictive. Model predictions can be tested, implying hypothesis evaluation. Then, comparison with observations can improve model construction. Modeling and testing define feedback loop guiding biological discovery. It is worth to notice that although we have seen the rise of Computational Systems Biology and it has established itself in the last two decades, attempts to build predictive models for gene transcriptional regulation started much earlier by the community of Biomathematics [5].

Biological activities are subject to a certain kind of control in many instances. In prokaryotic cells there are signaling pathways, regulation of gene transcription, and posttranscriptional, translational, and posttranslational control. In this work, we describe basic ideas of gene transcriptional regulation only and present a translation of these ideas into a predictive computational model. This text should be useful both for biologists who want to know about modeling processes and for modelers who want to apply their knowledge on biological problems.

2 Gene Transcriptional Regulation in Prokaryotes

2.1 Basics of Gene Transcriptional Regulation

Synthesis of mRNA molecules from a coding strand of DNA is carried out by RNA-polymerase protein complex, which is responsible for recognizing precisely the region where to bind at DNA (gene promoter region), physically separating coding from complementary DNA strands and synthesizing mRNA molecules. Transcrip-

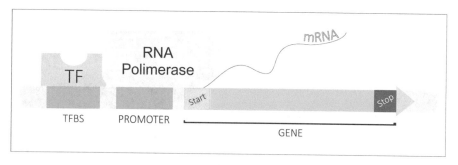

Fig. 1 Pictorial representation of a TF binding to DNA at its TFBS and an RNA-polymerase binding to its promoter site. In this figure, an interaction between TF and RNA-polymerase is also presented

tion does occur until a RNA-polymerase complex finds a transcript termination sign [6]. Other types of proteins are transcription factors (TFs). They also interact with DNA at specific sites in gene promoter regions, the transcription factor binding sites (TFBS). Transcriptional regulation is accomplished by interaction between TFs and RNA-polymerases, the first having the role of regulating the activity of the second (see Fig. 1).

A transcription factor can act as an activator, if its interaction with RNA-polymerase induces mRNA synthesis, or can be a repressor, if their interaction inhibits mRNA production. A single TF can have a large repertoire of interactions. Global regulators as the CRP in *E. coli* can have more than a hundred regulated transcription units (TUs) and can act as activator for a fraction of them and repressor for another [7].

The activity of a TF can be triggered by its interaction with a molecule called inducer. This interaction produces a protein complex TF-inducer that changes the original TF action mechanism. Activation and repression of transcription by a TF can be effective or not, depending on the presence/absence of the inducer molecule (see Fig. 2).

2.2 The Lac *Operon*

Transcriptional regulation of the *lac* operon has historical importance for biology. It was the first description of a genetic regulatory mechanism associated with a phenotype modification [2]. Jacob and colleagues observed that *E. coli* cultures exhibited a diauxic shift when left to grow in a media with glucose and lactose as carbon sources. The growth curves exhibit two phases. Both glucose and lactose are present in the medium during phase I, but only glucose is consumed. In this phase, lactose concentrations in medium are constant. Consumption of external lactose occurs only in phase II, after the consumption of all glucose available. There is

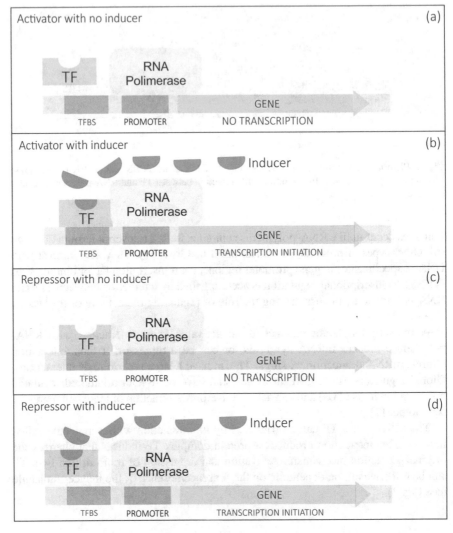

Fig. 2 This figure illustrates that TF inducer interaction can change TF action mode. We use two examples: (**a**) an activator without inducer is not effective, (**b**) TF binds to DNA and interacts with RNA-polymerase only if the inducer is present, (**c**) a repressor is active if the inducer is absent, and (**d**) TF releases DNA and repression ceases if the inducer is present

an adaptive plateau separating these two phases. Considering lactose and glucose, the latter is the preferable carbon source since its metabolism is more efficient energetically, since lactose metabolism needs first to break lactose into galactose and glucose.

The *lac* operon consists in three genes: the *lacZ* that codes β-galactosidase enzyme, responsible for lactose metabolism; *lacY* that codes a permease, respon-

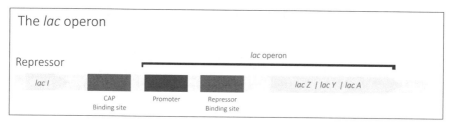

Fig. 3 Structure of *lac* operon

sible for lactose transport from extra- to intracellular medium; and *lacA* that codes for transacetylase. Close to the operon, there is also the gene *lacI*, encoding the repressor LacI of the *lac* operon (see Fig. 3).

Two transcription factors and their respective inducers are responsible for transcriptional regulation of lac operon: catabolite activator protein (CAP) and its inducer cAMP and repressor LacI, which have allolactose as inducer. CAP is an activator of transcription of *lac* operon, and to be effective, it needs to bind to the cAMP inducer, forming CAP-cAMP complex. This complex binds to the TFBS at DNA and activates the transcription initiation. Formation of this complex is directly associated with glucose, since inducer cAMP is produced only if this carbon source is absent. This is the same situation depicted in Fig. 2a, b. Another TF is repressor LacI, induced by allolactose. If allolactose is present in the medium, it binds to LacI repressor, which then releases the TFBS, enabling the transcription initiation. This is the same situation depicted in Fig. 2c, d. With the transcription of *lac* operon, permease transports external lactose into the cell, and β-galactosidase metabolizes it.

We can identify four situations for expression regulation of the lac operon:

1. In a medium with low glucose but with lactose available, the activator CAP binds to its TFBS, and the repressor LacI releases the DNA. The *lac* operon is fully expressed.
2. In a medium with low glucose and lactose unavailable, both the activator CAP and the repressor LacI bind to the DNA. The action of LacI prevents transcription.
3. In a medium with high glucose concentration and lactose unavailable, the activator CAP does not bind to its TFBS, and the repressor LacI binds to DNA. Operon transcription is then repressed.
4. In a medium with high glucose and lactose concentrations, CAP does not bind to its TFBS and does not activate transcription, but the repressor LacI releases DNA. In this case, there is a very low and basal level of the transcription of the *lac* operon.

Another important thing we learn with this mechanism is the following: An inducer is capable of triggering the synthesis of enzymes of its own metabolic pathway. This is important for carbon metabolism and many other important cellular processes, including bacterial multidrug resistance mechanisms, such as efflux pumps [1, 8, 9].

3 Computational Modeling of Transcriptional Regulation

This section presents some concepts about the modeling process and the formulation of a Boolean logic model for *lac* operon transcriptional regulation.

3.1 On the Modeling Approach

Modeling process means, in its essence, development of a representation of a phenomena or object. To start a representation process, we have to address two questions: What are the assumptions about the system, and what are the consequences of these assumptions? A second point is to find an adequate representation or formalism.

A representation of the transcriptional regulation, from a low-level perspective, can be extremely complex. In the previous section, we have ignored a number of phenomena that occurs in the cell. We did not address important details of the molecular transcription mechanism such as the structure and function of RNA-polymerase complex. We have also neglected that after *lacI* gene transcription and LacI repressor production, this protein needs to move and find its place at DNA binding site [10] and stay bound there, before the protein naturally degrades. We also did not see where and how cAMP is produced, neither its production control mechanisms. This was a matter of choice, since we are most interested in overall behavior of the system. This is typically one of the most important modeling choices. We need to define which observables and interactions will be included in the system and which observables are left outside. In terms of transcriptional regulation, natural observables are genes, metabolites, proteins, and mRNAs.

A second choice is about language. There are a number of mathematical and computational formalisms, which can be used to model gene transcriptional regulation [11–13]. We can divide them into three classes: continuous time models based on differential equations, for instance, which require real value (or estimated) parameters. This formalism is capable of directly describing and explaining experimental data quantitatively. We have also discrete-time models, such as logic models, which are the simplest modeling methodology, but based on orchestration of regulation, interactions can explain the overall behavior of a regulatory system. The most widespread methodology in this category is Boolean logic. A third category is the hybrid models, mixing continuous and discrete-time modeling techniques, which can appear in real phenomena. An example of hybrid model is the evolutionary trained neuro-fuzzy recurrent network (ENFRN), a methodology for regulatory network inference and dynamics, based on expression data [14].

A second stage of modeling process is the refinement cycle. Using model predictions and/or simulations, one can compare model behavior with biological phenomena under study and experimental data and from this comparison establish a direction for model improvement. Refined models can result in more accurate predictions and new comparisons with biological observations.

3.2 Boolean Logic for the Lac Operon

In this section, we describe one case of discrete-time, synchronous, and qualitative model using Boolean logic. This is the simplest formalism capable of describing the biological behavior and systemic properties of a regulatory system and has shown to be very effective for these purposes. For model construction, we will follow concepts described in [15] and will not be rigorous mathematically.

Consider $S = (s_1, s_2, ..., s_n)$ an n-tuple of state variables. A Boolean network $F = (F_1, F_2,..., F_n)$ is an n-tuple of functions F_i, defined over the set $\{0,1\}$. A function F_i determines a state transition for each variable s_i of the state vector S and is given in terms of logic operators $\bigvee$ (**OR**), $\bigwedge$ (**AND**), and $\neg$ (**NOT**). The possible values 0 and 1 define the possible states of variables s_i. We calculate the dynamics synchronously. Evaluation of all Boolean functions F_i on variables s_i occurs at the same time.

An n-tuple S, of values 0/1, defines a state of the system. The state space can be described as a collection of all possible n-tuples of 0/1 values, corresponding to all possible states in which the system can be found. A dynamics of the system is a succession of states. If we have a state **a**, applying a transition state to **a** that gives us **b**, that is, $F(\mathbf{a}) = \mathbf{b}$, then we can represent states **a** and **b** as nodes and F as an edge, building a directed graph. A path in this directed graph is a representation of the dynamics of the Boolean network.

In a regulatory system like the *lac* operon, we can define variables as mRNA, proteins, and carbon sources, and the state s_i of each variable corresponds to 0, if s_i is ABSENT/INACTIVE, and corresponds to 1, if s_i is PRESENT/ACTIVE.

We can now define a Boolean network for the transcriptional regulation of *lac* operon described in section "Gene Transcriptional Regulation in Prokaryotes". We identify the following set of variables (this is one of representations presented in [15]):

M = *lac* mRNA
P = *lac* permease
L = internal lactose
C = CAP
R = repressor LacI
B = β-galactosidase
A = allolactose

The external glucose G_e and external lactose L_e are parameters of the model.

Fig. 4 Boolean functions for the lac operon represented as an interaction graph. Narrow ends represent activation, and blocked lines represent repression

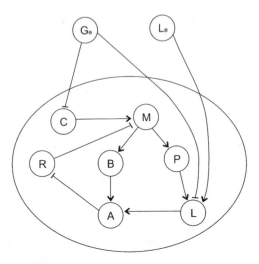

The derivation of Boolean functions follows directly from the knowledge about regulatory interactions. Here we describe Boolean functions for the *lac* operon network:

1. Boolean function for variable M: The mRNA of *lac* operon M will be PRESENT when CAP is PRESENT **AND** LacI repressor is ABSENT. The logic function reads $F_M = C$ **AND NOT** R.
2. Boolean function for L: Internal lactose L will be PRESENT when external lactose L_e is PRESENT **AND** permease P is PRESENT **AND** external glucose is ABSENT. This means $F_L = P$ **AND** L_e **AND NOT** G_e.
3. Boolean function for variable C: The CAP (C) is ACTIVE in the ABSENCE of glucose. Boolean function for C reads $F_C =$ **NOT** G_e and has exactly the opposite value of G_e.
4. Boolean function for R: The repressor LacI is ACTIVE when allolactose is ABSENT. Boolean functions reads $F_R =$ **NOT** A.
5. Both permease P and β-galactosidase B are PRESENT in the case where the transcript mRNA is also PRESENT. The corresponding Boolean functions are $F_B = M$ for β-galactosidase and $F_P = M$ for permease.
6. Boolean function for allolactose A: For allolactose to be PRESENT, we need internal lactose L PRESENT **AND** β-galactosidase B also PRESENT to metabolize L and generate A. Boolean functions read $F_A = L$ **AND** B.

We represent these Boolean functions as an interaction graph in Fig. 4.

The Boolean functions above can generate a dynamics on the state space. Any $S^0 = (s_M, s_P, s_B, s_C, s_R, s_A, s_L)$ can be chosen to be the initial state. Then we can apply the Boolean functions $F = (F_M, F_P, F_B, F_C, F_R, F_A, F_L)$ to S^0 to obtain the next state S^1. Application of F to S^0 is synchronous, and all variables s_i are updated at once. Iteration of this process generates a path on state space $P = (S^0, S^1, ..., S^m)$,

m being a positive integer. These dynamics can unveil important biological features of the system, as we will see below.

In [15], authors also make a comparison between the Boolean networks with other models represented by differential equations, and for this reason, they have extended the system representation, introducing the following three variables:

A_m = allolactose in medium-level concentration
L_m = internal lactose in medium-level concentration
R_m = repressor LacI in medium-level concentration

The idea is that external lactose, for instance, can occur in one of the three states: low (0), medium (1), or high (2). However, instead of changing the possible states of the system, they added a new variable L_m to keep only PRESENT (1) and ABSENT (0) as admissible state values.

A complete list of Boolean functions for this extended version of *lac* operon model is in [15]. Here we will only highlight their main result about dynamics, which is bistability. Bistability is a feature also observed with other modeling techniques [16, 17].

To describe bistability of *lac* operon dynamics, we consider what happens with all possible solutions obtained within state space.

For certain sets of parameter values which include external glucose $G_e = 1$ (glucose PRESENT), all possible paths in state space, which starts with any initial state S^0, eventually reaches a steady state corresponding to operon transcription OFF, that is, a state with the triplet (mRNA, permease, β-galactosidase) = (0, 0, 0).

For the case $G_e = 0$ (glucose ABSENT), final steady states depend on the values for external lactose L_e and external lactose in medium concentration L_{em}. We have the following sub-cases:

1. For external lactose ABSENT, $L_e = L_{em} = 0$, any possible initial state S^0 eventually reaches a steady state where the operon is OFF: (mRNA, permease, β-galactosidase) = (0, 0, 0).
2. For external lactose PRESENT, $L_e = L_{em} = 1$, any possible initial state S^0 eventually reaches a steady state where the operon is ON: (mRNA, permease, β-galactosidase) = (1, 1, 1).
3. For external lactose in high concentration ABSENT, $L_e = 0$, and external lactose in medium concentration PRESENT $L_{em} = 1$, there are two possible steady states: (mRNA, permease, β-galactosidase) = (0, 0, 0) and (mRNA, permease, β-galactosidase) = (1,1,1). Any possible initial state S^0 eventually reaches one of them.

Bistability is an important dynamical characteristic captured by the Boolean logic representation of *lac* operon described in [15], meaning that there is a set of model parameters for which the system can reach two distinct steady states, depending only on the choice of initial condition S^0.

Identification of multiple stable states in Boolean regulatory networks can lead to important results in larger scale problems. Transitions between two steady states of a regulatory network are associated with several biological processes such as cellular reprogramming and differentiation in eukaryotes [18].

3.3 Network Motifs

Regulatory networks can be mathematically represented by directed graphs. A node in such a graph represents a transcription unit, which can code for a transcription factor or a regulated operon. Arrows represent regulatory interactions. An arrow from X_i to Z_j means that X_i regulates the transcription of Z_j. In a genome-scale regulatory network, nodes and arrows are distributed in a particular manner. These distributions exhibit conspicuous characteristics such as the occurrence of regulatory motifs [19, 20]. Motifs are subgraphs with specific topologies and can be statistically overrepresented in the data, when compared with randomly distributed networks.

Randomization of a given a transcriptional regulation network (TRN) is made constructing a graph with the same number of nodes and edges, but with connections between nodes randomly distributed. We say a motif is overrepresented if its occurrence frequency in the TRN under study is significantly greater statistically than its occurrence in the equivalent randomized network. For *E. coli* TRN, there are three types of particularly overrepresented motifs [21]. There are feedforward loops (FFL), where a TF X_i regulates another TF Y_j and both regulate an operon Z_k; a second type of motifs is single input module (SIM) in which an autoregulated TF X_i regulates a set of operons $Z = \{Z_1, Z_2, ..., Z_n\}$; and there are dense overlapping regulons (DOR). In this case, a set of regulated operons $Z = \{Z_1, Z_2, ..., Z_m\}$ is regulated by a set of TFs $X = \{X_1, X_2, ..., X_n\}$ where each TF X_i regulates a subset of Z. They are highly connected subgraphs.

Motif identification described in [21] has important implications for the dynamics of *E. coli* TRN. A full description of motif repertoire, including simple autoregulatory interactions, is in [22].

The most common type of FFL is the coherent variation, having all regulatory interactions as activations. In this case, there is a TF X_i activating another TF Y_j, and both activate an operon Z_k. This motif acts as a filter in system dynamics, avoiding the activation of operon Z by a transient input signal. A signal for activation of operon Z_k must be persistent and last long as X_i activates the synthesis of Y_j, since Z_k requires activation from both X_i and Y_j to be fully transcribed.

A SIM motif has the effect of an orchestration of transcriptional regulation. A single TF X_i regulates a set of operons Z_k. This type of motif occurs when a group of operons code for products that will act coordinately as, for instance, as a protein complex. One example is flagella protein assembly.

There are multiple DOR motifs in *E. coli*. They are a compact form for processing multiple input signals by the TF layer $X = \{X_1, X_2, ..., X_n\}$, coordinately regulating their respective operons. They are associated to biological functions with ample extent, such as carbon utilization and anaerobic growth.

The occurrence of motifs involving regulatory circuits has a direct impact on dynamical properties of a network as a whole. It has been shown [23] that the presence of a feedback circuit is a sufficient condition for the occurrence of multiple stable states in a network. A model exhibiting multistability is a requirement to describe biological phenomena as cell differentiation and reprogramming [18]. Negative circuits also have their importance. Their occurrence is a sufficient condition for the presence of an attractive cycle in the network, and an attractive cycle is a property associated to system's homeostasis. A consequence of these results is that we can study dynamical properties of a network just identifying the presence of certain kinds of motifs, without solving the entire network dynamics.

4 Discussion

Boolean logic described in previous section is also useful for description of larger systems. This can give particularly interesting results when transcriptional regulation networks are coupled to other modeling aspects of the cell as metabolism and signaling pathways [24–26]. In [24], It is shown that these model couplings can reproduce many aspects of *E. coli* central metabolism. Many growth conditions are tested such as aerobic/anaerobic, diauxic shift, growth in presence/absence of amino acids, and complex medium. Growth curves can be reproduced in silico with Boolean modeling of transcriptional regulation. These models have further evolved for the concept of whole-cell model [27] that are capable of describing the behavior of a whole cell, integrating various distinct modeling paradigms as ordinary differential equations, Boolean logic, probabilistic, and constraint-based approach. In [27], authors integrate 28 cellular sub-models in order to build a whole-cell detailed model. Sub-models describe cellular processes such as DNA replication, damage, and repair; transcriptional regulation, RNA transcription, modification, and decay; translation, protein folding, modification, and decay; and many others. Pointing to this same direction, we have the book chapter [28], in this same volume, describing a strategy to build an integrated model for multidrug-resistant *Pseudomonas aeruginosa* CCBH4851, belonging to clone (ST277) endemic in Brazil. The objectives of this modeling effort are to study the mechanisms of multidrug resistance used by this strain and eventually identify new therapeutic targets for drug design.

There are an increasing number of transcriptional regulation networks [29–31], and the same is true for metabolic and signaling pathway reconstructions. Developing predictive models for these reconstructions can be of great value for the understanding biological phenomena in organisms as a whole.

In this chapter, we have described some introductory topics about the regulation of gene transcription and introduced some concepts about modeling process. We also described a Boolean logic model for the transcriptional regulation of the *lac* operon, discussing some results already presented in the literature. This text can be useful for those interested on having an overview of these subjects.

References

1. Hwang S, Kim CY, Ji SG, Go J, Kim H, Yang S, et al. Network-assisted investigation of virulence and antibiotic-resistance systems in *Pseudomonas aeruginosa*. Sci Rep. 2016;6:26223.
2. Jacob F, Monod J. Genetic regulatory mechanisms in the synthesis of proteins. J Mol Biol. 1961;3:318–56.
3. Watson JD, Crick FH. The structure of DNA. Cold Spring Harb Symp Quant Biol. 1953;18:123–31.
4. Kitano H. Systems biology: a brief overview. Science. 2002;295(5560):1662–4.
5. Leussler A, Van Ham P. Combinational systems. In: Thomas R, editor. Kinetic logic: a Boolean approach to the analysis of complex regulatory systems, vol. 29. Berlin: Springer Science & Business Media; 2013. p. 62–85.
6. Clancy S. DNA transcription. Nat Educ. 2008;1(1):41.
7. Zheng D, Constantinidou C, Hobman JL, Minchin SD. Identification of the CRP regulon using in vitro and in vivo transcriptional profiling. Nucleic Acids Res. 2004;32(19):5874–93.
8. Lister PD, Wolter DJ, Hanson ND. Antibacterial-resistant *Pseudomonas aeruginosa*: clinical impact and complex regulation of chromosomally encoded resistance mechanisms. Clin Microbiol Rev. 2009;22(4):582–610.
9. El Zowalaty ME, Al Thani AA, Webster TJ, El Zowalaty AE, Schweizer HP, Nasrallah GK, et al. *Pseudomonas aeruginosa*: arsenal of resistance mechanisms, decades of changing resistance profiles, and future antimicrobial therapies. Future Microbiol. 2015;10(10):1683–706.
10. Halford SE, Marko JF. How do site-specific DNA-binding proteins find their targets? Nucleic Acids Res. 2004;32(10):3040–52.
11. de Jong H. Modeling and simulation of genetic regulatory systems: a literature review. J Comput Biol. 2002;9(1):67–103.
12. Vijesh N, Chakrabarti SK, Sreekumar J. Modeling of gene regulatory networks: a review. J Biomed Sci Eng. 2013;6(02):223.
13. Le Novère N. Quantitative and logic modelling of molecular and gene networks. Nat Rev Genet. 2015;16(3):146–58.
14. Maraziotis IA, Dragomir A, Thanos D. Gene regulatory networks modelling using a dynamic evolutionary hybrid. BMC Bioinf. 2010;11:140.
15. Veliz-Cuba A, Stigler B. Boolean models can explain bistability in the *lac* operon. J Comput Biol. 2011;18(6):783–94.
16. Ozbudak EM, Thattai M, Lim HN, Shraiman BI, Van Oudenaarden A. Multistability in the lactose utilization network of *Escherichia coli*. Nature. 2004;427(6976):737–40.
17. Santillán M. Bistable behavior in a model of the lac operon in *Escherichia coli* with variable growth rate. Biophys J. 2008;94(6):2065–81.
18. Crespo I, Perumal TM, Jurkowski W, del Sol A. Detecting cellular reprogramming determinants by differential stability analysis of gene regulatory networks. BMC Syst Biol. 2013;7:140.
19. Mangan S, Alon U. Structure and function of the feed-forward loop network motif. Proc Natl Acad Sci U S A. 2003;100(21):11980–5.
20. Milo R, Shen-Orr S, Itzkovitz S, Kashtan N, Chklovskii D, Alon U. Network motifs: simple building blocks of complex networks. Science. 2002;298(5594):824–7.
21. Shen-Orr SS, Milo R, Mangan S, Alon U. Network motifs in the transcriptional regulation network of *Escherichia coli*. Nat Genet. 2002;31(1):64–8.
22. Alon U. Network motifs: theory and experimental approaches. Nat Rev Genet. 2007;8(6):450–61.
23. Remy E, Ruet P. From minimal signed circuits to the dynamics of Boolean regulatory networks. Bioinformatics. 2008;24(16):i220–i6.
24. Covert MW, Schilling CH, Palsson B. Regulation of gene expression in flux balance models of metabolism. J Theor Biol. 2001;213(1):73–88.

25. Covert MW, Xiao N, Chen TJ, Karr JR. Integrating metabolic, transcriptional regulatory and signal transduction models in *Escherichia coli*. Bioinformatics. 2008;24(18):2044–50.
26. Covert MW. Fundamentals of systems biology: from synthetic circuits to whole-cell models. Boca Raton: CRC Press; 2017.
27. Karr JR, Sanghvi JC, Macklin DN, Gutschow MV, Jacobs JM, Bolival B, et al. A whole-cell computational model predicts phenotype from genotype. Cell. 2012;150(2):389–401.
28. Silva FAB, Filho FM, Merigueti T, Giannini T, Brum R et al. Computational modeling of multidrug-resistant bacteria. In: Theoretical and applied aspects of systems biology. Springer; 2018.
29. Goelzer A, Bekkal Brikci F, Martin-Verstraete I, Noirot P, Bessières P, Aymerich S, et al. Reconstruction and analysis of the genetic and metabolic regulatory networks of the central metabolism of *Bacillus subtilis*. BMC Syst Biol. 2008;2:20.
30. Galán-Vásquez E, Luna B, Martínez-Antonio A. The regulatory network of *Pseudomonas aeruginosa*. Microb Inform Exp. 2011;1(1):3.
31. Ravcheev DA, Best AA, Sernova NV, Kazanov MD, Novichkov PS, Rodionov DA. Genomic reconstruction of transcriptional regulatory networks in lactic acid bacteria. BMC Genomics. 2013;14:94.

Cellular Reprogramming

Domenico Sgariglia, Alessandra Jordano Conforte, Luis Alfredo Vidal de Carvalho, Nicolas Carels, and Fabricio Alves Barbosa da Silva

Abstract With cellular reprogramming, it is possible to convert a cell from one phenotype to another without necessarily passing through a pluripotent state. This perspective is opening many interesting fields in the world of research and biomedical applications. This essay provides a concise description of the purpose of this technique, its evolution, mathematical models used, and applied methodologies. As examples, four areas in the biomedical field where cellular reprogramming can be applied with interesting perspectives are illustrated: diseases modeling, drug discovery, precision medicine, and regenerative medicine. Furthermore, the use of ordinary differential equations, Bayesian network, and Boolean network is described in these contexts. These strategies of mathematical modeling are the three main types that are applied in gene regulatory networks to analyze the dynamic interactions between their nodes. Ultimately, their application in disease research is discussed considering their benefits and limitations.

D. Sgariglia
Programa de Engenharia de Sistemas e Computação, COPPE-UFRJ, Rio de Janeiro, Brazil

A. J. Conforte · N. Carels
Laboratório de Modelagem de Sistemas Biológicos, Centro de Desenvolvimento Tecnológico em Saúde, Fundação Oswaldo Cruz, Rio de Janeiro, Brazil
e-mail: nicolas.carels@fiocruz.br

L. A. V. de Carvalho
Departamento de Medicina Preventiva, Faculdade de Medicina, UFRJ, Rio de Janeiro, Brazil
e-mail: luisalfredo@ufrj.br

F. A. B. da Silva (✉)
Laboratório de Modelagem Computacional de Sistemas Biológicos, Programa de Computação Científica, Fundação Oswaldo Cruz, Rio de Janeiro, Brazil
e-mail: fabricio.silva@fiocruz.br

© Springer International Publishing AG, part of Springer Nature 2018   41
F. A. B. da Silva et al. (eds.), *Theoretical and Applied Aspects of Systems Biology*,
Computational Biology 27, https://doi.org/10.1007/978-3-319-74974-7_3

1 Introduction

The concept of cellular reprogramming began in the 1960s with the idea of reversing the direction of cell differentiation, which was so far conceived only as occurring in a single irreversible direction. The differentiation of cellular state was schematically described through the Waddington epigenetic landscape [1, 2], where the metaphorical valleys represent states of cellular stability, and the hills around them represent the epigenetics barriers that prevent the transition from one state to another.

The goal of cellular reprogramming is to induce cells to overcome these barriers and move from one stable state (attractor) to another according to the simulations described in this chapter. Among the various scientific advances in this field, one may quote the work done by Takahashi and Yamanaka [3], concerning the generation of induced *pluripotent stem cells* (PSC), as an important reference in the progress of cellular reprogramming. The ability of a cell to reprogram itself from one attractor to another in the epigenetic landscape according to external and internal perturbations, or the overexpression of some key genes, has opened a huge field of investigation in the world of scientific research. Different strategies were followed with the aim of inducing phenotypic cell changes using the different mathematical and biological modeling techniques available.

Technological integration in different scientific areas such as biology, mathematics, statistics, and computational sciences is essential for the success in the simulation of cellular reprogramming. For this reason, the contribution of systems biology is determinant for the success of this emerging field.

This chapter first defines cellular reprogramming and its objective. Next, it provides a review of the methods used to achieve cellular reprogramming and the approaches to build the network models analyzed. Lastly, we discuss the applications of cellular reprogramming to diseases, highlighting the benefits and limitations of this technique and its potential application in different areas.

2 What Is Cellular Reprogramming?

2.1 Premise

We define cellular reprogramming as the conversion of one specific cell type to another one.

Eukaryote cells transit from one state to another through changes in gene expression and, consequently, protein levels in response to signals coming from the extracellular environment. The goal of cellular reprogramming is to artificially induce changes in a cell phenotype through perturbation of specific genes.

Until few years ago, cellular differentiation has long been thought of as "one-way traffic," without any possibility of returning to a previous cellular state. The idea that a cell could be induced to reverse its differentiated state toward a less specialized one was not even imagined.

The demonstration in 1963 [4] of cell dedifferentiation in culture of adult fibroblast through interaction with stem cells of a mouse teratocarcinoma [5] was a great step toward the concept that cellular differentiation is, indeed, reversible.

In 2006, Takahashi and Yamanaka induced PSCs from adult fibroblast cultures of mouse under the incubation with the transcriptional factors POU5F1, SOX2, KLF4, and MYC [3].

This remarkable discovery was a milestone for further advances and developments in the cellular reprogramming field. For the first time, it was shown to the scientific community that reversibility in the cell differentiation process was possible. Mature cells could be reverted to a previous pluripotent state, and it was possible to control the gene expression pattern with few transcription factors.

2.2 Meaning of Cellular Reprogramming

We begin with the mechanism of cell reprogramming by the definition of epigenetic given by Conrad Waddington (Fig. 1): "Epigenetic is the branch of biology that studies the causal interactions between genes and their products, which bring the phenotype into being" [6]. He conceived the epigenetic landscape as an inclined surface with a cascade of branches ridges, and valleys [1, 2, 7].

The goal of cellular reprogramming is to bring a cell (the ball of Fig. 1) from a valley of differentiation back to a state of pluripotency or to another differentiated state into a different valley passing a ridge.

Following the same logic, it becomes clear that inducing a cell to move from one specialized cell state to another without necessarily passing through the pluripotent state is also possible. Indeed, the transition from a differentiated state toward a progenitor state is referred to as *dedifferentiation*, while the transition between two differentiated states is called *transdifferentiation*.

Fig. 1 Waddington landscape representation of epigenetic space where the ball that can roll down from an undifferentiated cell state into a specialized state. The branches are the different potential states, and the ridges are the epigenetic barriers that prevent a cell from taking a different differentiation trajectory than the one in which it is already engaged

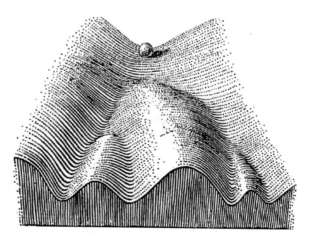

Keeping in mind the Waddington landscape representation described above, we might answer the following two questions:

(a) What are the barriers we must overcome to move from one cellular state to another?
(b) How can we induce cellular state transitions?

Answering the first question, we know that a stable cell state can be seen as a high-dimensional attractor of the gene regulatory network [8]. Attractors correspond to stable states associated with specific cell types [9].

In this context, cell fates are determined by gene expression and epigenetic patterns controlled by multiple factors [10], such as DNA methylation and histone modifications [11]. Both modifications can affect gene expression without inducing changes in DNA. DNA methylation involves the addition of methyl groups to the DNA molecule that usually results in the inhibition of eukaryotic gene transcription. Histone modifications are posttranslational processes that occur in the histone tails, which inhibit or induce local gene expression depending on the modification type [12].

After illustrating the role of the epigenetic activity that controls cellular states, the second question can be answered: How can we induce state transitions?

As outlined above, there are attractors corresponding to different cell fates and different epigenetic barriers that prevent transitions from one cell state to another. A stable cellular state is characterized by a given gene expression pattern. The perturbation of this pattern can induce cells to overcome these barriers by changing their steady state from one attractor to another in the epigenetic space [13]. This transition has the consequence of changing the cell phenotype.

As an example, we can cite the positive regulation of transcription factors responsible for the regulation of a gene expression pattern.

The scheme of Fig. 2 may represent both dedifferentiation and transdifferentiation processes. In general, we can think at epigenetic landscape as an energy configuration, where the cellular state is defined by the underlying transcriptional and epigenetic regulation [14].

2.3 Applications

Basically there are four main areas where cellular reprogramming are or could be applied in the biomedical research [15]:

(a) Disease modeling
(b) Drug discovery
(c) Precision medicine
(d) Regenerative medicine

With disease modeling (a), we may think about transforming a cell pathology into another desired cell condition, such as healthy, less aggressive phenotypes

PERTURBATION

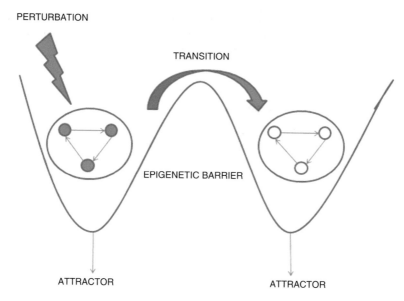

TRANSITION

EPIGENETIC BARRIER

ATTRACTOR   ATTRACTOR

Fig. 2 Schematic representation of the cellular transition from one attractor to another by overcoming an epigenetic barrier between two cell states as result of a specific perturbation

or even cell death. The benefit of this approach is to work with a human-specific representation that may not be available through cells coming from animal models. As an example, astrocytes dysfunction is related to several neurological and degenerative diseases, and their cellular reprogramming provides potential for the investigation of developmental and evolutionary features of the human brain. Exploring such potentialities, Dezonne et al. [16] successfully generated astrocytes from human cerebral organoids.

Concerning drug discovery (b), new drug targets can be inferred from a model representation and tested for cell reprogramming in vitro and in vivo before they reach clinical trials. For example, induced PSCs can be reprogramed into insulin-secreting pancreatic β cells, and their determinant genes could serve as targets for drug development. Also, induced PSCs from diabetes patients are being used to perform drug screening for new therapies against diabetes mellitus (DM) [17].

Precision medicine (c) aims to provide an individual treatment to patients and diseases. A key factor in this context is the pharmacogenomics that studies the influence of an individual's genetic characteristics in relation to its body's response to a drug. Succeeding in reprogramming a cell to a pluripotent state gives a chance to better understand the gentype-phenotype relationship at the individual level, which should allow the improvement of therapeutic efficacy [18].

Regenerative medicine (d) is the process of replacing, engineering, or regen-erating human cells, tissues, or organs to restore or establish normal function [19]. In therapies of cell replacement, the use of reprogrammed autologous cells can theoretically be a solution against the risk of graft rejection, due to cellular mismatch between the host and donor. In order to implement this idea in humans, nonhuman primates were studied regarding their potential to generate PSC cells through different cellular reprogramming techniques [20].

3 Reprogramming Methods

By *cell state*, one means the phenotype features of a cell as determined by the expression pattern of some of its key genes. Based on this definition, it is necessary to act on the expression of key genes to change a cell's phenotype features, which is the main purpose of cellular reprogramming. Consequently, one way to achieve such purpose is to modulate the regulation of the transcription factors that are responsible for the expression of those key genes. This method will be discussed below, together with other cellular reprogramming techniques that were also used [21].

3.1 Cellular Reprogramming Through the Overexpression of Transcription Factors

The discovery that it is possible to change cellular fate by overexpressing just four transcription factors [3] boosted the field of cellular reprogramming. After transfection, the cell was induced to a pluripotent state very much similar to that of embryonic stem cells; this similarity concerned morphology, phenotype, and epigenetics.

The switch from a somatic cell phenotype to induced PSCs through the modulation of transcription factor expression has an efficiency lower than 1% [22]. Once the genomic sequences of the original and reprogrammed cells are mostly identical, the reason for the low performance of cell reprogramming may be related to cell epigenetic factors, which indicates that induced PSCs have an epigenetic memory inherited from the previous cellular state [23].

Lineage reprogramming can also be obtained by cell reprogramming. As an example, Takahashi and Yamanaka [3] performed random gene integration at multiple DNA sites to obtain the overexpression of Oct4, Sox2, Klf4, and c-Myc transcription factors in adult fibroblasts, which caused their return to a pluripotent state. This transformation with retroviral vectors was performed for experimental purposes since the DNA integrates randomly at multiple sites and might promote the knockdown of essential genes and entail oncogenicity. To avoid such noxious risk, alternative transformation techniques were used, such as the combination of seven drug-like compounds that were able to generate iPSCs without the insertion of exogenous genes [24]. In addition to drug-like treatment, the repeated transfection of plasmids for transcriptional factor expression into mouse embryogenic fibroblast was also performed, but without any evidence of their genomic integration [25].

3.2 Somatic Cell Nuclear Transfer

Somatic cell nuclear transfer (SCNT) is a technique in which the nucleus of a donor somatic cell is transferred to another enucleated one called *egg cell*. After insertion, the somatic cell nucleus is reprogrammed by the egg cell. With this method it is possible to obtain embryonic stem cell (ESCs) [26] as well as to induce the differentiation of a cell phenotype into a different one [27].

3.3 Cell Fusion

It is possible to combine two nuclei within a same cell by the fusion of two cells. The dominant nucleus, the larger and more active one, imposes its pattern and consequently reprograms the somatic hybrid cell according to its dominant characteristics [28]. It is worth noting here that the cell fusing technique is not always efficient in achieving the desired result and the reprogramming is often incomplete.

4 Modeling Cellular Reprogramming

Reprogramming is obtained by resetting the regulation of gene expression in somatic cells, which depends on the knowledge of the key genes and proteins that may serve as target to induce this process, and the interactions between them.

The intracellular environment is continually subjected to stimuli from extracellular environment, such as nutrient availability, mechanical injury, cell competition, cooperation, etc. This type of stimulation affects the intracellular environment by changing the gene expression pattern in response to each stimulus. In this context, transcription factors are activated by external signals through transduction and promote the expression of specific genes and their respective pathways to set up a cellular response. This regulation process can be extended and include the induction of specific cell phenotypes.

Therefore, modeling the interaction between proteins in a living system and the transcription factors that regulate their expression is essential to carry out cellular reprogramming. As an approach to model such cellular systems, we may consider genes as variables and their activation state as "on" or "off." With these observations in mind, we may address some mathematical methodologies to represent the relationship between these state variables.

4.1 A Data-Oriented Approach

The development of new high-throughput technologies along with the growing amount of available data did promote computational frameworks based on protein interaction networks [29] integrated to different databases, such as (i) FANTOM consortium [30], which contains data on promoter characterization; (ii) STRING [31], which provides protein-protein interactions (PPI); and (iii) MARA (Motif Activity Response Analysis) [32], which provides interactions between proteins and DNA, to predict the reprogramming factors necessary to induce cell conversion.

In this context, Mogrify [29] is a predictive system that integrates gene expression data and regulatory network information. It searches for differentially expressed transcriptional factors that regulate most of the differentially expressed genes between two cell types. This methodology has been validated in vitro by inducing the transdifferentiations of dermal fibroblasts into keratinocytes and of keratinocytes into microvascular endothelial cells.

Basically, one may model a biological system through three different strategies (Fig. 3).

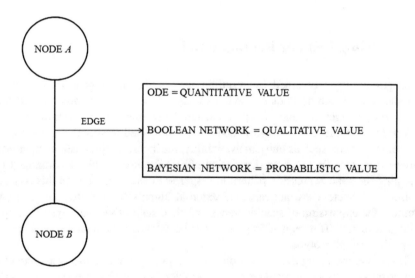

Fig. 3 Schematic representation showing the interpretation of an edge between two nodes by three different modeling methods: (i) ODE gives a quantitative description of the state of the connection by differential equation modeling, (ii) Boolean network gives a qualitative interpretation in terms of a connection being activated or not, and (iii) Bayesian network gives a probabilistic assessment of the connection state

4.2 Ordinary Differential Equation

In the context of a gene regulatory network, ordinary differential equations (ODEs) are used to describe the existing quantitative relationship between variables, i.e., nodes [33]. Theoretically, the use of ODE can provide a very accurate description of the existing interactions between system elements. In practice, the use of this technique, especially in complex networks, is difficult due to the high number of data and parameters involved in the process. The differential equation (formula 1) for each variable in the network is

$$\frac{dx_i}{dt} = f_i\left(x_{i_1}, x_{i_2}, \ldots, x_{i_l}\right) \tag{1}$$

where the right side of the equation represent all variable function linked to the gene x_i, and the left side is the variation in the gene x_i expression.

ODE can be used to model cellular reprogramming by determining the rate of change of a given substance concentration within the cell that determines a precise cellular state in response to some kind of cellular perturbation. For example, Mitra et al. [34] used ordinary differential equations to prove that time delays from chemical reactions are of crucial importance to understand cell differentiation and that it allows the introduction of a new system regime between two admissible steady states with sustained oscillations due to feedback loops in gene regulation circuits.

4.3 Bayesian Network

Bayesian network is an example of network analysis that takes into consideration the random behavior inherent to biological networks. Bayesian networks are acyclic graphs $G = (X, E)$, where X represents the network nodes and E the directed edges that represent the probabilistic relationship dependence between nodes. The relationship between the network's nodes is regulated by a conditional probability distribution (formula 2):

$$P\left(x_i \mid \text{Pa}\left(x_i\right)\right) \tag{2}$$

where $P_a(x_i)$ represent the antecessor nodes of the node x_i. A Bayesian network is a representation of a join probability distribution (formula 3):

$$P\left(x_1, x_2, \ldots, x_n\right) = \prod_{i=1}^{n} P\left(x_1 \mid \text{Pa}\left(x_1\right)\right) \tag{3}$$

It allows an intuitive visualization of the network conditional structural dependences between variables [35].

Bayesian networks that model sequences of variables varying over time are called *dynamical Bayesian networks* (DBNs). As proposed above, one may consider each protein in the network as being active or inactive. In this context, DBN allows the inference of the likelihood of each network node state, which is necessary to calculate the probability of each cell state [36] (an essential feature of cellular reprogramming). As an example, Chang et al. [36] established a cell-state landscape that allowed the search for optimal reprogramming combinations in human embryogenic stem cell (hESC) through the use of DBN.

4.4 Boolean Network

An alternative to differential equations and Bayesian network to describe variables' relationships in a gene regulatory network is the use of Boolean network. It is a qualitative dynamical model, describing a system change over time, which each network node being either "on" or "off." Its representation of the system is easier to derive that the one based on ordinary differential equations, since it does not require the inference of kinetic parameters and, consequently, it can process gene networks with a higher number of nodes.

A Boolean network is a directed graph $G(X,E)$ where X represent the nodes of the network and E are the edges between them. The vector of formula (4)

$$S(t) = (x_i(t), x_2(t), \ldots, x_n(t)) \tag{4}$$

describes the state of the network at any given time. The Boolean value, 1 or 0, of a node represents the state "on" or "off" of the gene considered, i.e., active or inactive, respectively.

The Boolean model is suitable to represent the evolution of biological systems over time and is relatively simple to implement and interpret. The greatest limitation of this type of network is that the state, 0 or 1, of a node is just an approximation of the reality. The state updating of all nodes across the entire system can be synchronous, asynchronous, or probabilistic depending on the modeling purpose and parameter's availability [37].

5 Cellular Reprogramming Using a Boolean Network

To address the problem of cellular reprogramming using the Boolean network in practice, one may use a modeling strategy of *gene regulatory network* (GRN) that warrants a relative simplicity in finding attractors. It should be noticed, however, that detailed information on the interactions within the elements of the network is not taken into account by this approach, since kinetic parameters or affinity terms may take different values according to different components of the network.

As seen above, gene interactions can be modeled based on the knowledge of the relationships between the genes of a set that should be modulated, activated, or inactivated, to achieve cellular reprogramming. Therefore, it is crucial to identify specific transcription factors that regulate these genes in order to enable a cell to perform a transition between its actual state and the wanted state.

Different cell types are defined as stable states, and a stable steady state is called an attractor. An attractor is characterized by a gene expression pattern that is specific of that attractor and whose perturbation can induce a transition from a stable cellular state to another [38]. It was shown that the number of genes to be modulated to reach attractor reprograming is relatively low, compared to the high number of genes differently expressed between two different cellular states [39].

Considering that the complexity of a gene regulatory network increases together with its number of nodes and that a phenotypic transition requires a low number of genes to be perturbed [40], different strategies are being used to reduce the number of network nodes to be analyzed. An iterative network pruning can be used to contextualize the network to the biological condition under which the expression data were obtained [41]. Pruning algorithms compare lists of genes and interactions from literature-based network with lists of genes differentially expressed from a bench experiment in two cellular phenotypes and then search for compatibility between both data sets. This comparison produces a score for each sample of pruned network in order to identify the genes to be perturbed according to the data pair that best matches the cell steady state regarded as a phenotype.

The topological relationship between the elements of a specific attractor in a network can be used to construct a protocol of cell reprogramming [40]. Based on data of topological configuration, it is possible to establish a hierarchical organization of *strongly connected components* (SCC), identify their respective *differentially expressed positive circuits* (DEPCs), and identify determinant genes able of promoting the transition from one stable cellular state to another.

The choice of genes to be perturbed can also be done based on dynamic simulation [38] through the combination of transcriptomic profiling and analyses of network stability in order to find the minimum number of DEPCs that needs to be perturbed to complete cellular transition.

6 Application of Cellular Reprogramming to Disease Control

All human diseases are intrinsic multifactorial and characterized by dysregulated processes in gene regulatory networks. The knowledge of GRN is important to understand how a molecular network robustness may lead malignant cells to overcome the inactivation of single protein targets by therapeutic treatment through alternative pathways or network propagation until a system accustoms to a new equilibrium [42]. Thus, network pharmacology and cellular reprogramming are promising methods for the identification of protein combinations with potential to disarticulate a key subnetwork that correlates with a disease and achieve an efficient therapeutic result [40, 43].

A very common problem is the bias in the modeling representation induced by reference to well-known pathways already described for the disease and the use of generic models that do not consider the specific features of the cell or tissue under consideration. The methods described in the previous section overcome this problem through the integration of gene expression data and regulatory networks, which allows the reconstruction of a network specific to the case under consideration. This specific network is more accurate, indicates specific aspects of the diseased cell or tissue, and may indicate genes related to dysregulated pathways responsible for the disease development [29, 38, 40].

The use of gene expression data from both ill and healthy cells is also important to identify the differentially expressed genes and target the ones preferentially expressed in ill cells in order to minimize the negative side effects of target inactivation to healthy cells.

The Mogrify methodology [29] considers all these features. However, it potentially may cause two types of negative effects if applied to patients in the context of a therapeutic treatment. First, with this methodology, one searches for differentially expressed transcription factors responsible for the regulation of genes related to the establishment of the disease phenotype. The problem is that transcription factors might be responsible for the regulation of hundreds of genes, and probably they are not all significantly more expressed in ill than in healthy cells. The perturbation of hundreds of genes, even if they are mostly differentially expressed in disease cells, may affect genes that are essential to cell maintenance and cause serious side effects. Second, this methodology requires the induction of gene expression through cell transfection. As already discussed above, the insertion of a plasmid into DNA occurs randomly and might knockdown some key genes, which increases the risk of oncogenicity. The most common approach applied in patients is the inhibition of a protein target with drugs. Even new innovative alternative patient therapies based on biopharmaceuticals as RNA interference, aptamer, peptides, or antibodies also target proteins with the aim to inactivate their function [44].

These limitations need to be considered when applying cellular reprogramming strategies in a disease context because they may exclude a number of possible alternative solutions. Once attractors for cell reprogramming have been considered, it is important to emphasize that focusing on the full reprogramming of a cell in order to reach a given steady state is not necessary. All stable attractors have a basin of attraction, in which trajectories spontaneously converge to the steady-state attractor [43]. The concept of basin of attraction should simplify the application of cellular reprogramming in diseases, since it reduces the number of required perturbations needed to achieve the desired stable state.

The perturbation capable of overcoming an epigenetic barrier and bringing a cell from a disease attractor to another desired one considered to match a healthy, or at least a less aggressive, condition for the patient needs to be carried out in a subspace where therapeutic options overlap with the basin of attraction.

As examples, we now propose putative applications of cellular reprogramming in two different diseases, cancer (cell disease) and malaria (infection disease).

Cancer cells accumulate malignant mutations during their development and, as result, present a different network topology if compared to healthy cells [45]. Due to mutations accumulation and its consequences on genome dysregulation, it would be impossible to control a cell in order to bring it back from its malignant attractor toward its healthy one. However, the key genes involved in the malignant attractor can be analyzed at the light of malignant features, such as continuous proliferation and escape from apoptosis or cell death. In addition, both malignant and healthy conditions can be analyzed in terms of differences according to their attractor phenotypes. This would allow the identification of key genes able to reprogram dysregulated cellular processes and achieve proliferation control and/or the induction of malignant cells to apoptosis.

The vaccines used against malaria uses live attenuated *salivary gland sporozoites* (SPZ) [46], and cannot be produced in large scale due to hurdles associated with SPZ obtainment. It is known that SPZ development occurs following three main stages according to the mosquito organs that are infected: midgut, hemolymph, and/or salivary gland. Therefore, if considering the salivary gland tissue, the cellular reprogramming analysis should allow the identification of key genes related to this tissue by comparison to the others two stages. The understanding of salivary gland SPZ genesis and maturation is crucial to develop a culture system in laboratory and produce SPZs in vitro for large-scale vaccine production.

Many advances were already made toward cell reprogramming, and it is effective for a number of purposes. However, much still need to be done in regard to diseases and patient treatment. A clear example is that, unfortunately, an efficient general method for identifying basins of attraction is still lacking [42].

7 Conclusion

The concept of cell reprogramming has evolved a lot during the last decade. The development of high-throughput technologies has also promoted more accurate applications of cell reprogramming through its integration with gene expression data. Currently, there is a great perspective of its application in multiple biomedical areas, such as drug screening and regenerative medicine. Nevertheless, there is still much to do in order to understand and predict the behavior of complex systems such as the biological ones.

Acknowledgment This study was supported by fellowships from *CAPES* to D.S. and from the Oswaldo Cruz Institute (https://pgbcs.ioc.fiocruz.br/) to A.C.

References

1. Waddington C. The strategy of the genes: a discussion of some aspects of theoretical biology. London: George Allen and Unwin; 1957. 262 pp
2. Waddington C. Organisers and genes. Cambridge: Cambridge University Press; 1940.
3. Takahashi K, Yamanaka S. Induction of pluripotent stem cells from mouse embryonic and adult fibroblast cultures by defined factors. Cell. 2006;126(4):663–76.
4. Siminovitch L, McCulloch EA, Till JE. The distribution of colony-forming cells among spleen colonies. J Cell Comp Physiol. 1963;62(3):327–36.
5. Martin GR. Isolation of a pluripotent cell line from early mouse embryos cultured in medium conditioned by teratocarcinoma stem cells. Proc Natl Acad Sci U S A. 1981;78(12):7634–8.
6. Waddington CH. Towards a theoretical biology. Nature. 1968;218(5141):525–7.
7. Waddington CH. An introduction to modern genetics. New York: The Macmillan Company; 1939.
8. Huang S, Eichler G, Bar-yam Y, Ingber DE. Cell fates as high-dimensional attractor states of a complex gene regulatory network. Phys Rev Lett. 2005;94(12):128701.
9. Huang S, Ernberg I, Kauffman S. Cancer attractors: a systems view of tumors from a gene network dynamics and developmental perspective. Semin Cell Dev Biol. 2009;20(7):869–76.
10. Lang AH, Li H, Collins JJ, Mehta P. Epigenetic landscapes explain partially reprogrammed cells and identify key reprogramming genes. Morozov AV., editor. PLoS Comput Biol. 2014;10(8):e1003734.
11. Seah Y, EL Farran C, Warrier T, Xu J, Loh Y-H. Induced pluripotency and gene editing in disease modelling: perspectives and challenges. Int J Mol Sci. 2015;16(12):28614–34.
12. Goldberg AD, Allis CD, Bernstein E. Epigenetics: a landscape takes shape. Cell. 2007;128(4):635–8.
13. Ding S, Wang W. Recipes and mechanisms of cellular reprogramming: a case study on budding yeast *Saccharomyces cerevisiae*. BMC Syst Biol. 2011;5(1):50.
14. del Sol A, Buckley NJ. Concise review: a population shift view of cellular reprogramming. Stem Cells. 2014;32(6):1367–72.
15. Mall M, Wernig M. The novel tool of cell reprogramming for applications in molecular medicine. J Mol Med. 2017;95(7):695–703.
16. Dezonne RS, Sartore RC, Nascimento JM, Saia-Cereda VM, Romão LF, Alves-Leon SV, et al. Derivation of functional human astrocytes from cerebral organoids. Sci Rep. 2017;7:45091.
17. Kawser Hossain M, Abdal Dayem A, Han J, Kumar Saha S, Yang G-M, Choi HY, et al. Recent advances in disease modeling and drug discovery for diabetes mellitus using induced pluripotent stem cells. Int J Mol Sci. 2016;17(2):256.
18. Hamazaki T, El Rouby N, Fredette NC, Santostefano KE, Terada N. Concise review: induced pluripotent stem cell research in the era of precision medicine. Stem Cells. 2017;35(3):545–50.
19. Mason C, Dunnill P. A brief definition of regenerative medicine. Regen Med. 2008;3(1):1–5.
20. Hemmi JJ, Mishra A, Hornsby PJ. Overcoming barriers to reprogramming and differentiation in nonhuman primate induced pluripotent stem cells. Primate Biol. 2017;4(2):153–62.
21. Halley-Stott RP, Pasque V, Gurdon JB. Nuclear reprogramming. Development. 2013;140(12):2468–71.
22. Takahashi K. Cellular reprogramming. Cold Spring Harb Perspect Biol. 2014;6(2):a018606.
23. D'urso A, Brickner J. Mechanisms of epigenetic memory. Trends Genet. 2014;30(6):230–6.
24. Hou P, Li Y, Zhang X, Liu C, Guan J, Li H, et al. Pluripotent stem cells induced from mouse somatic cells by small-molecule compounds. Science. 2013;341(6146):651–4.
25. Okita K, Nakagawa M, Hyenjong H, Ichisaka T, Yamanaka S. Generation of mouse induced pluripotent stem cells without viral vectors. Science. 2008;322(5903):949–53.
26. Byrne JA, Pedersen DA, Clepper LL, Nelson M, Sanger WG, Gokhale S, et al. Producing primate embryonic stem cells by somatic cell nuclear transfer. Nature. 2007;450(7169): 497–502.

27. Wakayama T, Tabar V, Rodriguez I, Perry AC, Studer L, Mombaerts P. Differentiation of embryonic stem cell lines generated from adult somatic cells by nuclear transfer. Science. 2001;292(5517):740–3.
28. Yamanaka S, Blau HM. Nuclear reprogramming to a pluripotent state by three approaches. Nature. 2010;465(7299):704–12.
29. Rackham OJL, Firas J, Fang H, Oates ME, Holmes ML, Knaupp AS, et al. A predictive computational framework for direct reprogramming between human cell types. Nat Genet. 2016;48(3):331–5.
30. Forrest ARR, Kawaji H, Rehli M, Kenneth Baillie J, de Hoon MJL, Haberle V, et al. A promoter-level mammalian expression atlas. Nature. 2014;507(7493):462–70.
31. Franceschini A, Szklarczyk D, Frankild S, Kuhn M, Simonovic M, Roth A, et al. STRING v9.1: protein-protein interaction networks, with increased coverage and integration. Nucleic Acids Res. 2012;41(D1):D808–15.
32. Suzuki H, Forrest ARR, van Nimwegen E, Daub CO, Balwierz PJ, Irvine KM, et al. The transcriptional network that controls growth arrest and differentiation in a human myeloid leukemia cell line. Nat Genet. 2009;41(5):553–62.
33. Cao J, Qi X, Zhao H. Modeling gene regulation networks using ordinary differential equations. Methods Mol Biol. 2012:185–97.
34. Mitra MK, Taylor PR, Hutchison CJ, McLeish TCB, Chakrabarti B. Delayed self-regulation and time-dependent chemical drive leads to novel states in epigenetic landscapes. J R Soc Interface. 2014;11(100):20140706.
35. Friedman N, Linial M, Nachman I, Pe'er D. Using Bayesian networks to analyze expression data. J Comput Biol. 2000;74(3):601–20.
36. Chang R, Shoemaker R, Wang W. Systematic search for recipes to generate induced pluripotent stem cells. PLOS Comput Biol Publ Libr of Sci. 2011;7(12):e1002300.
37. Xiao Y. A tutorial on analysis and simulation of Boolean gene regulatory network models. Curr Genomics. 2009;10(7):511–25.
38. Crespo I, Perumal TM, Jurkowski W, del Sol A. Detecting cellular reprogramming determinants by differential stability analysis of gene regulatory networks. BMC Syst Biol. 2013;7(1):140.
39. Lukk M, Kapushesky M, Nikkilä J, Parkinson H, Goncalves A, Huber W, et al. A global map of human gene expression. Nat Biotechnol. 2010;28(4):322–4.
40. Crespo I, del Sol A. A general strategy for cellular reprogramming: the importance of transcription factor cross-repression. Stem Cells. 2013;31(10):2127–35.
41. Crespo I, Krishna A, Le Bechec A, del Sol A. Predicting missing expression values in gene regulatory networks using a discrete logic modeling optimization guided by network stable states. Nucleic Acids Res. 2013;41(1):e8.
42. Cornelius SP, Kath WL, Motter AE. Realistic control of network dynamics. Nat Commun Nat Publ Group. 2013;4:1942.
43. Zickenrott S, Angarica VE, Upadhyaya BB, Sol A. Prediction of disease – gene – drug relationships following a differential network analysis. Cell Death Dis Nat Publ Group. 2016;7(1):e2040–12.
44. Tabernero J, Shapiro GI, LoRusso PM, Cervantes A, Schwartz GK, Weiss GJ, et al. First-in-humans trial of an RNA interference therapeutic targeting VEGF and KSP in cancer patients with liver involvement. Cancer Discov. 2013;3(4):406–17.
45. Jonsson PF, Bates PA. Global topological features of cancer proteins in the human interactome. Bioinformatics. 2006;22(18):2291–7.
46. Phillips MA, Burrows JN, Manyando C, van Huijsduijnen RH, Van Voorhis WC, TNC W. Malaria. Nat Rev Dis Prim. Macmillan Publishers Limited. 2017;3:17050.

Metabolic Models: From DNA
to Physiology (and Back)

Marcio Argollo de Menezes

Abstract Metabolic reconstructions constitute translations from genomic data to biochemical processes and serve as valuable tools to assess, along with mathematical models, the viability of organisms on different environments or the overproduction of industrially valuable metabolites following controlled manipulation of specific reaction rates. In the following, we review FBA, a constraint-based mathematical method which successfully predicts genome-wide metabolic fluxes, most notably the rate of accumulation of biomass precursors with stoichiometry determined by the cellular biomass composition. The practical implementation of the method on a synthetic metabolic model is offered as computer codes written for GNU-Octave, an open-source language with powerful numerical tools.

Systems biology is an emerging research field which integrates information from very distinct, well-established areas to deal with the (rather puzzling) question, "What is life and what underlies its agency?" [30], so that the innumerous molecular structures and procedures encoded therein can be exploited for diverse purposes [10, 22, 29], from drug design and crop yield optimization [12, 27] to tissue remodelling and winemaking [15, 28].

This chapter is a practical introduction to metabolic modelling, where one tests the flux capabilities of a given map of metabolic reactions with a mathematical representation of the map and computational techniques that solve numerical problems associated with tests of hypotheses, inference of behaviors, and generation of predictions [14, 22]. In the first section, I introduce the basic ideas and fundamental biological discoveries behind metabolic modelling, from map design to mathematical modelling. Next, I will describe the mathematical formulation and practical implementation of flux balance analysis (FBA) [25], one of the most successful computational techniques for organism-wide prediction of metabolic reaction fluxes and the effects of their modulation. We will find the maximal growth rate and determine the essentiality of genes/reactions for non-zero biomass

M. A. de Menezes (✉)
Instituto de Física, Universidade Federal Fluminense (UFF), Niterói, Brazil
e-mail: marcio@mail.if.uff.br

© Springer International Publishing AG, part of Springer Nature 2018   57
F. A. B. da Silva et al. (eds.), *Theoretical and Applied Aspects of Systems Biology*,
Computational Biology 27, https://doi.org/10.1007/978-3-319-74974-7_4

production [9, 12, 33, 36, 40] on an artificial metabolic model, with reactions derived from genes of a model cell and a pseudoreaction describing accumulation of biomass precursors. Numerical solutions will be implemented on *GNU Octave*, an open-source scripting language with many libraries for numerical calculus [6].

1 Metabolic Models

Advances on sequencing techniques and bioinformatics algorithms made it possible to reconstruct, from genomic sequence, the entire set of biochemical reactions and transport processes available to an organism [2, 5, 16, 24, 35].[1]

With this information one can build topological maps akin to metabolic pathways, where generation of a selected set of products, given a particular set of available substrates, is written in terms of reactions whose integrated interconversion of metabolites link the desired products to elements in the set of available substrates.

One can investigate, for instance, growth capacity in different environments, one of the central hypotheses behind the idea of life as self-replicating, autocatalytic sets [11, 34]: a cell must accumulate, from a basic food source, the set of metabolites which constitute its physical structure in amounts defined by the cellular composition, which is further (self) organized into a new (identical) cell [19]. This process can be incorporated in metabolic models in the form of a pseudoreaction (usually called biomass production reaction), where biomass precursor metabolites are the substrates with stoichiometric indices defined by their relative amounts in cellular composition [8].

Given enough precision in the reconstruction process [39], the physiological strategy for biomass generation, that is, the pathway design of metabolic reaction fluxes experimentally observed in living organisms [20, 38] should be one of the possible designs from the reconstructed map [8]. It is common sense to suppose that, under competition, evolution drives organisms toward maximization of fitness [31], which in simple prokaryotic organisms translates almost entirely to growth rate. This hypothesis is prone to be answered by mathematical and computational modelling [14, 26], which are very helpful tools, accelerating discovery and generating reproducible basic knowledge from biologically inspired hypotheses. Constraint-based analysis [3] is a mathematical methodology in which flux of metabolic reactions are predicted by systematic narrowing of the search space by the addition of biologically inspired constraints, the most notable being flux balance in freely dividing cells: when cells evolve with constant duplication time (constant growth rate), as expected in nutrient-rich media, their molecular composition remains unchanged after duplication. This steady state, easily reproducible in chemostats [21, 41], constrains metabolic reaction fluxes to values leading to balanced synthesis and consumption rates of every intracellular metabolite [17–19].

---

[1]Check http://systemsbiology.ucsd.edu/InSilicoOrganisms/OtherOrganisms or https://www.ebi.ac. uk/biomodels-main/ for an updated list.

Based on the premise that prokaryotes such as *Escherichia coli* have maximized their growth performance along evolution, flux balance analysis (FBA) [25] is used to predict the expected (physiological) metabolic phenotype of bacteria evolving in rich media (with constant growth rate) as the one, among all flux sets satisfying the stationarity constraint in the metabolic reconstruction, with the largest rate of biomass production (as noted by J. Monod [17], physical limits on uptake rates constrain growth rates to finite values). With simple mathematical formulation [26], FBA successfully connects cell's physiology to the capabilities of itś underlying metabolic network given flux constraints imposed by environmental nutrient composition and cellular state [7, 13, 23, 32].

2   FBA: Predicting Metabolic Phenotypes

Let's discuss the basic ideas behind FBA with its practical implementation on an artificial metabolic model containing $M = 7$ metabolites and $N = 12$ reactions (Fig. 1). Eleven reactions are gene-related (five occurring inside the cell and six transporting metabolites through cell boundary), and one is a pseudoreaction ($R12$, not gene-related) describing the accumulation of biomass precursor metabolites. Reversible reactions are split into the actual direct and reverse processes, and reaction fluxes are all positive-definite.

When reactions occur simultaneously in the intracellular medium, metabolite concentrations change in time as result of the difference between their rates of synthesis and consumption in all participating reactions. Writing the flux rate of reaction j as f_j and the stoichiometry of metabolite i in reaction j as S_{ij} (zero if not present in the reaction, negative if substrate and positive otherwise), the concentration of metabolite i evolves in time as

$$\frac{d\,[m_i]}{dt} = \sum_{j=1}^{N} S_{ij} f_j. \tag{1}$$

where $N = 12$ is the number of reactions and $M = 7$ the number of metabolites in the metabolic model. The stoichiometric matrix carries all the information contained

Fig. 1 Metabolic model from hypothetical cell, viewed as a list of reactions. Metabolites marked in red on the reactions list do not occur inside the cell

(R1,g1) m1_e → m1
(R2,g1) m1 → m1_e
(R3,g2) m2_e → m2
(R4,g2) m2 → m2_e
(R5,g3) m3_e → m3
(R6,g4) m4 → m4_e
(R7,g5) 2 m2+ m3 → m1 + m4
(R8,g6) m1+3 m3 → 2 m5
(R9,g7) m5 + m2 → m6
(R10,g7) m6 → m5 + m2
(R11,g8) m5 + 2 m6 → m7 + m1
(R12, ø) 0.3 m5 +0.54 m6 +0.16 m7 → Cell biomass

a)

$$\frac{d[m1]}{dt} = (1)f_1 + (-1)f_2 + (1)f_7 + (-1)f_8 + (1)f_{11}$$

$$\frac{d[m2]}{dt} = (1)f_3 + (-1)f_4 + (-2)f_7 + (-1)f_9 + (1)f_{10}$$

$$\frac{d[m3]}{dt} = (1)f_5 + (-1)f_7 + (-3)f_8$$

$$\frac{d[m4]}{dt} = (-1)f_6 + (1)f_7$$

$$\frac{d[m5]}{dt} = (2)f_8 + (-1)f_9 + (-1)f_{11} + (-0.3)f_{12}$$

$$\frac{d[m6]}{dt} = (1)f_9 + (-1)f_{10} + (-2)f_{11} + (-0.54)f_{12}$$

$$\frac{d[m7]}{dt} = (1)f_{11} + (-0.16)f_{12}$$

b)

$$\frac{d[mi]}{dt} = \sum_{j=1}^{N} S_{ij} f_j$$

$$S = \begin{pmatrix} 1 & -1 & 0 & 0 & 0 & 0 & 1 & -1 & 0 & 0 & 1 & 0 \\ 0 & 0 & 1 & -1 & 0 & 0 & -2 & 0 & -1 & 1 & 0 & 0 \\ 0 & 0 & 0 & 0 & 1 & 0 & -1 & -3 & 0 & 0 & 0 & 0 \\ 0 & 0 & 0 & 0 & 0 & -1 & 1 & 0 & 0 & 0 & 0 & 0 \\ 0 & 0 & 0 & 0 & 0 & 0 & 0 & 2 & -1 & 1 & -1 & -0.3 \\ 0 & 0 & 0 & 0 & 0 & 0 & 0 & 0 & 1 & -1 & -2 & -0.54 \\ 0 & 0 & 0 & 0 & 0 & 0 & 0 & 0 & 0 & 0 & 1 & -0.16 \end{pmatrix}$$

Fig. 2 Dynamics of metabolite concentrations, written in function of reaction fluxes

on the list of reactions and is depicted in Fig. 2, along with the differential equations describing time evolution of concentrations in our metabolic model.

Genome-wide metabolic reconstructions usually have hundreds to thousands of metabolites and reactions, and the solution of this set of coupled differential equations becomes unpractical. Nevertheless, in organisms growing in the stationary phase intracellular metabolite concentrations do not change with time, reducing Eq. 1 to

$$\sum_{j=1}^{N} S_{ij} f_j = 0 \quad \forall i \tag{2}$$

that can be written in a compact notation as

$$\mathbf{S}\vec{f} = \vec{0} \tag{3}$$

Equation 2 describes a set of coupled linear equations on fluxes, much simpler to solve than the set of coupled differential equations defined in (1). As there are more reactions than metabolites, multiple solutions exist for the problem [26], reflecting the multitude of strategies inscribed in metabolic networks.[2] This degeneracy can be lifted by the introduction of more constraints to solution. As stated in the introduction, prokaryotic cells growing in rich media should evolve toward maximization of growth rate. One can formulate this problem mathematically as

$$\text{MAX}\{f_{\text{bio}}\}$$

$$\text{Given} \sum_{j=1}^{N} S_{ij} f_j = 0 \quad \forall i \tag{4}$$

---

[2]Many organisms can, for instance, generate ATP either by respiration, fermentation, or both processes simultaneously [37].

which is a very popular problem in mathematics called linear programming [4] for which one finds public libraries implementing its solution with diverse algorithms. We choose to expose our examples in GNU Octave [6], an open-source scripting language with simple syntax and many libraries that solve a vast range of mathematical problems. It comes with an environment, where one can type commands that are interpreted on-the-fly.

2.1 Growth Prediction

To define the stoichiometric matrix of our metabolic model, just type, in the octave environment,

```
octave:1> A=[1 -1 0  0 0  0  1 -1  0  0  1  0;
             0  0 1 -1 0  0 -2  0 -1  1  0  0;
             0  0 0  0 1  0 -1 -3  0  0  0  0;
             0  0 0  0 0 -1  1  0  0  0  0  0;
             0  0 0  0 0  0  0  2 -1  1 -1 -0.3;
             0  0 0  0 0  0  0  0  1 -1 -2 -0.54;
             0  0 0  0 0  0  0  0  0  1 -0.16];
```

To find the maximum growth rate of our model cell, we use the linear programming library *glpk*. From the help function

```
octave:1> help glpk
 -- Function File: [XOPT, FMIN, ERRNUM, EXTRA] = glpk (C, A, B,
    LB, UB,
          CTYPE, VARTYPE, SENSE, PARAM)
    Solve a linear program using the GNU GLPK library.

    Given three arguments, 'glpk' solves the following
    standard LP:
         min C'*x
     subject to
         A*x  = b
          x >= 0

    Input arguments:
    C
          A column array containing the objective function
          coefficients.
    A
          A matrix containing the constraints coefficients.
    B
          A column array containing the right-hand side value
          for each constraint in the constraint matrix.
    LB
          An array containing the lower bound on each of the
          variables. If LB is not supplied, the default lower
          bound for the variables is zero.
```

UB

An array containing the upper bound on each of the
variables. If UB is not supplied, the default upper
bound is assumed to be infinite.

Since we want to maximize growth rate, the vector $\vec{C}$ must have a single non-zero element, $C[12]$, which we set to -1 to reflect the maximization of flux in the biomass production reaction $R12$.

```
octave:2> c = zeros(12,1);
octave:3> c(12)=-1;
```

The lower bound of all reactions is zero, and upper bounds, given no additional information on maximum reaction rates, are set to an arbitrary value (1 in our case).

```
octave:4> lb = zeros(12,1);
octave:5> ub = ones(12,1);
```

As no intracellular metabolite accumulates in time, all components of $\vec{B}$ are set to zero.

```
octave:6> B = zeros(7,1);
```

After setting all input parameters, *glpk* is evoked:

```
octave:7> [x0, FMIN, ERRNUM] = glpk(c,A,B,lb,ub);
```

If $ERRNUM = 0$, a valid solution is found, and the vector $\vec{X}0$ is returned with the respective reaction fluxes. In our case, $FMIN$ returns the growth rate. Typing the variable name in octave environment, one obtains its value

```
octave:8> ERRNUM
ERRNUM = 0
octave:9> x0
x0 =

   0.25253
   0.00000
   0.43434
   0.00000
   1.00000
  -0.00000
   0.00000
   0.33333
   0.43434
   0.00000
   0.08081
   0.50505
```

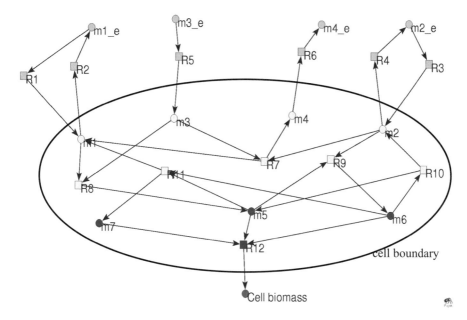

Fig. 3 Metabolic model from hypothetical cell viewed as a network. Metabolites marked in red on the reactions list do not occur inside the cell. In the network representation, metabolites and reactions are nodes (circles and squares, respectively), and directed links connect reaction substrates to their respective reactions and reactions to their products. Green nodes mark external metabolites, and transport reactions and biomass precursors are marked in blue

In order to visualize pathways involved in the strategy of optimal growth, we describe the metabolic model as a network (Fig. 3), with links connecting substrates to reactions and reactions to products. This representation evidences the molecular approach of physiology [19] in which growth is sustained by the uptake and sequential transformations of a small set of metabolites comprising the food source.

Since metabolite $m1$ is synthesized by an internal reaction, there should be another solution for the above problem given a medium without $m1$. In fact, if we set the upper bound of its uptake reaction to zero, we find another strategy for biomass generation (with smaller yield).

```
octave:10> ub(1,1)=0;
octave:11> [x0, FMIN, ERRNUM] = glpk(c,A,r,lb,ub);
octave:12> x0
x0 =

   0.00000
   0.00000
   0.75000
   0.00000
   1.00000
```

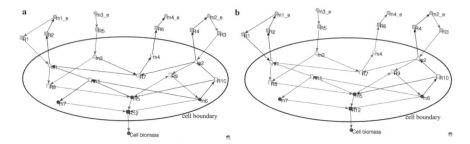

Fig. 4 Different strategies for biomass generation with metabolite $m1$ available as food source (left) and internally synthesized (right). (**a**) Pathway 1: Metabolite m1 available for uptake. (**b**) Pathway 2: Metabolite m1 not available for uptake

```
0.20161
0.20161
0.26613
0.34677
0.00000
0.06452
0.40323
```

The different pathways leading to biomass production are shown in Fig. 4, where active routes (links) are marked in red. In the left network, metabolite $m1$ is provided as a food source, while on the right network, it is not provided as a food source, but synthesized in the internal reaction $R7$ along with the by-product metabolite $m4$. This situation is analogous to the different pathways leading to ATP production in some organisms where the availability of oxygen in the environment determines whether respiration or fermentation takes place, with excretion of by-products evidencing the latter strategy [1].

2.2 Gene Essentiality

As stated previously, metabolic reactions are activated by the promotion of genes [1], and one can predict the essentiality of either reactions or genes for an organism's life by testing its capacity of producing biomass [33] in the metabolic reconstruction with upper bound for the selected reactions set to zero. Octave commands can be sequentially given to the interpreter as a script code. We let as a final exercise the interpretation of the code given below which output the critical genes which, when shut down, precludes biomass formation

```
marcio@sumbawa:~/cursos/fiocruz/redes metabolicas 2017/
book_chapter/codes$ ./fba1.m
######## Gene deletion studies ############################
Gene g2 [R3][R4] is critical
```

```
Gene g3 [R5] is critical
Gene g6 [R8] is critical
Gene g7 [R9][R10] is critical
Gene g8 [R11] is critical
marcio@sumbawa:~/cursos/fiocruz/redes metabolicas 2017/
book_chapter/codes$
```

2.3   Octave Code with FBA Analysis

```
#!/usr/bin/octave
printf("######## Metabolic reactions of model cell ########\n");
printf("# (R1, gene g1)        --> m1                       #\n");
printf("# (R2, gene g1)  m1 -->                             #\n");
printf("# (R3, gene g2)        --> m2                       #\n");
printf("# (R4, gene g2)  m2 -->                             #\n");
printf("# (R5, gene g3)        --> m3                       #\n");
printf("# (R6, gene g4)  m4 -->                             #\n");
printf("# (R7, gene g5)  2 m2 + m3 --> m1 + m4              #\n");
printf("# (R8, gene g6)  m1 + 3 m3 --> 2 m5                 #\n");
printf("# (R9, gene g7)  m5 + m2 --> m6                     #\n");
printf("# (R10, gene g7)  m6 --> m5 + m2                    #\n");
printf("# (R11, gene g8)  m5 + 2 m6 --> m7 + m1             #\n");
printf("# (R12, no gene)  0.3 m5 + 0.54 m6 + 0.16 m7 -->    #\n");
printf("###############################################\n");
printf("#R12 is a pseudo-reaction describing accumulation #\n");
printf("#of biomass precursors in proportions defined by  #\n");
printf("#cellular composition. Its flux mimics growth rate#\n");
printf("###############################################\n");

A=[1 -1 0  0 0  0  1 -1  0  0  1  0
   0  0 1 -1 0  0 -2  0 -1  1  0  0
   0  0 0 0  1  0 -1 -3  0  0  0  0
   0  0 0 0  0 -1  1  0  0  0  0  0
   0  0 0 0  0  0  0  2 -1  1 -1 -0.3
   0  0 0 0  0  0  0  0  1 -1 -2 -0.54
   0  0 0 0  0  0  0  0  0  1 -0.16];

biomass_reaction = 12;
a_uptake = 1;
b_uptake = 4;
c_uptake = 5;
d_excretion = 6;

[M,N]=size(A);
lb = zeros(N,1);
ub = ones(N,1);
r = zeros(M,1);
c = zeros(N,1);
c(biomass_reaction)=-1; # Maximization of biomass production
flux
```

```
[x0, FMIN, STATUS] = glpk(c,A,r,lb,ub);
if(STATUS==0)
  max_biomass=x0(biomass_reaction);
  printf("Maximum biomass production flux (pathway 1)=%f\n",x0
  (biomass_reaction));
  for i=1:N
    printf("Rxn %d flux=%f\n",i,x0(i));
  endfor
endif

ub(1,1)=0;
[x0, FMIN, STATUS] = glpk(c,A,r,lb,ub);
if(STATUS==0)
  max_biomass=x0(biomass_reaction);
  printf("Maximum biomass production flux (pathway 2)=%f\n",x0
  (biomass_reaction));
  for i=1:N
    printf("Rxn %d flux=%f\n",i,x0(i));
  endfor
endif

printf("######## Gene deletion studies #############\n");
# (R1, gene g1)      --> A
# (R2, gene g1)  A -->
# (R3, gene g2)      --> B
# (R4, gene g2)  B -->
# (R5, gene g3)      --> C
# (R6, gene g4)  D -->
# (R7, gene g5)  2B + C --> A + D
# (R8, gene g6)  A + 3C --> 2X
# (R9, gene g7)  X + B --> Y
# (R10, gene g7)  Y --> X + B
# (R11, gene g8) X + 2Y --> Z + A
ngenes=8;
nrxns_gene(1)=2;
nrxns_gene(2)=2;
nrxns_gene(3)=1;
nrxns_gene(4)=1;
nrxns_gene(5)=1;
nrxns_gene(6)=1;
nrxns_gene(7)=2;
nrxns_gene(8)=1;
gene_rxn(1,1)=1;
gene_rxn(1,2)=2;
gene_rxn(2,1)=3;
gene_rxn(2,2)=4;
gene_rxn(3,1)=5;
gene_rxn(4,1)=6;
gene_rxn(5,1)=7;
gene_rxn(6,1)=8;
gene_rxn(7,1)=9;
```

```
gene_rxn(7,2)=10;
gene_rxn(8,1)=11;


c = zeros(N,1);
c(biomass_reaction)=-1;
lb = zeros(N,1);
ub = ones(N,1);
for i=1:ngenes
  for j=1:nrxns_gene(i)
    k=gene_rxn(i,j);
    old_ub(k,1)=ub(k,1);
    old_lb(k,1)=lb(k,1);
    ub(k,1)=0;
    lb(k,1)=0;
  endfor
  [x0, FMIN, STATUS] = glpk(c,A,r,lb,ub);
  if(x0(biomass_reaction)<1e-6)
    printf("Gene g%d ",i);
    for j=1:nrxns_gene(i)
      k=gene_rxn(i,j);
      printf("[R%d]",k);
    endfor
    printf(" is critical\n");

  endif
  for j=1:nrxns_gene(i)
    k=gene_rxn(i,j);
    ub(k,1)=old_ub(k,1);
    lb(k,1)=old_lb(k,1);
  endfor
endfor
```

References

1. Alberts B, Johnson A, Lewis J, Morgan D, Raff M, Roberts K, Walter P. Molecular biology of the cell. 500 Tips. New York: Garland Science; 2014.
2. Bartell JA, Blazier AS, Yen P, Thgersen JC, Jelsbak L, Goldberg JB, Papin JA. Reconstruction of the metabolic network of Pseudomonas aeruginosa to interrogate virulence factor synthesis. Nat Commun. 2017;8:14631 EP. Article.
3. Bordbar A, Monk JM, King ZA, Palsson BO. Constraint-based models predict metabolic and associated cellular functions. Nat Rev Genet. 2014;15:107 EP. Review Article.
4. Cormen TH, Leiserson CE, Rivest RL, Stein C. Introduction to algorithms. MIT electrical engineering and computer science series. Cambridge: MIT Press; 2001.
5. Dias O, Rocha M, Ferreira EC, Rocha I. Reconstructing genome-scale metabolic models with merlin. Nucleic Acids Res. 2015l43(8):3899–910. 25845595[pmid].
6. Eaton JW, Bateman D, Hauberg S. GNU octave version 4.2.2 manual: a high-level interactive language for numerical computations. https://www.gnu.org/software/octave/doc/v4.2.2.

7. Feist AM, Palsson BO. The growing scope of applications of genome-scale metabolic reconstructions using escherichia coli. Nat Biotechnol. 2008;26:659–67.
8. Feist AM, Palsson BO. The biomass objective function. Curr Opin Microbiol. 2010;13(3):344–49. 20430689[pmid].
9. Joyce AR, Palsson B. Predicting gene essentiality using genome-scale in silico models. In: Osterman AL, Gerdes SY, editors. Microbial gene essentiality: protocols and bioinformatics. Totowa: Humana Press; 2008. p. 433–57.
10. Karr JR, Sanghvi JC, Macklin DN, Gutschow MV, Jacobs JM, Bolival B, Assad-Garcia N, Glass JI, Covert MW. A whole-cell computational model predicts phenotype from genotype. Cell. 2012;150:389–401.
11. Kauffman SA. Autocatalytic sets of proteins. J Theor Biol. 1986;119(1):1–24.
12. Kohanski MA, Dwyer DJ, Collins JJ. How antibiotics kill bacteria: from targets to networks. Nat Rev Microbiol. 2010;8(6):423–35. 20440275[pmid].
13. Mahadevan R, Palsson B, Lovley DR. In situ to in silico and back: elucidating the physiology and ecology of Geobacter spp. using genome-scale modelling. Nat Rev Microbiol. 2011;9:222 EP, Erratum.
14. Maranas CD, Zomorrodi AR. Optimization methods in metabolic networks. Hoboken: Wiley; 2016.
15. Mendoza SN, Can PM, Contreras N, Ribbeck M, Agosn E. Genome-scale reconstruction of the metabolic network in oenococcus oeni to assess wine malolactic fermentation. Front Microbiol. 2017;8:534. 28424673[pmid].
16. Monk J, Nogales J, Palsson BO. Optimizing genome-scale network reconstructions. Nat Biotechnol. 2014;32:447 EP.
17. Monod J. The growth of bacterial cultures. Annu Rev Microbiol. 1949;3(1):371–94.
18. Monod J. Recherches sur la croissance des cultures bactériennes. Actualités scientifiques et industrielles. Hermann; 1958.
19. Neidhardt FC. Bacterial growth: constant obsession with dN/dt. J Bacteriol. 1999;181(24):7405–08. 1365[PII].
20. Niedenfhr S, Wiechert W, Katharina NH. How to measure metabolic fluxes: a taxonomic guide for 13c fluxomics. Curr Opin Biotechnol. 2015;34(Supplement C):82–90. Systems biology Nanobiotechnology.
21. Novick A, Szilard L. Description of the chemostat. Science. 1950;112(2920):715–6.
22. O'Brien EJ, Monk JM, Palsson BO. Using genome-scale models to predict biological capabilities. Cell. 2015;161:971–87.
23. Oberhardt MA, Palsson BO, Papin JA. Applications of genome-scale metabolic reconstructions. Mol Syst Biol. 2009;5:1–15.
24. Oberhardt MA, Puchaka J, Martins dos Santos VAP, Papin JA. Reconciliation of genome-scale metabolic reconstructions for comparative systems analysis. PLoS Comput Biol. 2011;7(3):e1001116. 10-PLCB-RA-2544R2[PII].
25. Orth JD, Thiele I, Palsson B. What is flux balance analysis? Nat Biotechnol. 2010;28:245.
26. Palsson BØ. Systems biology: properties of reconstructed networks. Cambridge: Cambridge University Press; 2006.
27. Peyraud R, Dubiella U, Barbacci A, Genin S, Raffaele S, Roby D. Advances on plantpathogen interactions from molecular toward systems biology perspectives. Plant J. 2017;90(4):720–37.
28. Rajagopalan P, Kasif S, Murali TM. Systems biology characterization of engineered tissues. Annu Rev Biomed Eng. 2013;15(1):55–70.
29. Santos FB, Vos WM, Teusink B. Towards metagenome-scale models for industrial applicationsthe case of lactic acid bacteria. Curr Opin Biotechnol. 2013;24:200–6.
30. Schrödinger E. What is life? with mind and matter and autobiographical sketches. Cambridge paperback library. Cambridge: Cambridge University Press; 1992.
31. Shakiba N, Zandstra PW. Engineering cell fitness: lessons for regenerative medicine. Curr Opin Biotechnol. 2017;47(Supplement C):7–15. Tissue, cell and pathway engineering.

32. Smith CA, Neidhardt FC, Ingraham JL, Schaechter M. Physiology of the bacterial cell: a molecular approach. Sunderland: Sinauer Associates; 1990. p. 507; 43:95. ISBN: 0878936084; 2010;20:124–5.
33. Snitkin ES, Dudley AM, Janse DM, Wong K, Church GM, Segr D. Model-driven analysis of experimentally determined growth phenotypes for 465 yeast gene deletion mutants under 16 different conditions. Genome Biol. 2008;9(9):R140.
34. Sousa FL, Hordijk W, Steel M, Martin WF. Autocatalytic sets in E. coli metabolism. J Syst Chem. 2015;6(1):4. 9[PII].
35. Thiele I, Palsson B. A protocol for generating a high-quality genome-scale metabolic reconstruction. Nat Protoc. 2010;5:93–121.
36. Tobalina L, Pey J, Rezola A, Planes FJ. Assessment of FBA based gene essentiality analysis in cancer with a fast context-specific network reconstruction method. PLoS One. 2016;11(5):e0154583. PONE-D-15-35442[PII].
37. Vazquez A. Overflow metabolism: from yeast to marathon runners. Saint Louis: Elsevier Science; 2017.
38. Vinaixa M, Rodrguez MA, Aivio S, Capelládes J, Gmez J, Canyellas N, vis Stracker TH, Yanes O. Positional enrichment by proton analysis (pepa): a one-dimensional 1h-nmr approach for 13c stable isotope tracer studies in metabolomi cs. Angew Chem Int Ed. 2017;56(13):3531–5.
39. Walsh JR, Schaeffer ML, Zhang P, Rhee SY, Dickerson JA, Sen TZ. The quality of metabolic pathway resources depends on initial enzymatic function assignments: a case for maize. BMC Syst Biol. 2016;10:129. 369[PII].
40. Xavier JC, Patil KR, Rocha I. Integration of biomass formulations of genome-scale metabolic models with experimental data reveals universally essential cofactors in prokaryotes. Metab Eng. 2017;39:200–8.
41. Ziv N, Brandt NJ, Gresham D. The use of chemostats in microbial systems biology. J Vis Exp. 2013;14(80):50168. 50168[PII].

Analysis Methods for Shotgun Metagenomics

Stephen Woloszynek, Zhengqiao Zhao, Gregory Ditzler, Jacob R. Price, Erin R. Reichenberger, Yemin Lan, Jian Chen, Joshua Earl, Saeed Keshani Langroodi, Garth Ehrlich, and Gail Rosen

Abstract The development of whole metagenome shotgun sequencing (WGS) has enabled the precise characterization of taxonomic diversity and functional capabilities of microbial communities in situ while obviating organism isolation and cultivation procedures. WGS created with second- and third-generation sequencing technologies will generate millions of reads and tens (or hundreds) of gigabytes of information about the organisms under investigation. Despite

---

Author Contributions: SW, abstract, taxonomic binning, taxonomic classification, normalization, feature selection, feature extraction, distance-based approaches, neural network approaches, statistical inference, machine learning, drafted and ordered sub-sections, coordinated co-authors; ZZ, taxonomic classification, machine learning; GD, diversity metrics, feature selection, feature extraction; JRP, abstract, diversity metrics, distance-based approaches, diversity metrics; ERR, abstract, introduction; YL, functional annotation; JC, neural network approaches; JE, feature selection, feature extraction; SKL, taxonomic binning; GR, taxonomic classification, discussion, drafted sub-sections; all authors contributed to editing and revising.

S. Woloszynek · Z. Zhao · J. Chen · G. Rosen (✉)
Department of Electrical and Computer Engineering, Drexel University, Philadelphia, PA, USA
e-mail: gailr@coe.drexel.edu

G. Ditzler
Department of Electrical and Computer Engineering, University of Arizona, Tuscon, AZ, USA

J. R. Price · S. K. Langroodi
Department of Civil, Architectural, and Environmental Engineering, Drexel University, Philadelphia, PA, USA

E. R. Reichenberger
U.S. Department of Agriculture, Agricultural Research Service, Eastern Regional Research Center, Wyndmoor, PA, USA

Y. Lan
Department of Cell and Developmental Biology, Perelman School of Medicine, University of Pennsylvania, Philadelphia, PA, USA

J. Earl · G. Ehrlich
Department of Microbiology and Immunology, Drexel University College of Medicine, Philadelphia, PA, USA

© Springer International Publishing AG, part of Springer Nature 2018
F. A. B. da Silva et al. (eds.), *Theoretical and Applied Aspects of Systems Biology*, Computational Biology 27, https://doi.org/10.1007/978-3-319-74974-7_5

containing an immense amount of information, the reads are unorganized and unlabeled, leading to a significant challenge in discerning from which genome a read originated. Thus, analysis of WGS data necessitates first determining community structure and function from the raw reads before the focus can shift to making multi-sample comparisons. A typical WGS workflow consists of read assignment (taxonomic binning and classification), preprocessing techniques (normalization, dimensionality reduction), exploratory approaches (feature selection and extraction, ordination), statistical inference (regression, constrained ordination, differential abundance analysis), and machine learning. The following chapter provides an overview of these analytical approaches (including challenges and possible pitfalls that may be encountered by researchers) as well as steps toward their solutions. Relevant software packages and resources are also discussed.

1   Introduction to Metagenomics

The term "metagenome" originated with Handelsman et al. who defined it as a collection of genomes found in the microflora of soil and described an approach used to access the organisms living in this ecosystem [1]. Their motivation was influenced by a continual decline in the discovery of new compounds from an environment that had previously provided researchers and industry with chemicals that were antimicrobial or otherwise medicinal in nature. The paucity of newly discovered compounds followed the realization that many microbes were not culturable and that microbiologists had greatly underestimated both their numbers and diversity [1–7]. The reasons behind a microbe's resistance to culturing vary; their survival may be dependent upon compounds provided by other resident organisms, and/or the conditions (e.g., temperature, atmospheric pressure, gaseous elements (along with their amounts)) may be inadequate for their survival [2]. Regardless of the cause, it became apparent that the number of organisms that could not be cultured greatly surpassed the number of microorganisms that could be cultured [1, 2, 5, 6, 8–11]. Combined, these elements drove a new and oft-interdisciplinary field known as metagenomics – the study of uncultured genetic material acquired directly from environmental communities that contain a motley population of organisms. Ensuing from these developments was the inception of numerous large-scale metagenomic studies that investigated microbial communities in water, soil, and animals [12–16]. Information acquired from these studies have exposed the intricate influence and beauty of microbes on processes as vast as the geochemical to human health.

Although specimen isolation and cultivation are not required, sophisticated computational tools are a necessity in metagenomic analysis. This analysis has been aided greatly by advances in sequencing technology, which have yielded increased accuracy in base pair identification, longer reads, and decreases in sequencing costs. The reduction in sequencing pricing as well as faster computer processors have made metagenomic analysis more accessible to institutions and laboratories looking to investigate microbial communities. As such, clinical studies and research related to quorum sensing, antibiotic resistance, biofilms, bacteriophages, and food science along with other areas of interest have become far more common [17–23].

Knowledge obtained from these studies owe much not only to the individuals involved with undertaking the studies and improving sequencing technology but also those who have developed the algorithms and methods used to analyze next-generation sequencing (NGS) and metagenomic data [8, 24–31]. Metagenomic studies often revolve around determining what is in the sample (classification), how many organisms are in the sample (binning), and what they are doing (functional annotation). Additionally, researchers are interested in comparing samples (normalization, clustering, ordination) and determining similarities between samples (feature selection/extraction). These methods are often accomplished with machine learning techniques, and each of the aforementioned topics will be addressed in this chapter.

2 Sequence Quality and Identification

2.1 Introduction to Taxonomic Binning

High throughput whole metagenome shotgun sequencing (WGS) is a reliable technique used to characterize taxonomic diversity and function of microbial communities without cultivation of the microorganisms in a laboratory environment. After WGS, the primary goal is then to infer microbial community structure and function in the given microbiome from the millions of *unlabeled* genomic fragments (known as "reads") [32]. This is no easy task, however, since algorithmic approaches are necessary to discern taxonomic information. Extracting information from sequencing reads has accordingly been equated to simultaneously completing multiple puzzles with their pieces shuffled together [33]. While full-genome assembly is potentially an effective method for this purpose, constructing complete genomes from short reads often fails for many reasons including repetitive nucleotide patterns found within genomes, homologous regions of closely related regions, and conserved regions among different species [34, 35].

Binning is considered an alternative to full-length genome assembly [36]. Despite still relying on sequencing reads, binning is capable of approximating population composition and functional diversity of assigned genomes [37, 38]. There are two binning methods developed for disentangling metagenomic reads: "supervised" (taxa-dependent; classification) and "unsupervised" (taxa-independent; clustering). Supervised binning uses one or more phylogeny-based comparisons that involve aligning reads to reference genomes, assessing sequence composition properties such as GC content and oligonucleotide patterns (k-mers), and utilizing hybrid methods that leverage both alignment and sequence composition approaches [36, 39]. Supervised binning is often not effective for environmental samples or diverse microbial communities; however, due to bias with respect to previously sequenced or well-studied species, many of the reference databases in which supervised approaches rely are incomplete [33], which results in many reads going unassigned or being assigned incorrectly. Also, metagenomes with high interspecies diversity fail to be accurately classified by supervised binning tools [40].

Unsupervised binning, on the other hand, relies on discriminative nucleotides, sequence composition, and taxa abundance, which is inferred in terms of contig coverage [41, 42]. Binning techniques that rely on sequence composition assume that each taxon has a unique genomic signature, which is represented as k-mer frequency vectors (Fig. 1). Example tools include 2Tbinning, LikelyBin, Metawatt, SCIMM, self-organizing maps, and VizBin (Table 1). For low-abundance taxa, composition-based techniques are prone to incorrect taxon assignments since the generated clusters for these taxa tend to be poorly described [33]. In addition, they typically require high-quality reads or contigs that are over 1000 bp in length to achieve acceptable accuracy [43]. Abundance-based techniques are much better at handling low-abundance taxa and shorter reads. For single-sample studies, limitations associated with low-abundance taxa are mitigated by enforcing distributional assumptions (e.g., the Lander-Waterman model) to the k-mer abundance coverage profile. For multi-sample studies, the taxa abundance profiles are assumed to be correlated between samples [33]. Abundance-based techniques include AbundanceBin, Canopy, and MBBC. Lastly, hybrid techniques that utilize both sequence composition and taxa abundance include COCACOLA, CompostBin, CONCOCT, differential coverage binning, GroopM, MaxBin, MetaBAT, MetaCluster, and MyCC. For a detailed review of unsupervised binning approaches, see Sedlar et al. [33].

Selection of binning methods depends on the purpose of the metagenomic study, the computational requirements, as well as the time constraints. In supervised methods, the length of metagenomic reads, which is in turn dependent upon the sequencing platform, is also a factor [44–46]. In addition, read coverage must also be considered since greater coverage may capture rare species with more accurate results. On the other hand, unsupervised binning is effective for diverse microbiomes or low-coverage datasets [36]. To improve binning results, preprocessing (e.g., quality filtering of the sequencing reads) and post-processing techniques which use different reassembly approaches (e.g., mapping reads to the bins before reassembly) are options [47–49].

2.2   Taxonomic Classification

A variety of tools are currently available that perform taxonomic classification. These include methods that rely on a subset of marker genes (MetaPhlAn [50], MetaPhyler [51], mOTU [52], MicrobeCensus [53], GOTTCHA [54]), and those that use exploit the entire set of reads, using composition-based approaches, such as alignment (MEGAN [55]) or k-mer enumeration (CLARK [56], Kraken [57], LMAT [58], MetaFlow [59], NBC [60], and PhyloSift [61]) [62, 63]. Approaches

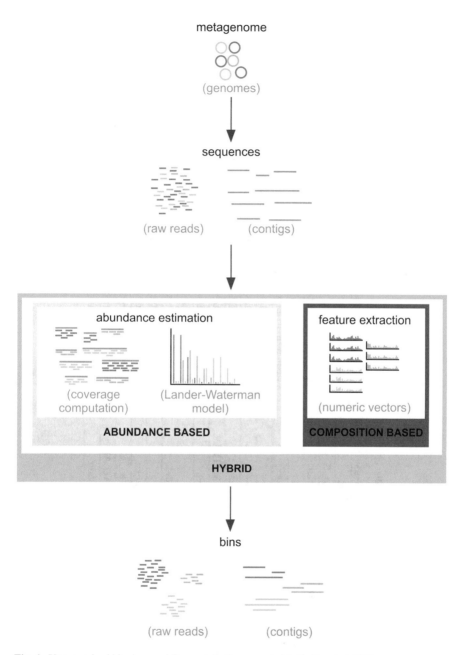

Fig. 1 Unsupervised binning workflow, originally presented in Sedlar et al. [33]

Table 1 Unsupervised binning tools. (Adapted from Sedlar et al. [33])

Method	Input data	Programming language	Type	Source
SOM	Raw reads or contigs	Perl	Composition	https://github.com/tetramerFreqs/Binning
LikelyBin	Raw reads	Perl, C	Composition	http://ecotheory.biology.gatech.edu/downloads/likelybin
SCIMM	Raw reads or contigs	Python	Composition	http://www.cbcb.umd.edu/software/scimm/
MetaWatt	Assembled contigs	Java	Composition	https://sourceforge.net/projects/metawatt/
VizBin	Contigs	Java	Composition	https://claczny.github.io/VizBin/
AbundanceBin	Raw reads	C++	Abundance	http://omics.informatics.indiana.edu/AbundanceBin/
Canopy	Gene abundance profiles	C++	Abundance	https://bitbucket.org/HeyHo/mgs-canopy-algorithm/wiki/Home
MBBC	Raw reads	Java	Abundance	http://eecs.ucf.edu/~xiaoman/MBBC/MBBC.html
CompostBin	Raw reads	C, MATLAB	Hybrid	https://sites.google.com/site/souravc/compostbin
MetaCluster	Raw reads (only pair-ends)	C++	Hybrid	http://i.cs.hku.hk/~alse/MetaCluster/index.html
DCB	Raw reads	R	Hybrid	https://github.com/MadsAlbertsen/multi-metagenome
CONCOCT	Contigs + BAM	Python	Hybrid	https://github.com/BinPro/CONCOCT
MaxBin	Contigs + (reads or abundance file)	Perl	Hybrid	https://sourceforge.net/projects/maxbin/
GroopM	Contigs + BAM	Python	Hybrid	http://ecogenomics.github.io/GroopM/
MetaBAT	Contigs + BAM	C++	Hybrid	https://bitbucket.org/berkeleylab/metabat
COCACOLA	Contigs + raw reads	MATLAB	Hybrid	https://github.com/younglululu/COCACOLA
MyCC	Contigs + BAM*optional	Python	Hybrid	https://sourceforge.net/projects/sb2nhri/files/MyCC/

that utilize sequences, while slower, are enticing in their ability to leverage additional information for assembly and contamination detection, for example [62].

Marker gene approaches are faster than composition-based approaches but are limited in the number of reads they can ultimately classify [62]. They vary from method to method mostly in terms of their marker gene database construction. Composition-based approaches, on the other hand, differ more algorithmically [63]. For example, CLARK performs classification by first identifying discriminative k-mers that uniquely characterize reference sequences, which it then uses to classify query reads based on the number of shared k-mers. Kraken is similar in that it uses the number of overlapping reads between query and reference to influence the classification; however, it leverages phylogenetic information during the mapping step, building a phylogenetic tree. The reference sequence is identified by determining the lowest common ancestor that contains the k-mer from the query. Other k-mer approaches that leverage phylogenetic information include PhyloSift and LMAT. MetaFlow treats classification as a query-to-reference matching problem, using a bipartite graph. Lastly, NBC is a metagenome fragment classification tool using k-mer frequency profiles. In short, this tool trains an NBC classifier based on the frequency of k-mers. Here, $\mathbf{X} = [x_1, x_2, \cdots, x_n]$ is the set of k-mers in a sequence. In the training phase, $p(x_i|C_k)$ is estimated by the total number of k-mers x_i occurring in all the training sequences that are labeled by C_k. In the testing phase, given a query sequence, the organism containing the sequence is predicted by the class that maximizes the posterior probability $P(C_k|\mathbf{X})$.

To evaluate the performance of the tools described above, McIntyre et al. designed an analysis involving 846 species across 67 simulated and datasets [62]. The performances were evaluated by each tool's ability to (1) identify taxa in a sample at genus, species, and strain levels, (2) estimate the relative abundances of taxa in a sample, and (3) classify individual reads at the species level. For taxa identification, all tools performed optimally at the genus level, but the performance dropped noticeably at the strain level. They also determined that the performance of k-mer-based tools could be improved by introducing an abundance threshold. Read depth was another important identified factor that had an effect on performance; they found a positive relationship between the number of recovered species and read depth. BLAST-MEGAN and PhyloSift were two exceptions, but this trend could be dampened with the addition of adequate filtering. On the other hand, read depth had little impact on marker gene-based tools. The authors also showed that an ensemble classifier that combined the results from the best performing tools could produce improved results in quantifying the number of species. Combining their approach with BLAST greatly improved performance; however, because BLAST is notoriously slow, a faster ensemble showed comparable performance. For relative abundance comparisons, the authors showed that most of the tools could predict the proportion of a particular species in a sample to within a few percentage points. CLARK slightly overestimated relative abundance, but had greater precision compared to other tools. k-mer-based methods achieved the highest recall with lower sequencing depth.

Long but low quality reads generated by newer sequencing platforms are becoming more readily available. For long, lower quality reads, CLARK and Diamond-MEGAN performed more robustly than other tools. For classifying individual reads, BLAST-MEGAN gave the best precision, whereas CLARK generally gave the best recall. The last considerations were runtime and memory. The authors benchmarked all tools under the same conditions and showed that MetaPhlAn, GOTTCHA, PhyloSift, and NBC used less memory; NBC and BLAST were the slowest; and CLARK, GOTTCHA, Kraken, MetaFlow, MetaPhlAn, Diamond-Megan, and LMAT were the fastest. The authors provided a decision tree summary of usage recommendations (Fig. 2).

It should be noted that despite the large study size, 846 species is only a small subset of all species that exist. Also, the ability of a given tool to identify "unknown" organisms was never evaluated. This is highlighted by the fact that as the read depth increased, most classifiers discovered more species – leaving a perplexing open problem in metagenomic taxonomic classification. Therefore, more research should be done to determine how database size affects classification, as well as how other parameters may affect classifier performance.

2.3   Functional Annotation

Unraveling the functional composition of metagenomes is crucial to understanding the microbe's metabolic dynamics and how they shape the environment or adapt to environmental changes. From either assembled individual genomes or the metagenome as an entity, protein-coding genes can be predicted by scanning the sequences for start/stop codons. However, gene prediction and the following functional profiling do not depend on full gene sequences. Functional profiling can be achieved using short reads directly, as they may be highly similar to gene sequence fragments or contain characteristic protein domains for recognition. As easy as it sounds, functional profiling of metagenomes remains challenging. One of the fundamental difficulties is that metagenomic sequences can be highly divergent in comparison to genes and proteins currently identified [64]. Therefore, profiling tools that rely on sequence similarity are subject to a tough dilemma between sensitivity and specificity. Another difficulty is that short sequencing reads may not contain sufficient information for us to accurately infer their functions. Therefore, increasing the number of annotated reads and improving the annotation accuracy remain top challenges for tools in development for functional profiling [65].

Recently, a lot of effort has been devoted to creating an accurate knowledge base of metagenomic functions and developing reliable and scalable profiling tools. These two types of efforts are tightly coupled, and in most cases, the choice of which database to profile against also decides which profiling software/tool should be used.

As of now, various databases have been constructed, and they represent different resolutions of metagenomic function. For example, NCBIs RefSeq database [66]

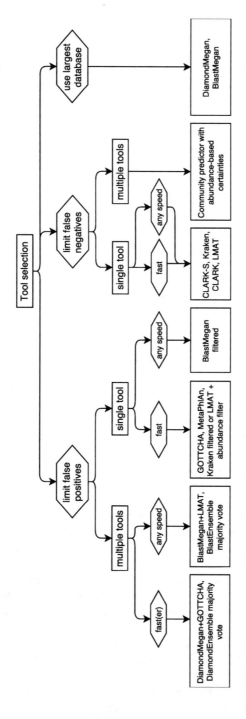

Fig. 2 Decision tree summary of usage recommendations, originally presented in McIntyre et al. [62]

and the UniProt database [67] are two of the largest reference sequence collections, both containing over 100 million annotated protein sequences. When provided with a reference database as comprehensive as these two, it is more likely to find annotated proteins that share high sequence similarity to an unknown read. However, one downside of big databases is that most of the "annotated" proteins in these databases were annotated automatically, i.e., the reference itself is subject to error. In many cases, these reference databases are adequate for profiling metagenomes and discovering significant changes between profiles. In other cases, however, one can opt for a reduced database with higher credibility, such as the Swiss-Prot database [68] which contains only a half-million annotated proteins but is manually curated and reviewed. There are also other reduced databases focusing on specific metagenomes (such as the UniMES database that hosts proteins inferred from environmental metagenomes) or datasets generated from specific metagenomic studies that can be used as reference databases of related metagenomes (such as the functional profiles generated from the Tara oceans project [69] of global ocean microbiomes and profiles generated from the Human Microbiome Project [70]).

As previously mentioned, the largest databases now contain up to 100 million annotated proteins. Although we may be able to annotate metagenomic reads with these proteins, it is not easy to interpret and understand a metagenomic profile without summarizing similar or relevant protein functions into groups. The gene ontology database is one of the many databases that strive to address this problem [71]. It annotates reference proteins with a carefully standardized vocabulary (called GO terms) and constructs a comprehensive relationship network between GO terms from the molecular level to larger pathways, as well as cellular and organismal-level systems. Therefore, we can use GO terms to profile metagenomes at molecular, pathway, or cellular levels. Besides gene ontology, several databases also summarize protein annotation into groups or hierarchical groups, such as the COG/EggNOG categorizations [72, 73], the KEGG pathways [74, 75], the MetaCyc pathways, and the SEED subsystems [76, 77]. The COG/EggNOG was generated by grouping orthologous proteins from numerous organisms into clusters, whereas KEGG, MetaCyc, and SEED group (or related) proteins are based on their related metabolic roles.

Annotating metagenomic reads using these databases – either large databases or reduced ones – relies on sequence similarity with reference proteins. Therefore, alignment-based methods such as BLAST search are often used for the functional profiling [78]. Additionally, numerous software tools were developed to make protein alignment and hence functional profiling computationally efficient. Besides individual tools, several large-scale pipelines have also been developed to annotate metagenomic data against multiple databases at once, such as IMG/M [79, 80], MG-RAST [81], MEGAN [82], and HUMAnN [83]. These pipelines stitch together multiple bioinformatics steps from raw metagenomic reads to functional profiling, making it easier for the user to interpret and compare the functional potential of different microbial communities.

Although proteins with similar functions may have evolved and become highly divergent in terms of nucleotide sequences, the protein domains they contain are more conserved and may function independently from the rest of the protein. Therefore, grouping proteins based upon their functional protein domains is yet another way of summarizing different protein sequences into a manageable profile. One example of such an approach is Pfam [84], which is a collection of protein sequence alignments and hidden Markov models (HMMs) and provides a good repository for identifying protein families, domains, and repeats. SMART is another example of a protein domain database, which is abundant with domains in signaling, extracellular, and chromatin-associated proteins. Other protein domain databases commonly used are PROSITE [85], PANTHER [86], HAMAP [87], ProDom [88], etc. Although each of these databases can be acquired separately and many of them have specific software that can profile metagenomes against it (such as HMMER for Pfam, ScanProsite for PROSITE, and HAMAP-Scan for HAMAP), it is worth mentioning that InterPro [89] has combined signatures from all of the aforementioned domain databases, as well as several others, into a single searchable resource for functional profiling. Therefore, InterProScan (developed for InterPro) can be a very handy software package to scan metagenomic reads against most domain databases.

2.4   Normalization

After sequencing DNA from microbial communities of interest and determining the abundances of genomes or genes present in the community, the next step is to perform comparisons between samples. However, to make these comparisons requires that the abundances first be normalized because raw metagenomic abundances fail to accurately represent the true configuration of the taxonomic community. Simply put, in a given sample from an environment of interest, the total number of sequenced reads does not accurately reflect the true amount of DNA present in the environment. This is primarily due to study-level variation in sample collection, DNA extraction, library preparation, and sequencing depth [90]. Obtaining true "absolute abundance" cannot be achieved with sequencing data alone; for example, quantitative PCR would have to be performed in tandem [90]. Thus, differences in library size is often mitigated by calculating relative abundances where each count is divided by the total abundance from its corresponding sample [91]. Specifically, given a vector of J raw abundances from sample i:

$$\left[x_{i,1}, x_{i,2}, \ldots, x_{i,J} \right] \tag{1}$$

the relative abundance for raw abundance j in sample i is given by

$$x_{i,j}^{*} = \frac{x_{i,j}}{\sum_{j} x_{i,j}} \tag{2}$$

However, this approach is considered inappropriate [92]: dividing a raw abundance by its sample's sum constrains it to the unit simplex (all values for that sample must sum to 1), thereby rendering the data compositional [93]. Increases in any gene or genomes abundance is coupled with a corresponding decrease in the relative abundance of all other genes or genomes. In other words, while the absolute abundance of a particular taxon may be constant between two communities, their relative abundances will be different if the abundances of other taxa differ. This complicates interpretability and introduces spurious correlations. Thus, techniques such as linear regression and Pearson's correlation are no longer appropriate [94, 95]. A better alternative is to perform a centered log-ratio transformation (CLR) [96]:

$$x_{i,j}^* = \log \frac{x_{i,j}}{g_i} \tag{3}$$

where g_i is the geometric mean for sample i given by

$$g_i = \sqrt[J]{x_{i,1} x_{i,2} \cdots x_{i,J}} \tag{4}$$

The CLR is free of the compositional artifacts described above, but is limited by a singular covariance matrix, which may limit its use in downstream modeling approaches [93]. In addition, sparse abundance data further complicates calculating the CLR due to a zero denominator and the calculation being done in log space. This necessitates stringent filtering or, more commonly, the addition of small non-zero values (pseudo-counts), which may introduce bias [97]. Also, if the pseudo-count is set to 1 and the dataset is very sparse, then each raw abundance will be divided by a geometric mean close to 1, drastically dampening any normalization effect, and use of smaller pseudo-counts does not remedy the situation [98]. Recent work has suggested using values based on percentiles in place of the geometric mean, but whether this approach is robust to highly sparse datasets is currently unknown [92, 98].

Silverman et al. [93] has introduced a phylogeny-based normalization approach (PhILR) that utilizes the isometric log-ratio transformation (ILR), which, unlike the CLR, returns an invertible covariance matrix. The ILR scales CLR transformed abundances by taxa-level weights $\boldsymbol{p}$ and a weight matrix ψ given by the binary partitioning of the phylogenetic tree:

$$x_{i,j}^* = \text{CLR}(x_{i,j}) \text{diag}(\boldsymbol{p}) \psi^T \tag{5}$$

The taxa-level weighting allows for soft-thresholding of low-abundance taxa and may dampen the bias resulting from use of a pseudo-count.

In addition to differences in sample read depth, there remain other potential biases – most notably from biological sources. These include a gene or genome's mappability and length. First, relative abundances are often overestimated since metagenomes are represented as the proportion of mapped reads present in the sample, and this ignores the variability of unmapped reads stemming from novel taxa or genes. Second, the probability of sequencing a read is a function of the length of the gene or genome from which the read originated. Correcting for gene-length permits gene-to-gene comparisons and is possible for well-described genes where the gene length is available; however, performing a genome-length correction is impractical due to the degree of diversity within a metagenome and the variability in genome lengths [90].

Accurately estimating the relative abundances of taxa in a metagenome can be accomplished via marker gene approaches, which circumvent issues with genome size since the marker genes themselves are well-characterized [90]. Marker gene approaches include MetaPhlAn [50], MetaPhyler [51], and MicrobeCensus [53]. Recent work has focused on calculating average genomic copy number, which corrects for the biases stemming from average genome size, genome mappability, and species richness. One approach, called MUSiCC, utilizes the median abundance of universal single-copy genes to normalize gene relative abundances. It is currently, however, only applicable to KEGG annotated data [91].

3 Comparative Analysis

3.1 Diversity Metrics and Distances

β-diversity allows us to examine the similarities and dissimilarities between multiple samples in a metagenomic study. Microbial ecologists begin by first computing a pair-wise distance matrix, $\mathbf{D} \in \mathbb{R}_+^{n \times n}$, where entry (i, j) is the distance between sample i and j with $i, j \in [n]$. One of the most important steps in this part of the analysis is the selection of the distance matrix. In general, microbial ecologists rarely, if ever, use the standard Euclidean distance to compare samples; rather, they use distances that are based on set theory or a distance between distributions.

The Jaccard index is a simple measure to determine the dissimilarity based solely on the presence or absence of a taxon in two samples. The index is given by

$$D_{\text{JAC}}(X_i, X_j) = 1 - \frac{|X_i \cap X_j|}{|X_i \cup X_j|} \tag{6}$$

where X_i and X_j represent a set of metagenomic features in sample i and j, respectively. One of the drawbacks to the Jaccard index is that it does not account for the magnitude of taxa presence, rather it only identifies whether the taxa were present in a sample. Bray-Curtis is another metric which, unlike the Jaccard index, has the abundances incorporated into the calculation. Formally, the Bray-Curtis dissimilarity is given by

$$D_{BC}(X_i, X_j) = \frac{2C_{ij}}{S_i + S_j} \tag{7}$$

where S_i and S_j are the total number of taxa counted at both sites and C_{ij} is the sum of the lesser value for only those species in common between samples. Note that because the triangle inequality does not hold, Bray-Curtis is a dissimilarity metric and not a distance metric.

The Hellinger distance is the distance between two probability distributions, and it has been used occasionally in microbial ecology. Also, similar to Jaccard and Bray-Curtis, the Hellinger distance is bounded. Let $\mathcal{P} := \{p_j : j \in [n]\}$ and $\mathcal{Q} := \{q_j : j \in [n]\}$ be the probability distributions over two different samples that are represented by n taxa. The Hellinger distance is defined as

$$D_{HEL}(\mathcal{P}, \mathcal{Q}) = \frac{1}{\sqrt{2}}\|\sqrt{\mathcal{Q}} - \sqrt{\mathcal{P}}\|_2 \tag{8}$$

where $\| \cdot \|_2$ is the ℓ^2-norm.

The aforementioned distances can all be found in traditional mathematical literature; however, given that microbial ecologists are using β-diversity in their studies, it should be the case that the distance measure being used in the analysis has some biological connection. The unique fraction metric (UniFrac) is perhaps the most widely used measure of distance in microbial ecology [99, 100]. UniFrac was proposed to measure the phylogenetic difference between microbial communities, as other measures such as Bray-Curtis, Hellinger, and Jaccard do not. The unweighted version of UniFrac, like the Jaccard index, only deals with the presence or absence of taxa. Unweighted UniFrac is implemented as follows: consider that you are provided two samples A and B, which are made up of metagenomic sequences, and build a phylogenetic tree using all available reads (see Zvelebi and Baum [101]). Color all the branches of the tree red where a path between two sequences in A exists, and perform the same operation for B but using a different color (e.g., blue). If a branch is colored both red and blue, then it is marked gray. The UniFrac distance is the ratio of the number of branches in the tree that are unique to either A or B to the total number of branches in the tree. Weighted UniFrac takes the concept of using this ratio to incorporate the frequency of the reads in the calculation.

3.2   Feature Representation and Dimensionality Reduction

Metagenomic datasets are often made up of thousands of features that represent abundances (i.e., the relative proportion of a protein family), and these datasets frequently have more features than the number of samples. A dataset with more features than samples is a challenging problem because the system is underdetermined. Furthermore, many of these features are often uncorrelated with sample data or even redundant with each other. For example, consider a metagenomic dataset that is being used to find the taxa that favor a high saline environment. This dataset has 50 samples from both a high and low saline environment, and each sample is made up of 5000 taxa. We refer to the high and low saline environments as the data that describe the sample classes. Feature selection and dimensionality reduction allow us to (1) represent these samples by either the bacteria that are relevant to differentiate between high and low environments and (2) visualize the 100 samples in a 2-D or 3-D space, respectively.

Feature selection is the process of identifying a subset of features that are relevant and possibly non-redundant with the class. This feature subset allows metagenomic research to identify the informative variables in a dataset, such that there is still significant predictive power in the reduced set. It is important to know that the feature subset has variables that still have physical meaning (e.g., bacteria, protein family, etc.). In contrast, feature extraction is an approach to transform the data into a new (lower) dimensional space, and the new features are typically combinations of all the other features; however, these new features no longer have physical meaning.

3.2.1 Feature Selection

Feature selection plays a central role in nearly all tasks of the data analysis; however, many popular feature selection algorithms do not scale well with a large metagenomic dataset. Therefore, computationally cheap methods are used to remove so-called low zero variance metagenomic features. This low zero variance is not the best method to use in every situation; however, it is a good place to start to remove complexity when faced with a high degree of dimensionality in the data. Related methods exist in information theory (i.e., measuring features for the amount of mutual information between a metagenomic feature and the class [102]). The objective is to eliminate metagenomic features with low mutual information and not redundant with the other features.

More sophisticated methods exist for performing feature selection, including ones utilizing more than just variance, in addition to other probabilistic quantities. For example, the Relief algorithm examines paired samples (based on Euclidean distance) and weights features based on the samples' proximity in Euclidean space. It updates a weight matrix by determining if the features belong to the same or different classes [103]. Correlation-based feature selection (CFS) identifies features that have a high correlation with the supplied class of the sample but low correlation with other features while being less computationally intensive than Relief [104]. Both of these approaches are known as filter-based feature selection since they are classifier independent. Brown et al. provide a comprehensive review of information-theoretic filter feature selection algorithms [105].

In addition to filter-based approaches, embedded feature selection algorithms jointly optimize the feature selector and classifier. The least absolute shrinkage and selection operator (Lasso) is an approach to feature selection that optimizes a model for linear regression [106]. However, unlike standard linear regression, Lasso adds a penalty on the ℓ_1-norm of the linear model. This penalty, which is shown in (9), forces the solution to be sparse (i.e., many entries in θ are zero), thus performing feature selection for the linear model $\theta^T\mathbf{x}$. Lasso is formally given by

$$\theta^* = \arg\min_{\theta \in \mathbb{R}^p} \frac{1}{2n} \sum_{i=1}^{n} (y_i - \theta^T\mathbf{x}_i)^2 + \lambda \sum_{j=1}^{p} |\theta_j| \tag{9}$$

where θ are the parameters of the linear model, $\mathbf{x}_i$ is the metagenomic feature vector, y_i is the class (± 1) or dependent variable, n is the number of samples, and p is the total number of metagenomic features. Bates and Tibshirani have recently adapted Lasso for compositional data, using log ratios as described above for CLR normalization [107]:

$$\theta^* = \arg\min_{\theta \in \mathbb{R}^p} \frac{1}{2} \sum_{i=1}^{n} \left(y_i - \mu - \sum_{1 \le j < k \le p} \theta_{j,k} \log \frac{x_{i,j}}{x_{i,k}} \right)^2 + \lambda \sum_{j=1}^{p} |\theta_j| \tag{10}$$

The log-ratio Lasso differs from (9) in that it aims to detect models composed of a sparse subset of ratios as opposed to models composed of a sparse subset of regression coefficients. Ditzler et al. have implemented an open-source feature selection software tool for analyzing metagenomic and 16S datasets [108]. Lasso and other sparse regression techniques are easily implemented in glmnet, available in R, MATLAB, and Python.

3.2.2 Feature Extraction

Feature selection reduces the set of metagenomic features to a subset that is informative – potentially non-redundant – and still maintains a physical interpretation. Feature extraction is a technique for dimensionality reduction that embeds the original set of features in a lower-dimensional space (e.g., apply a linear projection of the metagenomic data vectors from $\mathbb{R}^p$ to $\mathbb{R}^2$ where $p \gg 2$). Principal component analysis (PCA) is one of the more popular projections for feature extraction. In PCA, we seek to represent the p-dimensional data in a lower-dimensional space that maximizes the variance of the projections. It turns out these projections are the eigenvectors of the covariance matrix of the data that correspond to the largest eigenvalues. Note that PCA does not take the class into account when the projections are calculated. Sparse PCA can also be performed for feature extraction [109], where the difference between PCA is that the projection is made by adding a sparsity constraint on the input metagenomic features. Note that this form of feature extraction will result in a new set features that have a high variance; however, these features do not have any biological meaning because the new feature set is made up of linear combinations of all other features. There is also supervised principal component analysis (SPCA) that takes the classes into account to find the greatest degree of variance between classifications [110]. Linear discriminant analysis (LDA) is another linear transformation that reduces the dimensionality of the data that uses supervised information. The dimensionality of the reduced space is limited $C - 1$, where C is the number of classes, whereas supervised PCA does not have this limitation. LDA is also difficult to use on many metagenomic datasets because it suffers from the small sample size problem – thus, making supervised PCA a more appropriate feature extraction technique that takes into account the class separability as well.

Other related techniques include independent component analysis (ICA), non-negative matrix factorization (NMF), and canonical correlation analysis (CCA) [111]. ICA performs a linear transformation that, unlike PCA, which finds components that maximize the variance and identifies rotations that result in new, transformed features that are mutually statistically independent. In other words, each pair of features in this new feature space will have zero mutual information. NMF approximates a feature matrix X by $X \approx WH$, where each value $x_{i,j}$ is assumed to be Poisson distributed; hence, NMF is appropriate for nonnegative abundance data. CCA is another feature extraction technique. It uses a linear projection on a subset of features, then uses the correlation between the projections [112]. CCA can be applied to both continuous and discrete data, which is beneficial for analyzing not only the metagenomic features from abundance data but also the data associated with the samples. Finally, one of the advantages to CCA, as well as PCA, is that the projections can be computed efficiently using singular value decomposition (SVD).

Many datasets, even those in metagenomics, may not work well for data that lie on a nonlinear lower-dimensional manifold. t-Distributed Stochastic Neighbor Embedding (t-SNE) is a probabilistic nonlinear dimensionality reduction technique. It represents the similarity between any two points x_i and x_j as the conditional probability that x_i and x_j are neighbors, which is Gaussian distributed. It then attempts to learn a lower-dimensional embedding, where the similarities are now heavy tailed – that is, t-distributed. The Kullback-Leibler divergence between the estimated similarities in high- and low-dimensional space is minimized [113]. Visualizing the data with t-SNE can result in compact groups of classes (influenced by adjusted the "perplexity" parameter) in the lower-dimensional embedding. While the nonlinear embedding can be attractive to many metagenomic data analysis problems, there remain some drawbacks to t-SNE. Namely, t-SNE has a poor space complexity that can require a significant amount of memory to find the embedding.

3.2.3 Distance-Based Approaches for Feature Extraction

The remaining approaches are common in the statistical ecology and sequencing domains and are sometimes referred to as "unconstrained ordination" techniques (note that PCA described above can also be described as unconstrained ordination). These include principal coordinates analysis (PCoA) [114], otherwise known as metric multidimensional scaling (MDS), correspondence analysis (CA, or reciprocal averaging) [115], and nonmetric multidimensional scaling (NMDS) [116, 117]. PCoA is simply an eigenvalue decomposition of a distance matrix. If the chosen distance metric is Euclidean, then PCA and PCoA are equivalent, in which case the components are linear combinations of the original features. With an alternative metric, then the principal *coordinates* are governed by the distance function. CA aims to maximize the correspondence between column and row scores – or, equivalently, sample and feature scores – in a feature matrix with nonnegative elements. This approach is analogous to PCoA with χ^2 distance, a distance metric used to extract relationships between rows and columns. Lastly, NMDS maximizes the rank order between features and hence is less concerned with the underlying pair-wise distances [118].

3.2.4 Neural Network Approaches for Feature Extraction

Neural networks (NNs) are a popular machine learning model and have recently garnered heightened interest in the sequencing domains [119]. Essentially, NNs perform nonlinear adaptive regression. Unsupervised approaches in particular have garnered interest in their ability to extract meaningful features from unlabeled data. One architecture in particular is the denoising autoencoder (DAE), which has recently been shown to perform well when applied to high-dimensional gene expression datasets [120, 121]. Given an input matrix X, the DAE attempts to recover X after X has been corrupted with noise (Fig. 3). The noise enables the DAE to learn robust, potentially generalizable features while preventing it from simply learning the identity function.

Another NN approach involves applying word embeddings, a widely used strategy in the natural language processing domain, directly to sequencing reads. The word2vec model is one of the more popular word embedding models. It gives words continuous vector representation in a lower-dimensional space based on the frequency of pair co-occurrence in a context window of fixed length [122]. We can understand it as mapping each word to a point in a continuous high-dimensional space, such that the points of words with similar semantic meaning are closer to each other in terms of, for example, Euclidean distance. Ng utilized Skip-Gram word2vec to embed short DNA k-mers [123]. He demonstrated that the embedding space extracts useful properties. Specifically, k-mer pairs with high cosine similarity in the embedding space were consistent with high-scoring pairs identified via global sequence alignment.

Word2vec is a shallow, fully connected NN with one hidden layer (Fig. 4). The input and output layer have the same number of nodes which is the number of words in the vocabulary. The number of nodes in the hidden layer is the dimensionality of the embedding space – that is, the size of the reduced feature space. The first step is converting each word into a one-hot vector, thereby giving each word a unique index. Then, training can be performed in one of two varieties: (1) the Skip-Gram model, which uses a word to predict its context (i.e., neighboring words) and (2) the continuous bag-of-words model (CBOW), which uses the context to predict a

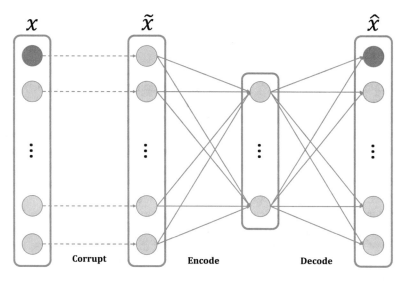

Fig. 3 Denoising autoencoder, where the input data x is corrupted with noise, producing $\tilde{x}$, which is then encoded into a lower-dimensional hidden layer. The hidden layer nodes are then decoded to produce $\hat{x}$, which has the same dimensionality as x. The distance between x and $\hat{x}$ is minimized such that the hidden layer is composed of features (nodes) capable of reconstructing x despite the addition of noise

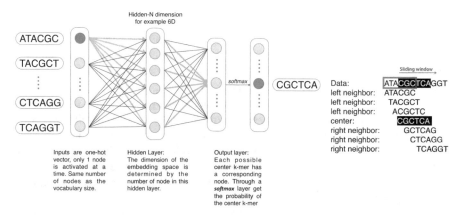

Fig. 4 (left) Neural network architecture for Skip-Graham word2vec. The training process requires the NN to predict the target word given the neighborhood. Words with similar context will activate similar nodes in the hidden layer. For the center k-mer "CGCTCA" and one of its neighbors "ATACGC," the corresponding node in the input and output layer is shown in red. Assuming "ATACGC" is the i-th word in the vocabulary, "CGCTCA" is then the $i + 1$-th word. The weights in blue connect between the input and hidden layers for the input word "ATACGC," i.e., the i-th column in weight matrix V. The weights in yellow connect the hidden and output layers for the output word "CGCTCA," i.e., the $i + 1$-th row of weight matrix U. (right) 6-mer neighborhood for word2vec training

word [123]. Because Skip-Gram does not average context vectors, but updates the weights for each word in the context separately, it can learn better representations for rare words compared to CBOW.

In the training process, the model will learn two vectors for each word w_i: (1) the i-th column of weight matrix V between the input and hidden layers and (2) the i-th row of weight matrix U between the hidden and output layers. We will refer to them as the input and output word vectors, respectively. The matrix product of V and U^T is the co-occurrence matrix. Levy and Goldberg described the word embedding as a matrix factorization of the co-occurrence matrix [124]. They also showed that a carefully constructed matrix factorization can produce word embeddings similar in quality to word2vec [125]. Also, Landgraf and Bellay showed that Skip-Gram word2vec is equivalent to weighted logistic PCA [126].

Training can be costly in terms of time and memory when the vocabulary is large. To accelerate the training process, Mikolov et al. utilized negative sampling [127]. Instead of updating the entire vocabulary each pass, they randomly sampled a subset of negative samples along with the context words to form a smaller vocabulary. Only the subset's weights are updated during a given pass. Another approach replaces the output weight matrix U with a Huffman tree [128].

4   Diversity Metrics and Constrained Ordination

After taxonomic or functional annotation has been performed, investigators are faced with the difficulty of quantitatively identifying and describing gradients, patterns, and variability within the dataset, particularly between individual samples or sample groups. Such analyses require simultaneous consideration of many, sometimes hundreds or thousands, distinct species or functions for each sample within the dataset. This effort is often further complicated by researchers who wish to include in their analysis information about the samples themselves or the sites from which they were collected, such as nutrient concentrations or availability, sample site location, host species (from which the samples were collected), vegetation composition or coverage, or watershed membership. The high degree of correlation expected between microbial community members and their environment requires the use of multivariate analytical methodologies.

Ordination is one of the most common analytical techniques used to explore the high-dimensional structure of microbial and molecular ecology datasets by using the distance matrix containing the similarity between metagenomic samples. Generally speaking, these methods attempt to identify the major ecological gradients or trends in high-dimensional datasets. Ordination methods can be largely categorized into two classes based upon the nature of the data to be used or the intent of the researcher. Unconstrained ordination methods (described above) employ only community data (i.e., the gene or taxon abundance table) in their calculations. Because unconstrained ordination relies only on species or functional abundances, the results expose or reveal the largest, and potentially most distinctive, gradients within the data. These methods are often used as a form of exploratory data analysis, where investigators may not possess well-conceptualized hypotheses or are interested in identifying unexpected gradients, patterns, or relationships between taxa, functions, or sample groups.

Constrained ordination methods exploit information about the samples themselves, or their environment, to hopefully explain the source of variation observed within the community dataset. This type of ordination is carried out by constraining the ordination object to be optimally correlated to the values of one or more predictor variables related to an ecologically relevant hypothesis under investigation. Constrained ordination methods can be viewed as analogs of regression models where data describing samples (e.g., sample types, source location, environmental data, etc.) are used directly to describe and subsequently interpret the structure of microbial communities. These methods are commonly used to directly test predetermined hypotheses, such as the effect of nutrient gradients or the impact of ecological disturbances. As with unconstrained ordination, there are many options available for carrying out constrained ordination. Some of the most commonly used are redundancy analysis (RDA) [129], distance-based redundancy analysis (db-RDA) [130], canonical correspondence analysis (CCA, which is distinct from canonical *correlation* analysis described above), and detrended canonical correspondence analysis (DCCA) [131].

With proper caution, constrained ordination methods may also be used during data exploration efforts, especially at the beginning of longer-term or larger-scale studies. Within this context, initial community results can be subjected to constrained ordination with explanatory variables being selected using stepwise variable selection methods such as those suggested by Blanchet et al. [132]. The resulting explanatory variable subset can be compared with results from other data exploration methods such as the BIOENV procedure proposed by Clarke and Ainsworth [133]. The end goal of these efforts is to enable the researcher to determine what explanatory variables may be the most important and will require further study, identify gaps in data collection, and improve or clarify the hypotheses driving the current study.

Ordination has traditionally been applied to manually collected taxa counts or coverage data as well as data describing environmental conditions. The emergence of sequencing technologies has led to the adoption of ordination to carry out similar analyses with both data resulting from both targeted amplicon and metagenomic sequencing. Metagenomic sequencing results are often annotated for both their taxonomic and functional content, providing investigators with two corresponding sources of information and reducing reliance on a single locus for taxonomic annotation and diversity estimates. In some cases side-by-side comparison of results obtained from ordinating the taxonomic and functional annotations have exposed interesting results [134].

Ordination provides a powerful way to probe large complex datasets, but as with any computational or statistical approach, an acute understanding is prerequisite for proper application and interpretation of results. Many decisions must be made regarding the proper choice of ordination method, the distance measures used (if any), whether or not the data should be transformed prior to calculation, as well as how to handle sample-level data. Further confusion may arise from the periodic development of new, perhaps better, ordination approaches that build upon the methods listed above. Thorough derivations and descriptions of ordination and clustering techniques are available in several well-written books [118, 135, 136] and review articles [137, 138], which may help educate investigators and students about appropriate approaches for answering ecological questions using ordination methods.

5 Statistical Inference

5.1 Multilevel Regression

Researchers may be interested in the relationship between a univariate statistic describing the taxonomic composition of a community (e.g., α-diversity, species richness, or evenness) and sample-level information such as site, temperature, time, or chemical concentration. Elucidating these relationships can be accomplished via linear regression:

$$y \sim N(X\beta, \sigma^2 I) \tag{11}$$

where y is a vector of length n, X is an $n \times p$ matrix of p sample-level covariates including an intercept term, β is a vector of regression coefficients of length p, and I is an $n \times n$ identity matrix [139]. The coefficients β and variance σ^2 can be estimated via least squares, where $\hat{\beta}$ represents the association between y and X.

Often, however, complex study designs necessitate the use of multilevel regression models, often referred to as mixed-effects models. As an example, suppose metagenomic samples are taken from ten sites, and α-diversity varies depending on which site the sample originated. One approach to model this data may involve coding each site with dummy variables, setting one arbitrary site as a "reference" level. The cost here is nine degrees of freedom, and we are limited in our ability to interpret the regression coefficients, since they can only be interpreted with respect to the reference level [140]. One can imagine that with even more sites, this approach becomes less practical.

An alternative strategy involves letting the intercept vary as a function of site (a random intercept model):

$$y_n = \mu + \alpha_{\text{site}[n]} + \epsilon_n \tag{12}$$

$$\alpha_{\text{site}} \sim N(\theta, \tau^2) \tag{13}$$

$$\epsilon \sim N(0, \sigma^2) \tag{14}$$

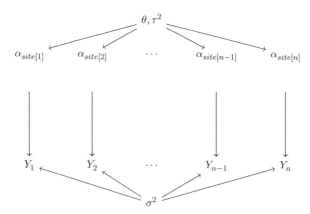

Fig. 5 Multilevel representation of sample-level means [141]. Each sample Y_n is influenced by its own between-site mean $\alpha_{\text{site}[n]}$. Each of these between-site means are generated by the same normal prior distribution with mean θ and variance τ^2. Within-site variability in Y is governed by σ^2. For sites with small sample sizes, $\alpha_{\text{site}[n]}$ will fall closer to θ, whereas sites with larger sample sizes will be represented more by their site-specific average (https://slack-files.com/T1VNV2ABW-F9YTS2K35-a06828b680)

where y_n is the alpha diversity for sample n, μ is the intercept, $\alpha_{\text{site}[n]}$ is the site-specific intercept for sample n, θ is the mean of site-specific intercepts, and τ^2 is their variance. Note that θ and τ^2 do not vary as a function of site. Figure 5 shows how the group-level means, α_{site}, distribute over the N samples. Let's now assume we believed that α-diversity varied as a function of temperature, but the degree of the relationship depended on the site. Here, we can let the slope between temperature and α-diversity vary:

$$y_n = \mu + \alpha_{\text{site}} \times \text{temp} + \epsilon_n \tag{15}$$

As more data become available, and study designs necessitate more complicated regression models, we can combine random effects and build complex multilevel regression models to help describe our community of interest. Such an approach is warranted because it allows an investigator to estimate the degree that specific effects vary by group (such as site) and not only with respect to a reference level. Moreover, because group-level effects share a common prior, a multilevel model can utilize group-level averages to "partially pool" information, thereby dampening the noisy contributions of underpowered group levels [139].

Multilevel models can easily be fit via the R package rstanarm. More sophisticated model designs can be implemented in Stan [142], which has interfaces in a variety of programming languages including R, Python, MATLAB, and Julia.

5.2 Multivariate Analysis

With univariate dependent variables, the regression approaches described above are an obvious choice. If, instead, we are interested in measuring the relationship between sample-level covariates and β-diversity, then we can turn to permutational MANOVA, which performs an analysis of variance on a distance matrix using sample-level covariates as predictors [143].

5.3 Differential Abundance Analysis

It is often of interest to detect which genes or taxa best differentiate two or more sample classes. A straightforward approach involves performing hypothesis testing for each variable in an abundance table and then correcting for the false discovery rate via the Benjamini-Hochberg procedure. Such an approach is limited in that it assumes specific assumptions are met prior to performing the analysis, which is typically not the case. These include assumptions regarding normality and mean-variance relationships.

More sophisticated strategies have been developed and applied to sequencing data of similar structure: edgeR and VOOM [144] for gene expression data, DESeq2 [145] for gene expression and 16S amplicon survey data [92], and MetagenomeSeq [146] also for 16S amplicon survey data. MetagenomeSeq, for example, applies either a zero-inflated Gaussian mixture model or a zero-inflated log-normal model to each feature separately. Account for zero inflation is thought to prevent overdispersed fits and also mitigate the detrimental effects of highly sparse sequencing data. DESeq2, on the other hand, first performs a variance-stabilization transformation, followed by fitting a negative binomial generalized linear model (GLM). For metagenomic data, under the right circumstances such that the abundance table consists of counts and hence has yet to be normalized in terms of sample library size imbalance, these approaches may prove viable. Still, however, they were developed with specific distributional assumptions in mind. Considering the plethora of normalization strategies available for metagenomic data, future work is necessary to demonstrate whether readily used differential abundance strategies remain appropriate after a particular normalization is performed and which metagenomic normalization procedures work well in tandem with which differential abundance strategies.

Approaches nevertheless exist that were designed with metagenomic data in mind. One such approach, LEfSe, performs the nonparametric Kruskal-Wallis sum-rank test to identify significant differences in abundances between genes or taxa belonging to a class of interest [147]. The Wilcoxon rank sum then disentangles pair-wise differences between sample subclasses. Linear discriminant analysis is applied last to estimate the effect size of the statistically significant features.

Johnsson et al. applied various differential abundance methods to metagenomic data and evaluated their performance in terms of statistical power, control of the false discovery rate, and uniformity of p-values given the null hypothesis [148]. They found that GLM-based models that combat potential overdispersion perform best. These include DESeq2, edgeR, and an overdispersed Poisson GLM. MetagenomeSeq generally performed well but was found to be inferior to simply performing t-tests to log-transformed features, suggesting that the zero-inflation mixture components have a negligible impact. Also, it was prone to highly biased p-values and consequently type 1 error. Also of note was that performing t-tests on square root transformed features was superior to utilizing non-parametric Wilcoxon rank-sum tests, which, as noted above, are used in LEfSe. The authors speculated this may be due to the latter's susceptibility to ties. It should be stressed that effect of different metagenomic-specific normalization approaches on differential abundance analysis was not explored.

6 Machine Learning and its Application to Metagenomics

6.1 Overview

Machine learning techniques are widely used in different steps in a metagenomic pipeline. For example, Naive Bayes has been applied for taxonomic classification, hidden Markov models (HMM) are often used for functional annotation, and random forest is readily utilized for phenotype prediction. From a research problem perspective, machine learning techniques are helpful in addressing the following questions:

- Who are there (what species are in a sample)?
- What are they doing (what functions are in a sample)?
- What can we infer from the sample (what is the state of the host/environment)?

In the following sections, we will talk about machine learning methods and tools that have been applied to metagenomics.

6.2 A General Machine Learning Review

One of the highly cited definitions of machine learning involves a computer program that is said to learn from experience E with respect to some class of tasks T and performance measure P, if its performance at tasks in T, as measured by P, improves with experience E [149]. Experience E usually refers to the data collected. Task T usually represents the decision or prediction we want to make. In a metagenomics context, E represents the samples or DNA sequences.

Machine learning methods can be classified into supervised, semi-supervised, and unsupervised learning approaches based on their dependency of labeled data. Supervised learning approaches (classification methods) require a training phase that utilizes the samples and their labels to minimize the cost function of the classifier. Unsupervised learning (clustering) methods, on the other hand, use a distance measure to group samples into clusters. Finally, semi-supervised learning methods act as a compromise between the two. They first use a subset of training data to train the classifier and then use unlabeled examples to improve performance. Table 2 lists the machine learning methods listed throughout this chapter.

In the context of metagenomics, machine learning techniques can be classified based on their application. The following sections will discuss the machine learning techniques that have been implemented in taxonomic classification, DNA binning, functional annotation, and phenotype prediction.

6.3   Taxonomic Classification and DNA Binning

One of the main challenges in metagenomics is the identification of microorganisms in clinical and environmental samples [150]. Taxonomic classification or DNA binning are helpful for researchers to determine the composition of their metagenomic samples. Taxonomic classification is a supervised learning problem, whereas DNA binning has traditionally been unsupervised, but could also be semi-supervised [151].

6.3.1   Naive Bayesian Classifier

A naive Bayes classifier (NBC) is a type of probabilistic classifier that exploits Bayes rule to perform classification. Naive refers to its assumption that features are independent from each other. Here $X = (x_1, x_2, \ldots, x_n)$ is an observation with n features. The probability of X coming from class k is

$$p(C_k | x_1, x_2, \ldots, x_n) = \frac{p(C_k) p(X | C_k)}{p(X)} = \frac{p(C_k) p(x_1 | C_k) p(x_2 | C_k) \cdots p(x_n | C_k)}{\sum_{k=1}^{|C|} p(C_k) p(x_1 | C_k) p(x_2 | C_k) \cdots p(x_n | C_k)} \tag{16}$$

So, the estimated class is

$$\hat{c} = \arg \max_{k \in 1, \ldots, K} p(C_k) \prod_{i=1}^{n} p(x_i | C_k) \tag{17}$$

NBC is easy to implement and has high accuracy when the features are independent. Rosen et al. proposed a metagenome fragment classification tool using k-mer frequency profiles [152, 153], which has proven to be fast and accurate when trained

Table 2 A list of machine learning algorithms and their implementations

Requires labels	Methods	Abbrev.	Implementation
N	Canonical correlation analysis	CCA	CCA(R), scikit-learn (Python)
N	Denoising autoencoder	DAE	keras
N	Independent component analysis	ICA	fastICA (R), scikit-learn (P)
Y	Linear discriminant analysis	LDA	MASS (R), scikit-learn (P)
N	Nonmetric multidimensional scaling	NMDS	vegan (R)
N	Nonnegative matrix factorization	NMF	nmf (R), scikit-learn (P)
N	Principal component analysis	PCA	prcomp (R), scikit-learn (P)
Y	Supervised principal component analysis	supervised PCA	superpc (R), supervisedPCA-Python (P)
N	Sparse principal component analysis	Sparse PCA	Elasticnet (R), scikit-learn (P)
N	Principal coordinate analysis	PCoA/MDS	vegan (R)
N	t-Distribution stochastic neighbor embedding	t-SNE	Rtsne (R), scikit-learn (P)
N	Word2vec	Word2vec	Gensim (P), keras
Y	Elastic net	EN	Glmnet (R, P)
Y	Hidden Markov model	HMM	HMM (R), Stan, hmmlearn (P)
Y	k-nearest neighbors	k-NN	Caret (R), scikit-learn (P)
Y	Naive Bayes	NB	Caret (R), scikit-learn (P), Stan, NBC[a]
Y	Support vector machine	SVM	Caret (R), scikit-learn (P)
Y	Random forest	RF	Caret (R), scikit-learn (P)
N	k-means/Medoids	k-means/Medoids	kmeans (R), scikit-learn (P), MetaCluster[b]

[a] http://nbc.ece.drexel.edu/
[b] http://i.cs.hku.hk/~alse/MetaCluster/

on large k-mers. Assuming $X = [x_1, x_2, \cdots, x_n]$ is a set of k-mers in a sequence, then, in the training phase, $p(x_i|C_k)$ is estimated by the total number of k-mers x_i that occurs in all the training sequences from class C_k. In the testing phase, the taxa that contains a given query sequence is predicted by the class that maximizes the posterior probability $P(C_k|X)$.

6.3.2 k-Nearest Neighbors

k-nearest neighbors (k-NN) classifies query samples $x_0 \in R^m$ based on distance. Given the labeled data, $T = \{x_i, y_i \mid i \in [1, N]\}$, $x_i \in R^m$, a new observation will be assigned to a class where the majority of the first K nearest labeled samples originate. The following is the classification workflow for 1-NN and K-NN:

1. 1-NN: (1) find the nearest instance in the training set $\min_{x_i}(\|x_o - x_i\|)$; (2) test data label is the same as the nearest instance via $y_0 = y_i$.
2. K-NN: (1) find the k nearest instances in the training set $\min_{x_1, \cdots, x_k}(\|x_o - x_i\|)$; (2) let the k nearest instances vote via $y_0 = \text{Mode}(y_1, \cdots, y_k)$.

Borozan et al. used K-NN to perform classification in their taxonomic lineage prediction tool, and they regarded K-NN as one of the simplest and most intuitive classification algorithms [154].

6.3.3 Clustering

The k-means clustering algorithm is used to partition N observations into k clusters. The observations' affiliations are determined by some distance measure, such as Euclidean distance. Hence, the observations that are close to each other will be grouped together, and the observations that are distant from each other will be assigned to different clusters. To converge to an optimum quickly, this clustering process utilizes an expectation-maximization (EM) procedure. This is an iterative refinement approach that assigns observations into k clusters by comparing the distance between the observations and k centroids (usually initialized randomly) and then updates the centroids with the new cluster assignment until convergence. The objective function is given by

$$\arg\min_C \sum_{i=1}^{k} \sum_{x \in C_i} ||x - \mu_i||^2 \qquad (18)$$

Many tools perform DNA binning by clustering sequences based on a predefined distance metric. Wang et al. used k-means to cluster sequences [155].

An alternative approach, CD-HIT, uses a greedy search algorithm to cluster the sequences [156]. First, it sorts sequences based on their length. The longest sequence will be representative of the first cluster formed. Then, the second longest sequence is compared to this cluster's representatives. It will be assigned to this cluster if the distance between it and the representative is within a user-selected distance threshold; otherwise, a new cluster will be created and the sequence becomes its representative. This process will be repeated for all remaining sequences until all sequences are assigned to either an existing cluster or a newly created one. In addition to CD-HIT, there are some other sequence clustering tools such as DNACLUST [157] and UCLUST [158]. These tools can be faster than CD-HIT under some circumstance. For example, UCLUST, by default, operates in an inexact mode that reduces the search space by only comparing the sequence to the representatives from a subset of all clusters.

6.4   Functional Annotation and Prediction

6.4.1   Hidden Markov Model

Hidden Markov models (HMMs) describe a sequential observation and their underlying latent states. The observed sequence of length L can be described as $\mathbf{x} = x_1, x_2, \cdots, x_L$ [159]. Its latent state sequence is $y = y_1, y_2, \cdots, y_L$. Each symbol x_n takes on a finite number of possible values from the set of observations $\mathbf{O} = O_1, O_2, \cdots, O_N$, and each state y_n takes one of the values from the set of states $\mathbf{S} = 1, 2, \cdots, M$, where N and M denote the number of distinct observations and states in the model. This model can be described by two matrices, the transition matrix and the emission matrix. One entry in the transition matrix is the probability of entering state j in the next time given in state i:

$$t(i, j) = P(y_{n+1} = j | y_n = i) \tag{19}$$

One entry in the emission matrix is the probability of observing x given state i:

$$e(x|i) = P(x_n = x | y_n = i) \tag{20}$$

The probability that an HMM will generate an observation $\mathbf{x}$ with underlying state sequence $\mathbf{y}$ is [159]:

$$p(x, y|t, e) = p(x|y, t, e) p(y|t, e) \tag{21}$$

where,

$$p(x|y, t, e) = p(x_1|y_1) p(x_2|y_2) \cdots p(x_L|y_L) \tag{22}$$

$$p(y|t, e) = p(y_0) p(y_1, y_2) p(y_2|y_3) \cdots p(y_{L-1}|y_L) \tag{23}$$

Rho et al. proposed to use HMMs to model nucleotide sequences to predict a given gene [160].

6.4.2 Logistic Regression

Logistic regression usually takes real value input and outputs a value between 0 and 1. This is accomplished by the logistic function:

$$\hat{y}(\mathbf{x}) = \frac{1}{1 + e^{-\sum_i^n \beta_i \cdot x_i}} \tag{24}$$

where $\mathbf{x} = [x_1, x_2, \cdots, x_n]$ is an input vector and $\boldsymbol{\beta} = [\beta_1, \beta_2, \cdots, \beta_n]$ are the weights estimated by the model. Since the output of this function is between 0 and 1, one can consider the output to be the probability of being classified into class C, i.e., $p(y = C|\mathbf{x}; \boldsymbol{\beta}) = 1/1 + e^{-\sum_i^n \beta_i \cdot x_i}$. Hence, we can determine the best parameter β using a maximum likelihood approach:

$$\max_{\beta} L(Y|\mathbf{X}; \boldsymbol{\beta}) \tag{25}$$

where $L(Y|\mathbf{X}; \boldsymbol{\beta})$ is the product of the probabilities that all labeled samples get classified into the correct class, i.e.,

$$L(Y|\mathbf{X}; \boldsymbol{\beta}) = \prod_{i=1}^{n} \hat{y}(\mathbf{x_i})^{y(\mathbf{x_i})} (1 - \hat{y}(\mathbf{x_i}))^{1-y(\mathbf{x_i})} \tag{26}$$

where $y(\mathbf{x_i})$ is the true label for sample $\mathbf{x_i}$. The optimal parameter to maximize the likelihood can be found using a gradient descent algorithm, which iteratively updates the parameters by the estimated derivative of the function given the current parameter such that the likelihood tends to increase after each update. In the testing phase, observations $\mathbf{x_i}$ will be classified into class C if the output $\hat{y}(\mathbf{x_i})$ is greater than 0.5 or a predefined threshold; otherwise, it will be classified as the alternative class $\neg C$.

Noguchi et al. used logistic regression to analyze the GC content of a given sequence and estimate the mono-codon and di-codon frequencies [161].

6.5 Phenotype Prediction

6.5.1 Random Forest

A decision tree is a supervised learning technique that looks at each feature individually to make a binary decision, thereby splitting samples into branches. The information gain is maximized during this process to help the classifier make an accurate decision. It is widely used because the decision process is interpretable, and the performance is often promising. Random forest (RF) is an ensemble learning extension of a decision tree where the decision is made by majority vote of many decision trees. Each tree is fit to only a subset of features, forcing the classifier to learn robust, potentially generalizable subsets of features, particularly when compared to simpler decision tree approaches. The following is the construction workflow of random forest [162]:

1. Draw $ntree$ bootstrap samples from the original data.
2. Grow a tree for each bootstrap dataset. At each node of the tree, randomly select $mtry$ variables for splitting. Grow the tree so that each terminal node has no fewer than $nodesize$ cases.
3. Aggregate information from the $ntree$ trees for new data prediction such as majority voting for classification.
4. Compute an out-of-bag (OOB) error rate by using the data not in the bootstrap sample.

RF has been applied in many metagenomic pipelines to predict phenotype given a high-dimensional abundance table. It naturally finds useful features and is robust to overfitting. Additional applications of RF can be found in Chen and Ishwaran [162].

6.5.2 Support Vector Machine

The support vector machine (SVM) finds a hyperplane or a set of hyperplanes that best separates labeled data in some geometric space. In the testing phase, samples are assigned to classes based on their location in this space. Normally, the linear SVM separates the space linearly, but when data are not linearly separable, the "kernel trick" enables the data to be projected into a higher dimensional space, thereby potentially rendering the data linearly separable. Hence, the core of the SVM model is a linear SVM algorithm. The following is an overview of the application of this model in a binary classification problem. Given some training data $\mathcal{D}$, a set of n points have the form

$$\mathcal{D} = \left\{ (\mathbf{x}_i, y_i) \mid \mathbf{x}_i \in \mathbb{R}^m, \ y_i \in \{-1, 1\} \right\}_{i=1}^{n} \tag{27}$$

where y_i is either 1 or -1, indicating the class to which the point $\mathbf{x}_i$ belongs. Each $\mathbf{x}_i$ is an m-dimensional real vector. We want to find the maximum-margin hyperplane that divides the points having $y_i = 1$ from those having $y_i = -1$. If the training data are linearly separable, we can select two hyperplanes that completely separate the data and then try to maximize the distance between the data and hyperplanes. The region bounded by them is called "the margin." These hyperplanes can be described by the equations

$$\begin{cases} \omega \cdot \mathbf{x} - b = 1 \\ \omega \cdot \mathbf{x} - b = -1 \end{cases} \tag{28}$$

It is an optimization problem to find these two hyperplane. Maximizing the distance between the data and hyperplane is equivalent to minimizing $||\boldsymbol{\omega}||$. While ensuring all positive and negative samples are separated, we have a constraint: $\forall(\mathbf{x}_i, y_i) \in \mathcal{D}$, $y_i(\boldsymbol{\omega} \cdot \mathbf{x}_i - b) \geq 1$. Again, in the testing phase, samples are assigned to classes by the subspaces segmented by the hyperplanes. Capriotti et al. developed a method based on kerneled SVMs to predict whether a new phenotype derived from a single nucleotide polymorphism can be related to a genetic disease in humans [163].

6.5.3 Elastic Net

A sparsity-promoting approach related to Lasso, but more robust for highly correlated features, is the elastic net (EN) [164]. It has been shown to perform well in both regression and classification, particularly with high-dimensional data where the number of features greatly outnumbers the number of samples [165]. Whereas the Lasso regularization penalty involves the L1-norm, EN compromises between the L1- and L2-norm (readers may recognize L2-norm penalized regression as ridge regression). EN is formally given by

$$\theta^* = \arg\min_{\theta \in \mathbb{R}^p} \frac{1}{2} \sum_{i=1}^{n} (y_i - \theta^\mathsf{T} \mathbf{x}_i)^2 + \lambda \left(\frac{1}{2}(1-\alpha)|\theta|_2^2 + \alpha|\theta|_1 \right) \qquad (29)$$

where $\alpha \in [0, 1]$ controls the relative contributions of the L1- and L2-norms. Notice that when $\alpha = 1$, the EN reduces to the Lasso, whereas when $\alpha = 0$, it reduces to ridge regression; thus, EN can be considered a generalization of the two regularization approaches.

7 Discussion and Conclusion

Techniques can generally be broken down into two main categories: (1) techniques that directly work with DNA/RNA sequences to classify attributes about them (taxonomy and function) and (2) techniques that facilitate comparative analyses. Some fundamental preprocessing steps – such as normalization, feature selection, and feature extraction – can be applied to single samples; however, most of these preprocessing steps are designed for multiple samples, as most studies use many samples and are usually limited by cost.

For the sequence identification problems, the longest-standing methods are those that extend read sequence length and identify its taxonomic origin and functional annotation. Assembly, with the most successful methods involving de Bruijn graphs, was one of the first algorithms to be developed because it was key in the Human Genome Project (even from the longer Sanger sequencing reads) [166]. In metagenomics, the problem is more complex since any read can come from any one of thousands of organisms in a sample, culminating in a demultiplexing (read binning) step before assembly. Taxonomic/functional binning and classification, with the rich history of k-mer-based clustering, alignment, and profile HMMs, have also been extensively studied but are still being investigated for their extensions to metagenomics (not just annotating whole genomes). The problem with metagenomic data is mainly that sequences come from a collection of possible species' origins and many sequences are from unknown (or not yet sequenced) species.

Comparative analysis is the most active area of development, notably the rich areas of statistical inference and machine learning, which are utilized to make cross-sample and even cross-study comparisons. Early attempts leveraged ordination, but the substantial growth of the machine learning field has provided researchers with an immense resource of potential tools, particularly classification algorithms, allowing one to apply discriminative and generative functions to discern groups of samples. Moreover, deep neural networks show much promise for learning complex relationships and hence are areas of active research.

We closed this chapter with a discussion of general machine learning approaches, since these techniques can be applied to not only sequence identification but also to comparative analysis and phenotype prediction. Machine learning and statistical inference can help researchers disentangle the complexity that make other models with strict assumptions fail. We show examples of where these algorithms have been applied. Still, when using learning algorithms, one must think about how much training data is available and whether supervised versus unsupervised learning is suitable. Also, sometimes there are many confounding factors, where feature selection or normalization may simplify and denoise the data. If there is inherent structure in data, hierarchical models which capture this structure should be used. If prediction is the goal, supervised approaches should be considered. Finally, no matter what method is used, researchers should be aware of class imbalance and model overfitting and try to mitigate these effects through carefully designing training/validation/testing regimes. There are many considerations that researchers should consider when analyzing complex metagenomic data, and these should be identified early and examined throughout analyses.

In this book chapter, important techniques in metagenomic analyses are reviewed. However, good benchmarking data and infrastructure is not available to ensure that future methods improve upon the state of the art. Therefore, there is not only much work to be done to improve metagenomic software, but there is need to standardize the way we assess these methods.

References

1. Handelsman J, Rondon M, Brady S, Clardy J, Goodman R. Molecular biological access to the chemistry of unknown soil microbes: a new frontier for natural products. Chem Biol. 1998;5(10):R245–9.
2. Handelsman J. Metagenomics: application of genomics to uncultured microorganisms. Microbiol Mol Biol Rev. 2004;68(4):669+.
3. Pace N, Stahl D, Lane D, Olsen G. The analysis of natural microbial-populations by ribosomal-RNA sequences. Adv Microb Ecol. 1986;9:1–55.
4. Simon C, Daniel R. Metagenomic analyses: past and future trends. Appl Environ Microbiol. 2011;77(4):1153–61.
5. Streit W, Schmitz R. Metagenomics – the key to the uncultured microbes. Curr Opin Microbiol. 2004;7(5):492–8.
6. Tringe SG, Hugenholtz P. A renaissance for the pioneering 16S rRNA gene. Curr Opin Microbiol. 2008;11(5):442–6.
7. Ward N. New directions and interactions in metagenomics research. FEMS Microbiol Ecol. 2006;55(3):331–8.
8. Lozupone C, Knight R. UniFrac: a new phylogenetic method for comparing microbial communities. Appl Environ Microbiol. 2005;71(12):8228–35.
9. Solden L, Lloyd K, Wrighton K. The bright side of microbial dark matter: lessons learned from the uncultivated majority. Curr Opin Microbiol. 2016;31:217–26.
10. Vieites JM, Guazzaroni ME, Beloqui A, Golyshin PN, Ferrer M. Metagenomics approaches in systems microbiology. FEMS Microbiol Rev. 2009;33(1):236–55.
11. Woese CR, Fox GE. Phylogenetic structure of the prokaryotic domain: the primary kingdoms. Proc Nat Acad Sci. 1977;74(11):5088–90. Available from: http://www.pnas.org/cgi/doi/10.1073/pnas.74.11.5088.
12. Eckburg PB, Bik EM, Bernstein CN, Purdom E, Dethlefsen L, Sargent M, et al. Diversity of the human intestinal microbial flora. Science. 2005;308(5728):1635–8.
13. Edwards RA, Rodriguez-Brito B, Wegley L, Haynes M, Breitbart M, Peterson DM, et al. Using pyrosequencing to shed light on deep mine microbial ecology. BMC Genomics. 2006;7:1–13.
14. Ley RE, Peterson Da, Gordon JI. Ecological and evolutionary forces shaping microbial diversity in the human intestine. Cell. 2006;124(4):837–48. Available from: http://www.ncbi.nlm.nih.gov/pubmed/16497592.
15. Turnbaugh PJ, Ridaura VK, Faith JJ, Rey FE, Knight R, Gordon JI. The effect of diet on the human gut microbiome: a metagenomic analysis in humanized gnotobiotic mice. Sci Trans Med. 2009;1(6):6ra14. Available from: http://www.pubmedcentral.nih.gov/articlerender.fcgi?artid=2894525&tool=pmcentrez&rendertype=abstract.
16. Venter J, Remington K, Heidelberg J, Halpern A, Rusch D, Eisen J, et al. Environmental genome shotgun sequencing of the Sargasso Sea. Science. 2004;304(5667):66–74.
17. Edwards RA, McNair K, Faust K, Raes J, Dutilh BE. Computational approaches to predict bacteriophage-host relationships. FEMS Microbiol Rev. 2016;40(2):258–72.
18. Forbes JD, Knox NC, Ronholm J, Pagotto F, Reimer A. Metagenomics: the next culture-independent game changer. Front Microbiol. 2017;8:1069. Available from: http://dx.doi.org/10.3389/fmicb.2017.01069.
19. Hurwitz BL, U'Ren JM, Youens-Clark K. Computational prospecting the great viral unknown. FEMS Microbiol Lett. 2016;363(10):1–12.
20. Kimura N. Metagenomic approaches to understanding phylogenetic diversity in quorum sensing. Virulence. 2014;5(3):433–42.
21. Mathieu A, Vogel TM, Simonet P. The future of skin metagenomics. Res Microbiol. 2014;165(2):69–76.
22. Sangwan N, Xia F, Gilbert JA. Recovering complete and draft population genomes from metagenome datasets. Microbiome. 2016;4:2–11.

23. Schmieder R, Edwards R. Insights into antibiotic resistance through metagenomic approaches. Future Microbiol. 2012;7(1):73–89.
24. Altschul S, Gish W, Miller W, Myers E, Lipman D. Basic local alignment search tool. J Mol Biol. 1990;215(3):403–10.
25. Caporaso JG, Kuczynski J, Stombaugh J, Bittinger K, Bushman FD, Costello EK, et al. QIIME allows analysis of high-throughput community sequencing data. Nat Methods. 2010;7(5):335–6.
26. Giardine B, Riemer C, Hardison R, Burhans R, Elnitski L, Shah P, et al. Galaxy: a platform for interactive large-scale genome analysis. Genome Res. 2005;15(10):1451–5.
27. Huson DH, Auch AF, Qi J, Schuster SC. MEGAN analysis of metagenomic data. Genome Res. 2007;17(3):377–86.
28. Li W, Godzik A. Cd-hit: a fast program for clustering and comparing large sets of protein or nucleotide sequences. Bioinformatics. 2006;22(13):1658–9.
29. Rosen GL, Reichenberger ER, Rosenfeld AM. NBC: the Naive Bayes classification tool web-server for taxonomic classification of metagenomic reads. Bioinformatics. 2011;27(1):127–9.
30. Schloss PD, Westcott SL, Ryabin T, Hall JR, Hartmann M, Hollister EB, et al. Introducing mothur: open-source, platform-independent, community-supported software for describing and comparing microbial communities. Appl Environ Microbiol. 2009;75(23):7537–41.
31. Wang Q, Garrity GM, Tiedje JM, Cole JR. Naive Bayesian classifier for rapid assignment of rRNA sequences into the new bacterial taxonomy. Appl Environ Microbiol. 2007;73(16):5261–67.
32. Tyson GW, Chapman J, Hugenholtz P, Allen EE, Ram RJ, Richardson PM, et al. Community structure and metabolism through reconstruction of microbial genomes from the environment. Nature. 2004;428(6978):37–43.
33. Sedlar K, Kupkova K, Provaznik I. Bioinformatics strategies for taxonomy independent binning and visualization of sequences in shotgun metagenomics. Computational and Structural Biotechnology Journal 2017;15:48–55. Available from: http://doi.org/10.1016/j.csbj.2016.11.005.
34. Mende DR, Waller AS, Sunagawa S, Järvelin AI, Chan MM, Arumugam M, et al. Assessment of metagenomic assembly using simulated next generation sequencing data. PLoS One. 2012;7(2):1–11.
35. Vázquez-Castellanos JF, García-López R, Pérez-Brocal V, Pignatelli M, Moya A. Comparison of different assembly and annotation tools on analysis of simulated viral metagenomic communities in the gut. BMC Genomics. 2014;15(1):37. Available from: http://bmcgenomics.biomedcentral.com/articles/10.1186/1471-2164-15-37.
36. Mande SS, Mohammed MH, Ghosh TS. Classification of metagenomic sequences: methods and challenges. Brief Bioinform. 2012;13(6):669–81.
37. Imelfort M, Parks D, Woodcroft BJ, Dennis P, Hugenholtz P, Tyson GW. GroopM: an automated tool for the recovery of population genomes from related metagenomes. PeerJ. 2014;2:e603. Available from: https://peerj.com/articles/603.
38. Ribeca P, Valiente G. Computational challenges of sequence classification in microbiomic data. Brief Bioinform. 2011;12(6):614–25.
39. Mohammed M, Ghosh TS, Singh NK, Mande SS. SPHINX – an algorithm for taxonomic binning of metagenomic sequences. Bioinformatics. 2010;27(1):22–30.
40. Cole ST, Brosch R, Parkhill J, Garnier T, Churcher C, Harris D, et al. Deciphering the biology of mycobacterium tuberculosis from the complete genome sequence. Nature. 1998, p. 537–544. Available from: http://dx.doi.org/10.1038/31159.
41. Albertsen M, Hugenholtz P, Skarshewski A, Nielsen KL, Tyson GW, Nielsen PH. Genome sequences of rare, uncultured bacteria obtained by differential coverage binning of multiple metagenomes. Nat Biotechno. 2013;31(6):533–8.
42. Alneberg J, Bjarnason BS, De Bruijn I, Schirmer M, Quick J, Ijaz UZ, et al. Binning metagenomic contigs by coverage and composition. Nat Methods. 2014;11(11):1144–6.
43. Miller IJ, Chevrette MG, Kwan JC. Interpreting microbial biosynthesis in the genomic age: biological and practical considerations. Marine Drugs. 2017, 1–24. Available from: http://dx.doi.org/10.3390/md15060165.

44. Lykidis A, Chen CL, Tringe SG, McHardy AC, Copeland A, Kyrpides NC, et al. Multiple syntrophic interactions in a terephthalate-degrading methanogenic consortium. ISME J. 2011;5(1):122–30.

45. Belda-Ferre P, Alcaraz LD, Cabrera-Rubio R, Romero H, Simón-Soro A, Pignatelli M, et al. The oral metagenome in health and disease. ISME J. 2012;6(1):46–56.

46. Qin J, Li R, Raes J, Arumugam M, Burgdorf KS, Manichanh C, et al. A human gut microbial gene catalogue established by metagenomic sequencing. Nature. 2010;464(7285):59–65.

47. Sangwan N, Xia F, Gilbert JA. Recovering complete and draft population genomes from metagenome datasets. Microbiome. 2016;4(1):8. Available from: http://www.microbiomejournal.com/content/4/1/8.

48. Mohammed MH, Ghosh TS, Reddy RM, Reddy CV, Singh NK, Mande SS. INDUS – a composition-based approach for rapid and accurate taxonomic classification of metagenomic sequences. BMC Genomics. 2011;12(Suppl 3). Available from: http://www.hubmed.org/display.cgi?uids=22369237.

49. Schmieder R, Edwards R. Fast identification and removal of sequence contamination from genomic and metagenomic datasets. PLoS One. 2011;6(3):1–11.

50. Segata N, Waldron L, Ballarini A, Narasimhan V, Jousson O, Huttenhower C. Metagenomic microbial community profiling using unique clade-specific marker genes. Nat Methods. 2012;9(8):811–4. Available from: http://www.pubmedcentral.nih.gov/articlerender.fcgi?artid=3443552&tool=pmcentrez&rendertype=abstract.

51. Liu B, Gibbons T, Ghodsi M, Pop M. MetaPhyler: taxonomic profiling for metagenomic sequences. In: Proceedings – 2010 IEEE international conference on bioinformatics and biomedicine, BIBM 2010; 2010, p. 95–100.

52. Sunagawa S, Mende DR, Zeller G, Izquierdo-Carrasco F, Berger Sa, Kultima JR, et al. Metagenomic species profiling using universal phylogenetic marker genes. Nat Methods. 2013;10(12):1196–9. Available from: http://www.ncbi.nlm.nih.gov/pubmed/24141494.

53. Nayfach S, Pollard KS. Average genome size estimation improves comparative metagenomics and sheds light on the functional ecology of the human microbiome. Genome Biol. 2015;16(1):51. Available from: http://genomebiology.com/2015/16/1/51.

54. Freitas TAK, Li PE, Scholz MB, Chain PSG. Accurate read-based metagenome characterization using a hierarchical suite of unique signatures. Nucleic Acids Res. 2015;43(10): e69(1–14).

55. Huson DH, Auch AF, Qi J, Schuster SC. MEGAN analysis of metagenomic data. Genome Res. 2007;17(3):377–86. Available from: http://www.hubmed.org/display.cgi?uids=17255551.

56. Ounit R, Wanamaker S, Close TJ, Lonardi S. CLARK: fast and accurate classification of metagenomic and genomic sequences using discriminative k-mers. BMC Genomics. 2015;16(1):236. Available from: http://www.biomedcentral.com/1471-2164/16/236.

57. Wood DE, Salzberg SL. Kraken: ultrafast metagenomic sequence classification using exact alignments. Genome Biol. 2014;15(3):R46. Available from: http://www.pubmedcentral.nih.gov/articlerender.fcgi?artid=4053813&tool=pmcentrez&rendertype=abstract.

58. Ames SK, Hysom DA, Gardner SN, Lloyd GS, Gokhale MB, Allen JE. Scalable metagenomic taxonomy classification using a reference genome database. Bioinformatics (Oxford, England). 2013;29(18):2253–60. Available from: http://www.ncbi.nlm.nih.gov/pubmed/23828782%5Cnhttp://www.pubmedcentral.nih.gov/articlerender.fcgi?artid=PMC3753567.

59. Sobih A, Tomescu AI, Mäkinen V. Metaflow: metagenomic profiling based on whole-genome coverage analysis with min-cost flows. In: Lecture Notes in Computer Science (including subseries Lecture Notes in Artificial Intelligence and Lecture Notes in Bioinformatics), vol. 9649; 2016. p. 111–121.

60. Rosen G, Garbarine E, Caseiro D, Polikar R, Sokhansanj B. Metagenome fragment classification using N-mer frequency profiles. Adv Bioinform. 2008;2008:205969. Available from: http://www.hubmed.org/display.cgi?uids=19956701.

61. Darling AE, Jospin G, Lowe E, Matsen FA, Bik HM, Eisen JA. PhyloSift: phylogenetic analysis of genomes and metagenomes. PeerJ. 2014;2:e243. Available from: https://peerj.com/articles/243.

62. McIntyre ABR, Ounit R, Afshinnekoo E, Prill RJ, Hénaff E, Alexander N, et al. Comprehensive benchmarking and ensemble approaches for metagenomic classifiers. Genome Biol. 2017;18(1):182. Available from: http://genomebiology.biomedcentral.com/articles/10.1186/s13059-017-1299-7.
63. Lindgreen S, Adair KL, Gardner PP. An evaluation of the accuracy and speed of metagenome analysis tools. Sci Rep. 2016;6:1–14. Available from: http://dx.doi.org/10.1038/srep19233.
64. Prakash T, Taylor TD. Functional assignment of metagenomic data: challenges and applications. Brief Bioinform. 2012;13(6):711–27. Prakash, Tulika Taylor, Todd D eng Research Support, Non-U.S. Gov't Review England 2012/07/10 06:00 Brief Bioinform. 2012;13(6):711–27. https://doi.org/10.1093/bib/bbs033.Epub2012Jul6.
65. Carr R, Borenstein E. Comparative analysis of functional metagenomic annotation and the mappability of short reads. PLoS One. 2014;9(8):e105776. Carr, Rogan Borenstein, Elhanan eng DP2 AT007802/AT/NCCIH NIH HHS/ P30 DK089507/DK/NIDDK NIH HHS/ DP2 AT007802-01/AT/NCCIH NIH HHS/Comparative Study Research Support, N.I.H., Extramural 2014/08/26 06:00 PLoS One. 2014;9(8):e105776. https://doi.org/10.1371/journal.pone.0105776. eCollection 2014.
66. O'Leary NA, Wright MW, Brister JR, Ciufo S, Haddad D, McVeigh R, et al. Reference sequence (RefSeq) database at NCBI: current status, taxonomic expansion, and functional annotation. Nucleic Acids Res. 2016;44(D1):D733–45. O'Leary, Nuala A Wright, Mathew W Brister, J Rodney Ciufo, Stacy Haddad, Diana McVeigh, Rich Rajput, Bhanu Robbertse, Barbara Smith-White, Brian Ako-Adjei, Danso Astashyn, Alexander Badretdin, Azat Bao, Yiming Blinkova, Olga Brover, Vyacheslav Chetvernin, Vyacheslav Choi, Jinna Cox, Eric Ermolaeva, Olga Farrell, Catherine M Goldfarb, Tamara Gupta, Tripti Haft, Daniel Hatcher, Eneida Hlavina, Wratko Joardar, Vinita S Kodali, Vamsi K Li, Wenjun Maglott, Donna Masterson, Patrick McGarvey, Kelly M Murphy, Michael R O'Neill, Kathleen Pujar, Shashikant Rangwala, Sanjida H Rausch, Daniel Riddick, Lillian D Schoch, Conrad Shkeda, Andrei Storz, Susan S Sun, Hanzhen Thibaud-Nissen, Francoise Tolstoy, Igor Tully, Raymond E Vatsan, Anjana R Wallin, Craig Webb, David Wu, Wendy Landrum, Melissa J Kimchi, Avi Tatusova, Tatiana DiCuccio, Michael Kitts, Paul Murphy, Terence D Pruitt, Kim D eng Intramural NIH HHS/ Research Support, N.I.H., Intramural England 2015/11/11 06:00 Nucleic Acids Res. 2016;44(D1):D733–45. https://doi.org/10.1093/nar/gkv1189. Epub 8 Nov 2015.
67. UniProt Consortium. Reorganizing the protein space at the universal protein resource (UniProt). Nucleic Acids Res. 2012;40:D71–5.
68. Gasteiger E, Jung E, Bairoch A. SWISS-PROT: connecting biomolecular knowledge via a protein database. Curr Issues Mol Biol. 2001;3(3):47–55. Gasteiger, E Jung, E Bairoch, A Eng Review England 2001/08/08 10:00 Curr Issues Mol Biol. 2001;3(3):47–55.
69. Alberti A, Poulain J, Engelen S, Labadie K, Romac S, Ferrera I, et al. Viral to metazoan marine plankton nucleotide sequences from the Tara Oceans expedition. Sci Data. 2017;4:170093. Alberti, Adriana Poulain, Julie Engelen, Stefan Labadie, Karine Romac, Sarah Ferrera, Isabel Albini, Guillaume Aury, Jean-Marc Belser, Caroline Bertrand, Alexis Cruaud, Corinne Da Silva, Corinne Dossat, Carole Gavory, Frederick Gas, Shahinaz Guy, Julie Haquelle, Maud Jacoby, E'krame Jaillon, Olivier Lemainque, Arnaud Pelletier, Eric Samson, Gaelle Wessner, Mark Acinas, Silvia G Royo-Llonch, Marta Cornejo-Castillo, Francisco M Logares, Ramiro Fernandez-Gomez, Beatriz Bowler, Chris Cochrane, Guy Amid, Clara Hoopen, Petra Ten De Vargas, Colomban Grimsley, Nigel Desgranges, Elodie Kandels-Lewis, Stefanie Ogata, Hiroyuki Poulton, Nicole Sieracki, Michael E Stepanauskas, Ramunas Sullivan, Matthew B Brum, Jennifer R Duhaime, Melissa B Poulos, Bonnie T Hurwitz, Bonnie L Pesant, Stephane Karsenti, Eric Wincker, Patrick eng Research Support, Non-U.S. Gov't England 2017/08/02 06:00 Sci Data. 2017;4:170093. https://doi.org/10.1038/sdata.2017.93.
70. The Human Microbiome Project Consortium. Structure, function and diversity of the healthy human microbiome. Nature. 2012;486:207–14.

71. Ashburner M, Ball CA, Blake JA, Botstein D, Butler H, Cherry JM, et al. Gene ontology: tool for the unification of biology. Nat Genetics. 2000;25(1):25–9.

72. Tatusov RL, Fedorova ND, Jackson JD, Jacobs AR, Kiryutin B, Koonin EV, et al. The COG database: an updated version includes eukaryotes. BMC Bioinform. 2003;4:41–7.

73. Grigoriev IV, Nordberg H, Shabalov I, Aerts A, Cantor M, Goodstein D, et al. The Genome portal of the department of energy joint Genome Institute. Nucleic Acids Res. 2012;40: D26–32.

74. Kanehisa M, Goto S, Kawashima YSM, Furumichi M, Tanabe M. Data, information, knowledge and principle: back to metabolism in KEGG. Nucleic Acids Res. 2014;42: D199–205.

75. Kanehisa M, Goto S. KEGG: kyoto encyclopedia of genes and genomes. Nucleic Acids Res. 2000;28:27–30.

76. Caspi R, Altman T, Dale JM, Dreher K, Fulcher CA, Gilham F, et al. The MetaCyc database of metabolic pathways and enzymes and the BioCyc collection of pathway/genome databases. Nucleic Acids Res. 2010;38:D473–9.

77. Overbeek R, Begley T, Butler RM, Choudhuri JV, Chuang HY, Cohoon M, et al. The subsystems approach to genome annotation and its use in the project to annotate 1000 genomes. Nucleic Acids Res. 2005;33:5691–702.

78. Altschul S, Gish W, Miller W, Myers E, Lipman D. Basic local alignment search tool. J Mol Biol. 1990;215:403–10.

79. Markowitz VM, Ivanova NN, Szeto E, Palaniappan K, Chu K, Dalevi D, et al. IMG/M: a data management and analysis system for metagenomes. Nucleic Acids Res. 2008;36:D534–8.

80. Markowitz V, Chen IM, Palaniappan K, Chu K, Szeto E, Grechkin Y, et al. IMG: the integrated microbial genomes database and comparative analysis system. Nucleic Acids Res. 2012;40:D115–22.

81. Aziz RK, et al. The RAST server: rapid annotations using subsystems technology. BMC Genomics. 2008;9(75):1–15.

82. Huson DH, Auch AF, Qi J, Schuster SC. MEGAN analysis of metagenomic data. Genome Res. 2007;17(3):377–86.

83. Abubucker S, Segata N, Goll J, Schubert AM, Izard J, Cantarel BL, et al. Metabolic reconstruction for metagenomic data and its application to the human microbiome. PLoS Comput Biol. 2012;8(6):1–17.

84. Finn RD, Coggill P, Eberhardt RY, Eddy SR, Mistry J, Mitchell AL, et al. The Pfam protein families database: towards a more sustainable future. Nucleic Acids Res. 2016;44(D1):D279–85. Finn, Robert D Coggill, Penelope Eberhardt, Ruth Y Eddy, Sean R Mistry, Jaina Mitchell, Alex L Potter, Simon C Punta, Marco Qureshi, Matloob Sangrador-Vegas, Amaia Salazar, Gustavo A Tate, John Bateman, Alex eng 108433/Z/15/Z]/Wellcome Trust/United Kingdom BB/L024136/1/Biotechnology and Biological Sciences Research Council/United Kingdom Howard Hughes Medical Institute/ Research Support, Non-U.S. Gov't England 2015/12/18 06:00 Nucleic Acids Res. 2016;44(D1):D279–85. https://doi.org/10.1093/nar/gkv1344. Epub 15 Dec 2015.

85. Sigrist CJA, de Castro E, Cerutti L, Cuche BA, Hulo N, Bridge A, et al. New and continuing developments at PROSITE. Nucleic Acids Res. 2013;41(D1):E344–7. 062BE Times Cited:260 Cited References Count:14.

86. Mi H, Muruganujan A, Thomas PD. PANTHER in 2013: modeling the evolution of gene function, and other gene attributes, in the context of phylogenetic trees. Nucleic Acids Res. 2013;41(Database issue):D377–86. Mi, Huaiyu Muruganujan, Anushya Thomas, Paul D eng GM081084/GM/NIGMS NIH HHS/ Research Support, N.I.H., Extramural Research Support, Non-U.S. Gov't England 2012/11/30 06:00 Nucleic Acids Res. 2013;41(Database issue):D377–86. https://doi.org/10.1093/nar/gks1118. Epub 27 Nov 2012.

87. Pedruzzi I, Rivoire C, Auchincloss AH, Coudert E, Keller G, de Castro E, et al. HAMAP in 2013, new developments in the protein family classification and annotation system. Nucleic Acids Res. 2013;41(Database issue):D584–9. Pedruzzi, Ivo Rivoire, Catherine Auchincloss, Andrea H Coudert, Elisabeth Keller, Guillaume de Castro, Edouard Baratin, Delphine Cuche,

Beatrice A Bougueleret, Lydie Poux, Sylvain Redaschi, Nicole Xenarios, Ioannis Bridge, Alan eng 5R01GM080646-07/GM/NIGMS NIH HHS/ 8P20GM103446-12/GM/NIGMS NIH HHS/ 5G08LM010720-03/LM/NLM NIH HHS/ 2P41 HG02273/HG/NHGRI NIH HHS/ 3R01GM080646-07S1/GM/NIGMS NIH HHS/ SP/07/007/23671/British Heart Foundation/United Kingdom 1 U41 HG006104-03/HG/NHGRI NIH HHS/ Research Support, N.I.H., Extramural Research Support, Non-U.S. Gov't Research Support, U.S. Gov't, Non-P.H.S. England 2012/11/30 06:00 Nucleic Acids Res. 2013 Jan;41(Database issue):D584–9. https://doi.org/10.1093/nar/gks1157. Epub 27 Nov 2012.

88. Bru C, Courcelle E, Carrere S, Beausse Y, Dalmar S, Kahn D. The ProDom database of protein domain families: more emphasis on 3D. Nucleic Acids Res. 2005;33(Database issue):D212–5. Bru, Catherine Courcelle, Emmanuel Carrere, Sebastien Beausse, Yoann Dalmar, Sandrine Kahn, Daniel eng Research Support, Non-U.S. Gov't England 2004/12/21 09:00 Nucleic Acids Res. 2005;33(Database issue):D212–5. https://doi.org/10.1093/nar/gki034.

89. Hunter S, Apweiler R, Attwood TK, Bairoch A, Bateman A, Binns D, et al. InterPro: the integrative protein signature database. Nucleic Acids Res. 2009;37(Database issue):D211–5. Hunter, Sarah Apweiler, Rolf Attwood, Teresa K Bairoch, Amos Bateman, Alex Binns, David Bork, Peer Das, Ujjwal Daugherty, Louise Duquenne, Lauranne Finn, Robert D Gough, Julian Haft, Daniel Hulo, Nicolas Kahn, Daniel Kelly, Elizabeth Laugraud, Aurelie Letunic, Ivica Lonsdale, David Lopez, Rodrigo Madera, Martin Maslen, John McAnulla, Craig McDowall, Jennifer Mistry, Jaina Mitchell, Alex Mulder, Nicola Natale, Darren Orengo, Christine Quinn, Antony F Selengut, Jeremy D Sigrist, Christian J A Thimma, Manjula Thomas, Paul D Valentin, Franck Wilson, Derek Wu, Cathy H Yeats, Corin eng BB/F010508/1/Biotechnology and Biological Sciences Research Council/United Kingdom 087656/Wellcome Trust/United Kingdom GM081084/GM/NIGMS NIH HHS/ Wellcome Trust/United Kingdom BB/F010435/1/Biotechnology and Biological Sciences Research Council/United Kingdom Research Support, N.I.H., Extramural Research Support, Non-U.S. Gov't England 2008/10/23 09:00 Nucleic Acids Res. 2009;37(Database issue):D211–5. https://doi.org/10.1093/nar/gkn785. Epub 21 Oct 2008.

90. Nayfach S, Pollard KS. Toward accurate and quantitative comparative metagenomics. Cell. 2016;166(5):1103–16. Available from: http://dx.doi.org/10.1016/j.cell.2016.08.007.

91. Manor O, Borenstein E. MUSiCC: a marker genes based framework for metagenomic normalization and accurate profiling of gene abundances in the microbiome. Genome Biol. 2015;16(1):53. Available from: http://www.ncbi.nlm.nih.gov/pubmed/25885687%5Cnhttp://www.pubmedcentral.nih.gov/articlerender.fcgi?artid=PMC4391136.

92. McMurdie PJ, Holmes S. Waste not, want not: why rarefying microbiome data is inadmissible. PLoS Comput Biol. 2014;10(4):1–11.

93. Silverman JD, Washburne AD, Mukherjee S, David LA. A phylogenetic transform enhances analysis of compositional microbiota data. eLife. 2017;6:1–20.

94. Li H. Microbiome, metagenomics, and high-dimensional compositional data analysis. Ann Rev Stat Appl. 2015;2(1):73–94. Available from: http://www.annualreviews.org/doi/abs/10.1146/annurev-statistics-010814-020351?journalCode=statistics.

95. Kurtz ZD, Mueller CL, Miraldi ER, Littman DR, Blaser MJ, Bonneau RA. Sparse and compositionally robust inference of microbial ecological networks. PLoS Comput Biol. 2015;11(5):1–25.

96. Gloor GB, Reid G. Compositional analysis: a valid approach to analyze microbiome high throughput sequencing data. Can J Microbiol. 2016;703(April):2015–0821. Available from: http://www.nrcresearchpress.com/doi/abs/10.1139/cjm-2015-0821#.VxVj4pMrJIX.

97. Kumar MS, Slud EV, Okrah K, Hicks SC, Hannenhalli S, Corrada Bravo H. Analysis and correction of compositional bias in sparse sequencing count data. bioRxiv. 2017;1–34. Available from: http://www.biorxiv.org/content/early/2017/05/27/142851?%3Fcollection=.

98. Tsilimigras MCB, Fodor AA. Compositional data analysis of the microbiome: fundamentals, tools, and challenges. Ann Epidemiol. 2016;26(5):330–5. Available from: http://dx.doi.org/10.1016/j.annepidem.2016.03.002.

99. Lozupone C, Knight R. UniFrac: a new phylogenetic method for comparing microbial communities. App Environ Microbiol. 2005;71(12):8228–8235.
100. Lozupone C, Hamady M, Kelley S, Knight R. Quantitative and qualitative β diversity measures lead to different insights into factors that structure microbial communities. Appl Environ Microbiol. 2007;73(5):1576–1585.
101. Zvelebil M, Baum J. Understanding bioinformatics. New York: Garland Science; 2008.
102. Cover TM, Thomas JA. Elements of information theory. New York: Wiley-Interscience; 2006.
103. Kira K, Rendell L. A practical approach to feature selection. In: National conference on artificial intelligence; 1992.
104. Hall MA. Correlation-based feature selection for discrete and numeric class machine learning. In: Proceedings of the seventeenth international conference on machine learning; 2000, p. 359–366. Available from: http://www.ime.unicamp.br/~wanderson/Artigos/correlation_based_feature_selection.pdf.
105. Brown G, Pocock A, Zhao MJ, Luján M. Conditional likelihood maximisation: a unifying framework for information theoretic feature selection. J Mach Learn Res. 2012;13:27–66.
106. Tibshirani R. Regression shrinkage and selection via the lasso. J R Stat Soc. 1996;58(1):267–88.
107. Bates S, Tibshirani R. Log-ratio Lasso: scalable, sparse estimation for log-ratio models. 2017;1–24. Available from: http://arxiv.org/abs/1709.01139.
108. Ditzler G, Morrison JC, Lan Y, Rosen G. Fizzy: feature selection for metagenomics. BMC Bioinform. 2015;16(358):1–8.
109. Zou H, Hastie T, Tibshirani R. Sparse principal component analysis. J Comput Graph Stat. 2006;15(2):262–86.
110. Blair E, Hastie T, Paul D, Tibshirani R. Prediction by supervised principal components. J Am Stat Assoc. 2006;101(473):119–37.
111. Hastie T, Tibshirani R, Friedman J. The elements of statistical learning. Elements. 2009;1:337–87. Available from: http://www.springerlink.com/index/10.1007/b94608.
112. Hotelling H. Relations between two sets of variates. Biometrika. 1936;28(3):321–77.
113. van der Maaten L, Hinton GE. Visualizing high-dimensional data using t-SNE. J Mach Learn Res. 2008;9:2579–605.
114. Gower JC. Some distance properties of latent root and vector methods used in multivariate analysis. Biometrika. 1966;53(3/4):325. Available from: http://www.jstor.org/stable/2333639?origin=crossref.
115. Hirschfeld HO. A connection between correlation and contingency. Math Proc Camb Philos Soc. 1935;31(4):520–24. Available from: http://journals.cambridge.org/action/displayAbstract?fromPage=online&aid=1737020%5Cnhttp://journals.cambridge.org/action/displayFulltext?type=1&fid=2109508&jid=&volumeId=&issueId=04&aid=1737020&bodyId=&membershipNumber=&societyETOCSession=.
116. Kenkel NC, Orloci L. Applying metric and nonmetric multidimensional scaling to ecological studies: some new results. Ecology. 1986;67(4):919–928.
117. Kruskal JB. Nonmetric multidimensional scaling: a numerical method. Psychometrika. 1964;29(2):115–29.
118. Legendre P, Legendre L. Numerical ecology. Amsterdam: Elsevier Science; 2008.
119. Ching T, Himmelstein DS, Beaulieu-Jones BK, Kalinin AA, Do BT, Way GP, et al. Opportunities and obstacles for deep learning in biology and medicine. bioRxiv. 2017;142760. Available from: https://www.biorxiv.org/content/early/2017/05/28/142760.full.pdf+html.
120. Tan J, Doing G, Lewis KA, Price CE, Chen KM, Kyle C, et al. System-wide automatic extraction of functional signatures in Pseudomonas aeruginosa with eADAGE. bioRxiv. 2016, p. 1–25.
121. Xie R, Wen J, Quitadamo A, Cheng J, Shi X. A deep auto-encoder model for gene expression prediction. BMC Genomics. 2017;18(S9):845. Available from: https://bmcgenomics.biomedcentral.com/articles/10.1186/s12864-017-4226-0.
122. Mikolov T, Chen K, Corrado G, Dean J. Distributed representations of words and phrases and their compositionality. CrossRef Listing of Deleted DOIs. 2000;1:1–9. Available from: http://www.crossref.org/deleted_DOI.html.

123. Ng P. dna2vec: consistent vector representations of variable-length k-mers. 2017;1–10. Available from: http://arxiv.org/abs/1701.06279.
124. Levy O, Goldberg Y. Neural word embedding as implicit matrix factorization. Adv Neural Inf Process Syst (NIPS). 2014;2177–85. Available from: http://papers.nips.cc/paper/5477-neural-word-embedding-as-implicit-matrix-factorization.
125. Levy O, Goldberg Y, Dagan I. Improving distributional similarity with lessons learned from word embeddings. Trans Assoc Comput Linguist. 2015;3:211–25. Available from: https://tacl2013.cs.columbia.edu/ojs/index.php/tacl/article/view/570.
126. Landgraf AJ, Bellay J. word2vec skip-gram with negative sampling is a weighted logistic PCA. 2017;1–5. Available from: http://arxiv.org/abs/1705.09755.
127. Mikolov T, tau Yih W, Zweig G. Linguistic regularities in continuous space word representations. In: North American Chapter of the Association for Computational Linguistics. 2015.
128. Mikolov T, Chen K, Corrado G, Dean J. Efficient estimation of word representations in vector space. CoRR. 2013;abs/1301.3781. Available from: http://arxiv.org/abs/1301.3781.
129. Rao C. The use and interpretation of principal component analysis in applied research; 1964. Available from: http://www.jstor.org/stable/25049339.
130. Legendre P, Andersson MJ. Distance-based redundancy analysis: Testing multispecies responses in multifactorial ecological experiments. Ecol Monogr. 1999;69(1):1–24.
131. ter Braak CJ. Canonical correspondence analysis: a new eigenvector technique for multivariate direct gradient analysis. Ecology. 1986;67(5):1167–79.
132. Blanchet G, Legendre P, Borcard D. Forward selection of spatial explanatory variables. Ecology. 2008;89(9):2623–32.
133. Clarke KR, Ainsworth M. A method of linking multivariate community structure to environmental variables. Marine ecology progress series. 1993;92:205–219.
134. MacKelprang R, Waldrop MP, Deangelis KM, David MM, Chavarria KL, Blazewicz SJ, et al. Metagenomic analysis of a permafrost microbial community reveals a rapid response to thaw. Nature. 2011;480(7377):368–71.
135. Borcard D, Gillet F, Legendre, Legendre P. Numerical ecology with R. Springer. 2011.
136. McCune B, Grace JB. Analysis of ecological communities. Gleneden Beach: MjM Software Design; 2002.
137. Ramette A. Multivariate analyses in microbial ecology. Fems Microbiology Ecology 2007;62(2):142–160. Available from: http://doi.org/10.1111/j.1574-6941.2007.00375.x.
138. Ter Braak CJF. Canonical community ordination. Part I: basic theory and linear methods. Ecoscience. 1994;1:127–40.
139. Gelman A, Stern H. The difference between significant and not significant is not itself statistically significant. Am Stat. 2006;60(4):328–31.
140. Zuur AF, Ieno EN, Elphick CS. A protocol for data exploration to avoid common statistical problems. Methods Ecol Evol. 2010;1(1):3–14. Available from: http://doi.wiley.com/10.1111/j.2041-210X.2009.00001.x.
141. Hoff PD. A first course in Bayesian statistical methods, vol. 64; 2009. Available from: http://books.google.com/books?id=9tv0taI8l6YC%5Cnhttp://www.amazon.com/Bayesian-Statistical-Methods-Springer-Statistics/dp/0387922997.
142. Team SD. Stan modeling language. User's guide and reference manual. 2017; p. 1–488. Available from: http://mc-stan.org/manual.html%5Cnpapers2://publication/uuid/C0937B19-1CC1-423C-B569-3FDB66090102.
143. Paliy O, Shankar V. Application of multivariate statistical techniques in microbial ecology. Mol Ecol. 2016;25(5):1032–57.
144. Law CW, Chen Y, Shi W, Smyth GK. Voom: precision weights unlock linear model analysis tools for RNA-seq read counts. Genome Biol. 2014;15(2):R29. Available from: http://www.pubmedcentral.nih.gov/articlerender.fcgi?artid=4053721&tool=pmcentrez&rendertype=abstract.
145. Love MI, Anders S, Huber W. Differential analysis of count data – the DESeq2 package, vol. 15; 2014. Available from: http://biorxiv.org/lookup/doi/10.1101/002832%5Cnhttp://dx.doi.org/10.1186/s13059-014-0550-8.

146. Paulson J. MetagenomeSeq: statistical analysis for sparse high-throughput sequencing. BioconductorJp. 2014;1–20. Available from: http://bioconductor.jp/packages/2.14/bioc/vignettes/metagenomeSeq/inst/doc/metagenomeSeq.pdf.

147. Segata N, Izard J, Waldron L, Gevers D, Miropolsky L, Garrett WS, et al. Metagenomic biomarker discovery and explanation. Genome Biol. 2011;12:R60(1–18).

148. Jonsson V, Österlund T, Nerman O, Kristiansson E. Statistical evaluation of methods for identification of differentially abundant genes in comparative metagenomics. BMC Genomics. 2016;17(1):78. Available from: http://www.pubmedcentral.nih.gov/articlerender.fcgi?artid=4727335&tool=pmcentrez&rendertype=abstract.

149. Mitchell TM. Machine learning. 1st ed. New York: McGraw-Hill, Inc.; 1997.

150. McIntyre ABR, Ounit R, Afshinnekoo E, Prill RJ, Hénaff E, Alexander N, et al. Comprehensive benchmarking and ensemble approaches for metagenomic classifiers. Genome Biol. 2017;18(1):182. Available from: https://doi.org/10.1186/s13059-017-1299-7.

151. Chatterji S, Yamazaki I, Bai Z, Eisen J. CompostBin: a DNA composition-based algorithm for binning environmental shotgun reads. ArXiv e-prints. 2007 Aug.

152. Rosen G, Garbarine E, Caseiro D, Polikar R, Sokhansanj B. Metagenome fragment classification using N-mer frequency profiles. Adv Bioinform. 2008;2008(205969):1–12164: e79(1–11).

153. Rosen GL, Reichenberger ER, Rosenfeld AM. NBC: the Naive Bayes classification tool webserver for taxonomic classification of metagenomic reads. Bioinformatics. 2011;27(1):127–9. Available from: +http://dx.doi.org/10.1093/bioinformatics/btq619.

154. Borozan I, Watt S, Ferretti V. Integrating alignment-based and alignment-free sequence similarity measures for biological sequence classification. Bioinformatics. 2015;31(9):1396–404.

155. Wang Y, Leung H, Yiu S, FY C. MetaCluster 4.0: a novel binning algorithm for NGS reads and huge number of species. J Comput Biol. 2012;19(2):241–9.

156. Li W, Godzik A. Cd-hit: a fast program for clustering and comparing large sets of protein or nucleotide sequences. Bioinformatics. 2006;22(13):1658–9. Available from: +http://dx.doi.org/10.1093/bioinformatics/btl158.

157. Ghodsi M, Liu B, Pop M. DNACLUST: accurate and efficient clustering of phylogenetic marker genes. BMC Bioinformatics. 2011;12(1):271. Available from: https://doi.org/10.1186/1471-2105-12-271.

158. Edgar RC. Search and clustering orders of magnitude faster than BLAST. Bioinformatics. 2010;26(19):2460–1. Available from: +http://dx.doi.org/10.1093/bioinformatics/btq461.

159. Yoon BJ. Hidden Markov models and their applications in biological sequence analysis. Curr Genomics. 2009;10(6):402–15.

160. Rho M, Tang H, Ye Y. FragGeneScan: predicting genes in short and error-prone reads. Nucleic Acids Res. 2010;38(20):e191. Available from: +http://dx.doi.org/10.1093/nar/gkq747.

161. Noguchi H, Park J, Takagi T. MetaGene: prokaryotic gene finding from environmental genome shotgun sequences. Nucleic Acids Res. 2006;34(19):5623–30. Available from: +http://dx.doi.org/10.1093/nar/gkl723.

162. Chen X, Ishwaran H. Random forests for genomic data analysis. Genomics. 2012;99(6):323–9.

163. Capriotti E, Calabrese R, Casadio R. Predicting the insurgence of human genetic diseases associated to single point protein mutations with support vector machines and evolutionary information. Bioinformatics. 2006;22(22):2729–34. Available from: +http://dx.doi.org/10.1093/bioinformatics/btl423.

164. Hastie T, Tibshirani R, Wainwright M. Statistical learning with sparsity: the Lasso and generalizations. Boca Raton: CRC; 2015; p. 362.

165. Hughey JJ, Butte AJ. Robust meta-analysis of gene expression using the elastic net. Nucleic Acids Res. 2015;43(12):e79(1–11). Available from: http://doi.org/10.1093/nar/gkv229.

166. Venter JC, Adams MD, Myers EW, Li PW, Mural RJ, Sutton GG, et al. The sequence of the human genome. Science (N Y). 2001;291(5507):1304–51. Available from: http://www.ncbi.nlm.nih.gov/pubmed/11181995.

ANOCVA: A Nonparametric Statistical Test to Compare Clustering Structures

Alexandre Galvão Patriota, Maciel Calebe Vidal, Davi Augusto Caetano de Jesus, and André Fujita

Abstract Clustering is an important tool in biological data investigation. For example, in neuroscience, one major hypothesis is that the presence or not of a disorder can be explained by the differences in how brain's regions of interest cluster. In molecular biology, genes may cluster in a different manner in controls and patients or also among different stages or grades of a certain disease (e.g., cancer). Therefore, it is important to statistically test whether the properties of the clusters change between groups of patients and controls. To this end, we describe a nonparametric statistical test called analysis of cluster structure variability (ANOCVA). ANOCVA is based on two well-established ideas: the silhouette statistic to measure the variability of the clustering structures and the analysis of variance. The advantages of ANOCVA are the following: (i) it allows the comparison of the clustering structures of multiple groups simultaneously; (ii) it identifies features that contribute to the differential clustering; (iii) it is fast and easy to implement; and (iv) it can be applied in combination with a wide variety of clustering algorithms. Finally, we describe an R implementation of ANOCVA, freely available at https://CRAN.R-project.org/package=anocva (package `anocva`).

1 Introduction

The number of biological datasets has been increasing over the past decades, leading to an era of information-driven science [11] and allowing previously unimaginable breakthroughs. In this scenario, statistical methods that are capable of mining and identifying important characteristics in these large datasets are necessary.

A. G. Patriota · D. A. C. de Jesus
Department of Statistics, Institute of Mathematics and Statistics, University of São Paulo, São Paulo, Brazil
e-mail: patriota@ime.usp.br; dejesus@ime.usp.br

M. C. Vidal · A. Fujita (✉)
Department of Computer Science, Institute of Mathematics and Statistics, University of São Paulo, São Paulo, Brazil
e-mail: calebe@ime.usp.br; fujita@ime.usp.br

© Springer International Publishing AG, part of Springer Nature 2018
F. A. B. da Silva et al. (eds.), *Theoretical and Applied Aspects of Systems Biology*,
Computational Biology 27, https://doi.org/10.1007/978-3-319-74974-7_6

Among several exploratory methods, one widely used approach is the clustering of items to identify patterns that may aid in generating more specific hypotheses. In other words, biomedical researchers are interested in identifying features that cluster in a different manner between two (or more) conditions. For example, in molecular biology, it is important to recognize the genes that present different clustering patterns between controls and tumor tissues in terms of gene expression [5, 13]. Another example is in neuroscience, where it is hypothesized that the brain is organized in clusters of neurons with different major functionalities and that deviations from the typical clustering pattern can lead to a disease condition [6]. Therefore, to better understand diseases, it is necessary to compare the clustering structures among different populations. This leads to the problem of how to statistically test the equality of the clustering structures of two or more populations followed by the identification of features that are not equally clustered.

With this motivation, we describe a method to statistically test whether the items of two or more populations are equally clustered, namely, analysis of cluster structure variability (ANOCVA) [3]. Formally, the problem that ANOCVA solves can be described as follows: given k populations $T_1, T_2, \ldots, T_k$, where each population T_j $(j = 1, \ldots, k)$ is composed of n_j subjects and each subject has N items that are clustered, we would like to verify whether the clustering structures of the k populations are equal and, if not, which items are differently clustered (i.e., which items contribute to the differential clustering).

ANOCVA is based on two well-established concepts: the silhouette statistic [10] and the classic analysis of variance (ANOVA). Essentially, ANOCVA uses the silhouette statistic to measure the "variability" of the clustering structure in each population. Next, it compares the silhouette ("variability") among populations, which is similar to the procedure of ANOVA. The intuitive idea behind this approach is that populations with the same clustering structures also have the same "variability." This simple idea leads to a powerful statistical test for the equality of clustering structures, which (i) can be applied to a variety of clustering algorithms, (ii) allows comparing the clustering structures of multiple groups simultaneously, (iii) is fast and easy to implement, and (iv) identifies the items (features) that significantly contribute to the differential clustering [3].

We will first define the "clustering variability," and then we describe the statistics related with the hypothesis tests used in ANOCVA. Finally, we describe its R implementation.

2 The Silhouette Statistic

The silhouette statistic is a measure of how well an item is clustered given a clustering algorithm. In other words, this statistic can also be interpreted as a measure of "clustering variability" [10]. Formally, let $\chi = \{x_1, \ldots, x_N\}$ be the N items of one subject that are grouped into $\mathbb{C} = \{C_1, \ldots, C_r\}$ clusters by a clustering algorithm according to an optimal criterion. Note that $\chi = \cup_{q=1}^{r} C_q$. Denote the dissimilarity (e.g., the normalized Euclidean distance) between items x and y as

$d(x, y)$. Let $|D|$ be the number of items of cluster D. Then, define the average dissimilarity of x to all items of $D \subset \chi$ (or $D \in \mathbb{C}$) as

$$d(x, D) = \frac{1}{|D|} \sum_{y \in D} d(x, y).$$

Denote $E_q \in \mathbb{C}$ as the cluster to which x_q has been assigned by the clustering algorithm. Define the within dissimilarity of x_q as

$$a_q = d(x_q, E_q),$$

and the smallest between dissimilarity of x_q as

$$b_q = \min_{E_q \neq F_q} d(x_q, F_q),$$

for $q = 1, \ldots, N$.

Then, we can measure how well each item x_q has been clustered by analyzing the silhouette statistic given by [10]:

$$s_q = \begin{cases} \frac{b_q - a_q}{\max\{a_q, b_q\}} & \text{if} |E_q| > 1 \\ 0 & \text{if} |E_q| = 1. \end{cases} \tag{1}$$

The silhouette statistic s_q assumes values between -1 and $+1$, and its interpretation is as follows [10]. If $s_q \approx 1$, then $a_q \ll b_q$. This situation occurs when item x_q has been assigned to an appropriate cluster and the second-best choice cluster is not as close as the actual cluster (Fig. 1a). If $s_q \approx 0$, then $a_q \approx b_q$, which means that item x_q is equally far away from the cluster to which it was assigned to and from the second closest cluster. Consequently, it is not clear whether item x_q should have been assigned to the actual cluster or to the second-best choice cluster (Fig. 1b). If $s_q \approx -1$, then $a_q \gg b_q$, and clearly item x_q was poorly clustered because it is closer to the second-best choice cluster than to the cluster to which it was assigned to by the clustering algorithm (Fig. 1c). In summary, the silhouette statistic s_q can be interpreted as a measure of the goodness of fit of item x_q to the cluster that it was assigned to by the clustering algorithm. This measure will be useful in Sect. 3 for developing a statistic to evaluate the clustering structure variability.

3   ANOCVA

In this section, we describe ANOCVA [3].

Let N be the number of items and $\mathbf{Q} = \{d(x_l, x_q)\}$ be the $(N \times N)$ matrix of dissimilarities. Since $\mathbf{Q}$ is a matrix of dissimilarities, it is symmetric and has zero diagonal elements. Let $\mathbf{l} = (l_1, l_2, \ldots, l_N)$ be the labels obtained by applying a clustering algorithm to the dissimilarity matrix $\mathbf{Q}$, that is, the labels represent the cluster to which each item belongs to. Note that the dissimilarity matrix $\mathbf{Q}$ and the vector of labels $\mathbf{l}$ are sufficient for computing the silhouettes $s_1, \ldots, s_N$ (see Sect. 2).

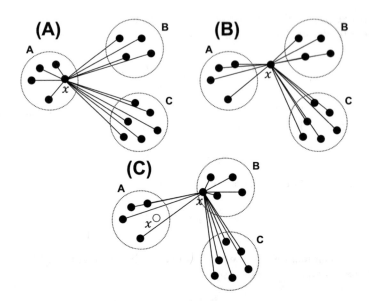

Fig. 1 An illustration for interpreting the silhouette statistic when item x is assigned to cluster A by the clustering algorithm. (**a**) Situation where item x is closer to cluster A (the actual cluster) than to cluster B (the second-best choice cluster), i.e., $a_x \ll b_x$. (**b**) Situation where the item x is equally far away from both the actual and the second-best choice clusters, i.e., $a_x \approx b_x$. (**c**) Situation where item x should be assigned to the second-best choice cluster (cluster B) because it lies much closer to it than to the cluster that it was assigned to (cluster A), i.e., $a_x \gg b_x$

To avoid notational confusion, we adopt the notation $s_q^{(\mathbf{Q},\mathbf{l})}$ to denote the qth item's silhouette obtained by using the dissimilarity matrix $\mathbf{Q}$ and the vector of labels $\mathbf{l}$ for $q = 1, \ldots, N$.

Let $T_1, T_2, \ldots, T_k$ be k types of populations. For the jth population, n_j subjects are collected for $j = 1, \ldots, k$. The items of the ith subject taken from the jth population are represented by the matrix $\mathbf{X}_{i,j} = (\mathbf{x}_{i,j,1}, \ldots, \mathbf{x}_{i,j,N})$, where each item $\mathbf{x}_{i,j,q}$ $(q = 1, \ldots, N)$ is a vector containing the features. Thus, the steps to compute the statistics of ANOCVA are described in Algorithm 1.

The statistics $s_q^{(\bar{\bar{\mathbf{A}}}_j, \mathbf{l}_{\bar{\mathbf{A}}})}$ and $s_q^{(\bar{\mathbf{A}}_j, \mathbf{l}_{\bar{\mathbf{A}}})}$ will be used in Sect. 4 to test whether the k populations are equally clustered.

4 Statistical Tests

ANOCVA is composed of one joint and possibly N marginal statistical tests, which are described as follows. The joint null hypothesis states that all N items from all k populations are equally clustered (they all present the same clustering structure). The qth marginal null hypothesis states that item q is equally clustered among all k populations. To test these null hypotheses, joint and marginal test statistics are derived from the silhouette statistics by regarding their behavior under the respective null hypotheses.

Algorithm 1: ANOCVA

Input: the k types of populations $T_1, \ldots, T_k$, a dissimilarity metric, and a clustering algorithm.

Output: the statistics $s_q^{(\bar{\bar{\mathbf{A}}}_j, \mathbf{l}_{\bar{\bar{\mathbf{A}}}})}$ and $s_q^{(\bar{\mathbf{A}}_j, \mathbf{l}_{\bar{\bar{\mathbf{A}}}})}$.

---

1 Define the $(N \times N)$ matrix of dissimilarities among items of each matrix $\mathbf{X}_{i,j}$ by
  $\mathbf{A}_{i,j} = \{d(\mathbf{x}_{i,j,q}, \mathbf{x}_{i,j,q'})\}$, for $i = 1, \ldots, n_j$ and $j = 1, \ldots, k$.

2 Let $n = \sum_{j=1}^{k} n_j$; then, define the following average matrices of dissimilarities:

3

$$\bar{\mathbf{A}}_j = \frac{1}{n_j} \sum_{i=1}^{n_j} \mathbf{A}_{i,j} = \frac{1}{n_j} \sum_{i=1}^{n_j} \{d(\mathbf{x}_{i,j,q'}, \mathbf{x}_{i,j,q'})\}, \tag{2}$$

4 and

5

$$\bar{\bar{\mathbf{A}}} = \frac{1}{n} \sum_{j=1}^{k} n_j \bar{\mathbf{A}}_j. \tag{3}$$

6 To determine the clustering labels $\mathbf{l}_{\bar{\bar{\mathbf{A}}}}$, apply the clustering algorithm to the matrix of dissimilarities $\bar{\bar{\mathbf{A}}}$.

7 Compute the silhouette statistic of the qth item based on the dissimilarity matrix $\bar{\bar{\mathbf{A}}}$ and the vector of labels $\mathbf{l}_{\bar{\bar{\mathbf{A}}}}$, i.e., $s_q^{(\bar{\bar{\mathbf{A}}}_j, \mathbf{l}_{\bar{\bar{\mathbf{A}}}})}$ for $q = 1, \ldots, N$ (Eq. 1).

8 Compute the silhouette statistic of the qth item based on the dissimilarity matrix $\bar{\mathbf{A}}_j$ and the vector of labels $\mathbf{l}_{\bar{\bar{\mathbf{A}}}}$, i.e., $s_q^{(\bar{\mathbf{A}}_j, \mathbf{l}_{\bar{\bar{\mathbf{A}}}})}$ for $q = 1, \ldots, N$ (Eq. 1).

For the joint null hypothesis, if all N items from all populations $T_1, \ldots, T_k$ are equally clustered, then the silhouette statistics $s_q^{(\bar{\bar{\mathbf{A}}}, \mathbf{l}_{\bar{\bar{\mathbf{A}}}})}$ and $s_q^{(\bar{\mathbf{A}}_j, \mathbf{l}_{\bar{\bar{\mathbf{A}}}})}$ are expected to be close for all $j = 1, \ldots, k$ and $q = 1, \ldots, N$. Therefore, let

$$\mathbf{S} = \left(s_1^{(\bar{\bar{\mathbf{A}}}, \mathbf{l}_{\bar{\bar{\mathbf{A}}}})}, \ldots, s_N^{(\bar{\bar{\mathbf{A}}}, \mathbf{l}_{\bar{\bar{\mathbf{A}}}})} \right)^{\top}, \quad \mathbf{S}_j = \left(s_1^{(\bar{\mathbf{A}}, \mathbf{l}_{\bar{\bar{\mathbf{A}}}})}, \ldots, s_N^{(\bar{\mathbf{A}}, \mathbf{l}_{\bar{\bar{\mathbf{A}}}})} \right)^{\top},$$

and

$$\delta \mathbf{S}_j = \mathbf{S} - \mathbf{S}_j.$$

Then, the joint test statistic is

$$\Delta \mathbf{S} = \sum_{j=1}^{k} \delta \mathbf{S}_j^{\top} \delta \mathbf{S}_j, \tag{4}$$

for $j = 1, \ldots, N$.

The joint null hypothesis can be defined in terms of the proposed test statistic, given the clustering algorithm: $H_0 : \mathbb{E}(\delta \mathbf{S}_1) = \ldots = \mathbb{E}(\delta \mathbf{S}_k) = 0$. Under the null hypothesis, all N items are equally clustered along the k populations, i.e., $s_q^{(\bar{\bar{\mathbf{A}}}, \mathbf{l}_{\bar{\bar{\mathbf{A}}}})} \approx s_q^{(\bar{\mathbf{A}}_j, \mathbf{l}_{\bar{\bar{\mathbf{A}}}})}$ for all $q = 1, \ldots, N$; thus, we expect a small $\Delta \mathbf{S}$. Conversely, a large $\Delta \mathbf{S}$ suggests a rejection of the null hypothesis.

For the marginal null hypothesis, Fujita et al. [3] defined the marginal test statistic as

$$\Delta s_q = s_q^{\left(\bar{\bar{\mathbf{A}}}, \mathbf{l}_{\bar{\bar{\mathbf{A}}}} \right)} - \frac{1}{k} \sum_{j=1}^{k} \left(s_q^{\left(\bar{\mathbf{A}}_j, \mathbf{l}_{\bar{\bar{\mathbf{A}}}} \right)} \right), \tag{5}$$

for $q = 1, \ldots, N$.

Later, de Jesus [2] improved Eq. 5 as follows:

$$\Delta s_q = \sum_{j=1}^{k} \left(s_q^{\left(\bar{\bar{\mathbf{A}}}, \mathbf{l}_{\bar{\bar{\mathbf{A}}}} \right)} - s_q^{\left(\bar{\mathbf{A}}_j, \mathbf{l}_{\bar{\bar{\mathbf{A}}}} \right)} \right)^2, \tag{6}$$

for $q = 1, \ldots, N$.

In Sect. 6, we show the advantages of Eq. 6 over Eq. 5.

Then, the N marginal null hypotheses can also be written in terms of the proposed statistic given the clustering algorithm: $H_{0q} : \text{“} \mathbb{E}\left(s_q^{(\bar{\bar{\mathbf{A}}}, \mathbf{l}_{\bar{\bar{\mathbf{A}}}})} \right) = \mathbb{E}\left(s_q^{(\bar{\mathbf{A}}_j, \mathbf{l}_{\bar{\bar{\mathbf{A}}}})} \right)$, $j = 1, \ldots, k$."

Under the null hypothesis, we expect a small Δs_q. Conversely, a large Δs_q suggests a rejection of the null hypothesis.

The exact or asymptotic distributions of both $\Delta \mathbf{S}$ and Δs_q are difficult to derive. Therefore, to compute the empirical distributions of ΔS and Δs_q under the null hypothesis, Fujita et al. [3] proposed a bootstrap procedure, as described in Algorithm 2.

The entire ANOCVA pipeline is illustrated in Fig. 2.

Algorithm 2: Bootstrap

Input: the dataset $\{T_1, T_2, \ldots, T_k\}$, the dissimilarity metric, the clustering algorithm, and the clustering labels for $\bar{\bar{\mathbf{A}}}$, i.e., $\mathbf{l}_{\bar{\bar{\mathbf{A}}}}$.

Output: the p-values for $\Delta \hat{S}$ and $\Delta \hat{S}_q$.

1 To construct bootstrap samples T_j^*, for $j = 1, \ldots, k$, resample with replacement n_j subjects from the entire dataset $\{T_1, T_2, \ldots, T_k\}$.

2 Use Algorithm 1 with T_j^* as input to compute $\bar{\mathbf{A}}_j^*$, $\bar{\bar{\mathbf{A}}}^*$, $\mathbf{l}_{\bar{\bar{\mathbf{A}}}}^*$, $s_q^{(\bar{\bar{\mathbf{A}}}^*, \mathbf{l}_{\bar{\bar{\mathbf{A}}}}^*)*}$, and $s_q^{(\bar{\mathbf{A}}_j^*, \mathbf{l}_{\bar{\bar{\mathbf{A}}}}^*)*}$, for $j = 1, \ldots, k$ and $q = 1, \ldots, N$.

3 Calculate $\Delta \hat{S}^*$ (Eq. 4) and $\Delta \hat{s}_q^*$ (Eq. 5).

4 Repeat steps 1 to 3 until the desired number of bootstrap replicates is obtained.

5 The p-values for the bootstrap tests based on the observed statistics $\Delta \hat{S}$ and $\Delta \hat{S}_q$ are the fraction of replicates of $\Delta \hat{S}^*$ and $\Delta \hat{s}_q^*$ on the bootstrap dataset T_j^*, respectively, that are at least as large as the observed statistics on the original dataset.

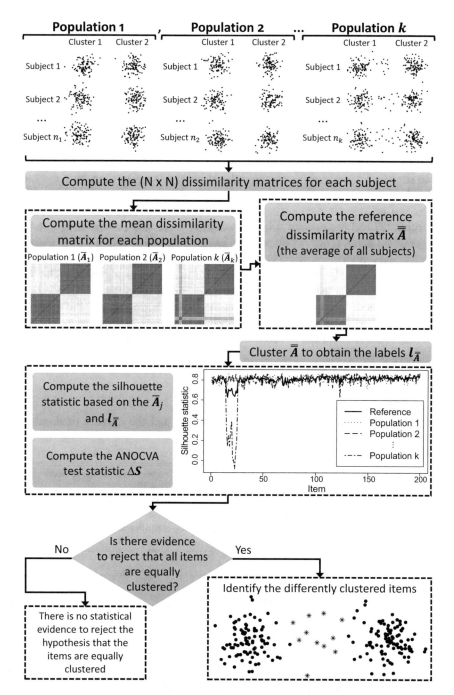

Fig. 2 The ANOCVA pipeline

5 Implementation

ANOCVA is implemented in R and is freely available at the R project website https://CRAN.R-project.org/package=anocva (package `anocva`) [12].

This implementation requires as input the dissimilarity matrices among items, a vector of labels describing which individual belongs to which group, the number of clusters, and the number of bootstrap samples.

Internally, the `anocva` implementation uses the spectral clustering algorithm [8, 9], which is described in Algorithm 3.

In step 4 of Algorithm 3, we generally apply the k-means algorithm to cluster the points (y_i). However, the `anocva` package uses the k-medoids procedure (described in Algorithm 4) because it is more robust to outliers than the k-means algorithm [7] and because its R implementation presents a deterministic solution, i.e., given an input, it always outputs the same clustering structure due to a greedy search at the initialization step [1].

The k-medoids algorithm is very similar to the k-means algorithm. The main difference is that the k-medoids algorithm uses the items as centers (medoids) or representatives of the clusters rather than the centroids. In general, the medoid is the item with the minimum sum of distances to all items in the cluster. The most common realization of the k-medoids algorithm is the partition around medoid (PAM) algorithm [7].

Algorithm 3: Spectral clustering algorithm
Input: the dissimilarity matrix $\mathbf{W}$ of a graph G and the number of desired clusters k.
Output: Clusters $C_1, \ldots, C_k$.

1 Let $\mathbf{D}$ be a diagonal matrix with degrees $d_1, \ldots, d_n$ of the vertices $v_1, \ldots, v_n$, respectively, on the diagonal. Then, compute the Laplacian matrix $\mathbf{L} = \mathbf{D} - \mathbf{W}$.
2 Compute the k eigenvectors $\mathbf{u}_1, \ldots, \mathbf{u}_k$ of $\mathbf{L}$ associated with the k largest eigenvalues.
3 Let $\mathbf{U} \in \mathbb{R}^{n \times n}$ be the matrix containing the eigenvectors $\mathbf{u}_1, \ldots, \mathbf{u}_k$ as columns.
4 Cluster the points (y_i) $i = 1, \ldots n$ in $\mathbb{R}^k$ with a clustering algorithm into clusters $C_1, \ldots, C_k$.

Algorithm 4: K-medoids
Input: the n items and the number of clusters k
Output: the k clusters.

1 Randomly select k items to be the medoids.
2 Assign the items to the cluster whose medoid is nearest.
3 For each medoid $l = 1, 2, \ldots, k$ and item i, swap the lth medoid and the ith item, and compute the "cost" (the sum of the distances of the medoid to all items in the cluster). If the "cost" of this new configuration increased, then undo the swap.
4 Go to step 3 while the cost decreases.

If the input number of clusters is unknown *a priori*, then the anocva package provides two options for estimating it: the silhouette [10] and the slope criteria [4]. The silhouette criterion is described as follows. First, given the k clusters obtained by a clustering algorithm, define the overall silhouette statistic as

$$s(k) = \frac{1}{N} \sum_{q=1}^{N} s_q. \tag{7}$$

The silhouette criterion consists of selecting the number of clusters k that maximizes Eq. 7, i.e., $\hat{k} = \arg\max_{k \in 2,...,n-1} s(k)$. Similar to the silhouette criterion, the slope criterion also uses the silhouette statistic, but it selects the number of clusters k as $\hat{k} = \arg\max_{k \in \{2,...,n-1\}} -(s(k+1) - s(k))s(k)^p$, where p is a positive integer value. The tuning parameter p is useful for interpolating between a criterion where the gap $(s(k+1) - s(k))$ is more important (small p) and a criterion where the silhouette value has more weight (large p). In other words, the slope criterion is the difference of silhouette statistics as a function of the number of clusters. The difference between the silhouette and slope can be understood by noting the following fact: in maximizing the silhouette statistic, as described by Rousseeuw [10], the number of clusters is estimated correctly only when the within-cluster variances are equal or very similar. If the within-cluster variances are unequal (e.g., in the presence of a dominant cluster), then the slope criterion is more robust than the silhouette criterion.

The output of anocva consists of one p-value, which represents whether there is at least one group that clusters in a different manner, and a vector of size N containing the p-values for each item.

An example of the application of anocva to synthetic data is as follows:

```
library(anocva)

set.seed(39487049)

# Generate simulated data under H0
dataset = array(NA, c(90, 50))
data.dist = array(NA, c(90, 50, 50))
# 90 subjects, each one with 50 items divided into two clusters
for (i in 1:90) {
   dataset[i,] = c(rnorm(25, mean = 0, sd = 1), rnorm(25,
                  mean = 10, sd = 1))
   # Calculate dissimilarities (Euclidian distance)
   data.dist[i,,] = as.matrix(dist(dataset[i,]))
}

# Separate the 90 subjects into three populations
id = c(rep(1, 30), rep(2, 30), rep(3, 30))

# ANOCVA under H0 (populations equally clustered)
# Call ANOCVA statistical test with 1,000 bootstrap replicates
resh0 = anocva(data.dist, id, replicates = 1000, r = NULL)
```

```
# There are no statistical evidences to reject H0 at a p-value
# threshold of 5%
sprintf(''P-value: %.4f,'' resh0$pValueDeltaS)
[1] ''P-value: 0.7433''

# P-values for each item
resh0$pValueDeltaSq
 [1] 0.86513487 0.34465534 0.61138861 0.63336663 0.99400599 0.68631369
 [7] 0.88911089 0.12387612 0.80019980 0.13186813 0.33066933 0.83316683
[13] 0.12987013 0.29370629 0.79620380 0.34965035 0.54445554 0.76923077
[19] 0.62337662 0.53946054 0.85914086 0.39960040 0.76823177 0.69630370
[25] 0.07192807 0.68931069 0.51248751 0.94705295 0.73726274 0.63836164
[31] 0.16283716 0.43056943 0.65934066 0.91108891 0.27172827 0.04795205
[37] 0.75424575 0.17182817 0.28171828 0.58741259 0.59840160 0.92507493
[43] 0.88211788 0.31668332 0.61738262 0.30669331 0.82217782 0.65834166
[49] 0.45254745 0.29470529


# Simulated data under H1: items 49..50 of population 3
# are not equally clustered
for (i in 61:90) {
  dataset[i,49:50] = dataset[i,49:50] + rnorm(1, 1.3, 1.2)
  data.dist[i,,] = as.matrix(dist(dataset[i,]))
}


# ANOCVA UNDER H1 (populations are not equally clustered)
resh1 = anocva(data.dist, id, replicates = 1000, r = NULL)

# H0 rejected at a p-value threshold of 5%
sprintf(''P-value: %.4f,'' resh1$pValueDeltaS)
[1] ''P-value: 0.0250''

# P-values for each item
resh1$pValueDeltaSq
 [1] 0.98801199 0.39160839 0.85514486 0.45454545 0.84315684 0.46453546
 [7] 0.84115884 0.05694306 0.66633367 0.38061938 0.41358641 0.57542458
[13] 0.13186813 0.30769231 0.92807193 0.40059940 0.46653347 0.92807193
[19] 0.31768232 0.75324675 0.47952048 0.56143856 0.72127872 0.82617383
[25] 0.05194805 0.58441558 0.39360639 0.54345654 0.66833167 0.39060939
[31] 0.07492507 0.20779221 0.84215784 0.63536464 0.59940060 0.06993007
[37] 0.85514486 0.08491508 0.60339660 0.31868132 0.84015984 0.81218781
[43] 0.81418581 0.21678322 0.74625375 0.21578422 0.30369630 0.48751249
[49] 0.00000000 0.00000000
```

6   Numerical Results

In this section, we present some simulation results to illustrate the performance of ANOCVA and compare the use of Eqs. 5 and 6.

The simulations were designed as follows: two clusters are generated by bivariate normal distributions with means $(0, 0)$ (cluster 1) and $(2, 0)$ (cluster 2) and with covariance matrix $\sigma^2 \mathbf{I}$. Clusters 1 and 2 are composed of $N + M$ and N items, respectively. All items are well clustered except for the M items of cluster 1 that "move" from position $(0, 0)$ (center of cluster 1) in direction to position $(2, 0)$ (center of cluster 2). The distance between these M items and the center of its original cluster $(0, 0)$ is given by Δ. We performed 100 Monte Carlo realizations of this scenario for all combinations of parameters ($\sigma^2 = 10^{-5}, 10^{-3}, 10^{-1}, 1$; $M = 1, 2, 5$, $\Delta = 0, 0.25, 0.50, \ldots, 2$) and number of subjects ($n_1 = n_2 = 10, 20, 30, 40$). Here, we only present the cases for $N = 10$, $M = 2, 5$, and $\sigma^2 = 10^{-1}, 1$ since all other settings present similar behavior.

Figure 3 panels A and B show the difference between the use of Eqs. 5 and 6 (by considering $N = 10$, $M = 5$, and $\sigma^2 = 1$). Whereas Eq. 5 presents non-monotonous behavior with respect to Δ (Fig. 3a), the test statistic proposed by de Jesus (2017) [2] (Eq. 6) presents the desired behavior (Fig. 3b).

Figure 3 panels C and D show the difference between the tests when we use the Euclidean distance and the normalized Euclidean distance (by considering $N = 10$, $M = 2$, and $\sigma^2 = 10^{-1}$). The Euclidean distance causes high levels of rejection rates for items under the null hypothesis (Fig. 3c). This problem is not observed when normalized Euclidean distance is used (Fig. 3d).

7 Final Remarks

Fujita et al. [3] conducted thorough simulations to show that ANOCVA controls the type I error and has sufficient power to capture small clustering differences among populations under unity variance and Gaussian noise. De Jesus [2] improved Eqs. 5 and 6 for the case of non-unity variance and proposed the test statistic for the marginal hypotheses together with the normalized Euclidean distance. This marginal test statistic correctly identifies discrepancies from the null hypotheses when the one proposed by Fujita et al. (2014a) [3] could not. Moreover, the normalized Euclidean distance is required under non-unity variance samples. Therefore, we recommend using Eq. 6 in ANOCVA. It is already implemented in the new version of anocva freely available at https://CRAN.R-project.org/package=anocva [12].

ANOCVA succeeded in identifying the features (items) that cluster in a different manner among groups. One advantage of ANOCVA is that it is possible to simultaneously test two or more groups, similar to what is performed by ANOVA for means. It allows the application of ANOCVA to diseases that present subclassifications, for example, different grades of cancer.

Applications of ANOCVA were illustrated in large (order of hundreds of individuals) functional magnetic resonance imaging data and shown to be useful for identifying brains' regions of interest (ROIs) that are differently clustered between controls and subjects diagnosed with attention deficit hyperactivity disorder [3] or autism spectrum disorder [4].

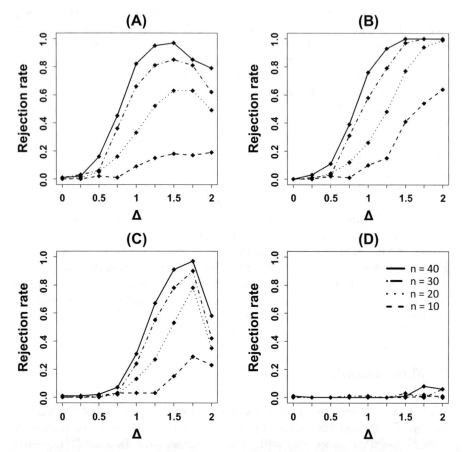

Fig. 3 Simulation results. (**a**) Rejection rate of the null hypothesis with the method proposed by [3] for an item under the null hypothesis with $N = 10$, $M = 5$, and $\sigma^2 = 1$. (**b**) Rejection rate of the null hypothesis with the method proposed by [2] for an item under the null hypothesis with $N = 10$, $M = 5$, and $\sigma^2 = 1$. (**c**) Rejection rate of the null hypothesis with the proposed test for an item under the alternative hypothesis with $N = 10$, $M = 2$, and $\sigma^2 = 10^{-1}$ and Euclidian distance. (**d**) Rejection rate of the null hypothesis with proposed test for an item under the alternative hypothesis with $N = 10$, $M = 2$, and $\sigma^2 = 10^{-1}$ and standardized Euclidean distance

The flexibility of ANOCVA that allows the application of the test on several populations simultaneously (rather than being limited to pairwise comparisons), along with its performance, makes it applicable to many areas where clustering structure is of interest.

Acknowledgements A.F. was partially supported by the São Paulo Research Foundation (2015/01587-0, 2016/13422-9), CNPq (304876/2016-0), CAPES, NAP eScience-PRP-USP, and the Alexander von Humboldt Foundation.

References

1. Aggarwal CC, Reddy CK. Data clustering: algorithms and applications. Bosa Roca: Chapman and Hall/CRC; 2013.
2. Caetano de Jesus DA. Evaluation of ANOCVA test for cluster comparison through simulations. Master Dissertation. Institute of Mathematics and Statistics, University of São Paulo; 2017.
3. Fujita A, Takahashi DY, Patriota AG, Sato JR. A non-parametric statistical test to compare clusters with applications in functional magnetic resonance imaging data. Stat Med. 2014a;33:4949–62.
4. Fujita A, Takahashi DY, Patriota AG. A non-parametric method to estimate the number of clusters. Comput Stat Data Anal. 2014b;73:27–39.
5. Furlan D, Carnevali IW, Bernasconi B, Sahnane N, Milani K, Cerutti R, Bertolini V, Chiaravalli AM, Bertoni F, Kwee I, Pastorino R, Carlo C. Hierarchical clustering analysis of pathologic and molecular data identifies prognostically and biologically distinct groups of colorectal carcinomas. Mod Pathol. 2011;24:126–37.
6. Grossberg S. The complementary brain: unifying brain dynamics and modularity. Trends Cogn Sci. 2000;4:233–46.
7. Kaufman L, Rousseeuw P. Clustering by means of medoids. North-Holland: Amsterdam; 1987.
8. Von Luxburg U. A tutorial on spectral clustering. Stat Comput. 2007;17:395–416.
9. Ng A, Jordan M, Weiss Y. On spectral clustering: analysis and an algorithm. In: Dietterich T, Becker S, Ghahramani Z, editors. Advances in neural information processing systems. Cambridge, MA: MIT Press; 2002, vol. 14, p. 849–56.
10. Rousseeuw PJ. Silhouettes: a graphical aid to the interpretation and validation of cluster analysis. J Comput Appl Math. 1987;20:53–65.
11. Stein LD. Towards a cyber infrastructure for the biological sciences: progress, visions and challenges. Nat Rev Genet. 2008;9:678–88.
12. Vidal MC, Sato JR, Balardin JB, Takahashi DY, Fujita A. ANOCVA in R: a software to compare clusters between groups and its application to the study of autism spectrum disorder. Front Neurosci. 2017;11:1–8.
13. Wang YK, Print CG, Crampin EJ. Biclustering reveals breast cancer tumour subgroups with common clinical features and improves prediction of disease recurrence. BMC Genomics. 2013;14:102.

Part II
Applied Systems Biology

Modeling of Cellular Systems: Application in Stem Cell Research and Computational Disease Modeling

Muhammad Ali and Antonio del Sol

Abstract The large-scale development of high-throughput sequencing technologies has allowed the generation of reliable omics data at different regulatory levels. Integrative computational models enable the disentangling of complex interplay between these interconnected levels of regulation by interpreting these large quantities of biomedical information in a systematic way. In the context of human diseases, network modeling of complex gene-gene interactions has been successfully used for understanding disease-related dysregulations and for predicting novel drug targets to revert the diseased phenotype. Furthermore, these computational network models have emerged as a promising tool to dissect the mechanisms of developmental processes such as cellular differentiation, transdifferentiation, and reprogramming. In this chapter, we provide an overview of recent advances in the field of computational modeling of cellular systems and known limitations. A particular attention is paid to highlight the impact of computational modeling on our understanding of stem cell biology and complex multifactorial nature of human diseases and their treatment.

1 Introduction to Systems Biology

Systems biology is the integration of computational and experimental research to study the mechanisms underlying complex biological processes as integrated systems of many interacting components. Systems biology offers a holistic rather than reductionist approach for understanding and controlling biological complexity,

M. Ali
Computational Biology Group, Luxembourg Centre for System Biomedicine (LCSB), University of Luxembourg, Luxembourg City, Luxembourg
e-mail: muhammad.ali@uni.lu

A. del Sol (✉)
Computational Biology Group, Luxembourg Centre for System Biomedicine (LCSB), University of Luxembourg, Luxembourg City, Luxembourg

Moscow Institute of Physics and Technology, Dolgoprudny, Moscow, Russia Federation
e-mail: antonio.delsol@uni.lu

© Springer International Publishing AG, part of Springer Nature 2018   129
F. A. B. da Silva et al. (eds.), *Theoretical and Applied Aspects of Systems Biology*,
Computational Biology 27, https://doi.org/10.1007/978-3-319-74974-7_7

which arises due to the interconnected components working together in a synchronized fashion to maintain the phenotype of an organism. Systems biology-based approaches help us in exploring these systems at the level of a cell, tissue, organ, organism, as well as a population and an ecosystem. Characterization of these systems in their full complexity allows us to better understand the properties of the components involved and their static as well as the dynamic behaviors.

During the last decade, various experimental techniques have enabled the large-scale generation of high-throughput (HT) biological data across different levels of regulation. Among them, the ones which have been extensively used for modeling biological systems are mutation detection by single nucleotide polymorphism (SNP) genotyping [1], gene expression quantification by messenger ribonucleic acid sequencing (RNA-seq) [2], identification of protein interactions with deoxyribonucleic acid (DNA) via chromatin immunoprecipitation sequencing (ChIP-seq) technique [3], and quantification of different metabolite levels in the organism by HT metabolic screening [4]. This plethora of data has enabled the development of computational models, allowing the dissection of the complex mechanism underlying different biological processes at different regulatory levels. This vast amount of data across different levels of a biological system has also opened a new gateway to integrate data from these different but interconnected layers to gain a deeper system-level understanding.

2 Computational Modeling of Cellular Systems

The complexity of biological systems can be broken down to an individual molecule or atom, but to study their overall effect on the system, we need to understand their interactions with each other and with other ongoing processes or pathways in the system. This is even crucial for understanding their role in the onset or progression of the diseases such as cancer and Alzheimer's disease. Mathematical models of biological systems, which use efficient algorithms and data structures, enable researchers to investigate how complex regulatory processes are intertwined and how any perturbation in these processes can lead to the development of disease. Recent advancement in computational resources and large-scale generation of so-called "omics" data sets has led to model, visualize, and rationally perturb systems at different levels such as modeling and designing from atomic resolution to cellular pathways and analysis of guided alterations in the system and their propagation.

A computational model of a complex system can help us in understanding the behavior of that system by simulating its dynamics. Numerous computational models have been developed to address different kinds of processes – for example, flight simulator models [5], protein folding models [6], and artificial neural network models [7]. Moreover, computational modeling has emerged as a powerful and promising approach to investigate and manipulate biological systems. In particular, different categories of cellular processes have been modeled by using the computational models, such as gene regulation, signaling pathways, and metabolic processes

[8]. However, modeling the biological system at a cellular level is a convoluted problem involving the challenging task of understanding the cellular dynamics and characterizing the underlying biological principles. Gaining a system-level under-standing of these intertwined cellular processes and their complex interconnections may serve as a critical foundation for developing therapeutic fronts where we anticipate that computational cellular modeling approaches will make an impact.

2.1 Gene Regulatory Networks

It is increasingly recognized that complex biological systems cannot be described in a reductionist view. To understand the behavior of such complex system, a deeper understanding of the different components of this system and their interactions with each other is required. This knowledge can help us in viewing the system under study as a network of components, which has a certain topology. This topological information is fundamental in constructing a realistic model to unlock the functions of the network. There are various types of biological networks, which have been extensively studied by the researchers, such as gene regulatory networks (GRNs), protein-protein interaction (PPI) networks, signal transduction networks, and metabolic networks. In particular, GRNs are the on-off switches of a cell operating at the gene levels where two genes are connected to each other if the expression of one gene modulates expression of another one by either activation or inhibition. A GRN can be represented by a directed graph where nodes represent the genes and directed edges among these nodes represent gene-gene interactions. As a simple example of a GRN, Fig. 1 depicts the schematic illustration of core pluripotency transcription factors (TFs) that maintain the pluripotency potential of stem cells. *Oct4*, *Sox2*, and *Nanog* have a positive self-regulation, while they also positively regulate each other.

Fig. 1 Schematic representation of the transcriptional regulation of core pluripotency factors

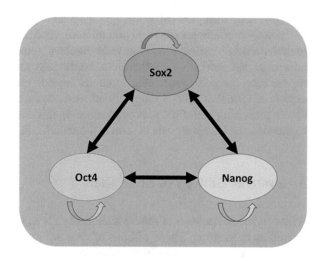

Genes in a GRN are not independent from each other; rather they regulate each other and act collectively. This collective behavior can be observed by mRNA quantification obtained from a microarray or mRNA-seq experiment where some genes are significantly upregulated, while others are downregulated, suggesting that upregulated genes might be the one inhibiting the downregulated genes. The connections among all the genes in a GRN cannot be inferred correctly by just relying on their mRNA levels or simple gene expression correlation-based methods but by obtaining the literature-based information stored in the repositories. These repositories, such as MetaCore database from Thomson Reuters and gene pathway studio [9], contain the experimental evidences of gene-gene interactions where one gene regulates the expression of its target genes.

The topological analysis of a GRN can help in identifying some important genes in the network, such as those involved in network motifs. Network motifs are topological patterns that occur in real networks significantly more often than in randomized networks [10]. These patterns have been preserved over evolutionary time scale against mutations that can randomly change edges. Similarly, the detection of elementary circuits, which is the path starting from and ending in the same gene visiting each intermediate gene only once, has been associated with the stability of GRNs [11, 12]. These circuits can either have an even number of inhibitions hence called positive circuits (positive feedback loops) or an odd number of inhibitions, therefore called negative circuits. Moreover, the genes in the strongly connected components (SCCs) of a network – a subnetwork in which every gene is reachable from every other genes in that subnetwork through a directed path – are interconnected positive and negative circuits and usually considered to be the pivotal genes, maintaining the network phenotype.

GRNs play an important role in unraveling the molecular mechanisms underlying a particular biological process, such as cell cycle, apoptosis, and cell differentiation. A paramount problem in modeling a GRN is to understand the dynamical interactions among the genes in the GRN, which collectively govern the behavior of the cell. Several methods have been proposed so far to infer GRN from gene expression and epigenetic data [13–16]. Although, the goal is the same – to model biological processes – available methods rely on different modeling formalisms, for example, logical models have been used to infer Boolean networks, probabilistic Boolean networks, and petri nets. Furthermore, continuous models were also introduced for the same purpose; prominent examples include continuous linear models and models of TF activity [17]. Computational methods for modeling GRNs have proved to be a promising bioinformatics application. In this chapter, we tried to explore the applications of GRNs models in stem cell research and disease modeling.

3   Systems Biology of Stem Cells

A human body comprises of different kinds of cells that are distinct in their structure as well as in function. These trillions of cells are expected to have same genomic material and comprise of only a limited number of distinct cell types, which are estimated to be approximately 400 [18]. Cell identity specification is considered to be determined by cell-specific gene expression program. Regulation of cell-specific gene expression program is a complex process tightly controlled at the transcriptional and epigenetics regulatory levels. In order for a cell-specific gene to be expressed, the DNA corresponding to this gene and its distal regulatory elements must be in an accessible and active state. In this context, cell-specific epigenetics landscape is hypothesized to be an explanation for the differences between heterogeneous cell fates. The different types of cells in the body and their structure perfectly suit the role they perform. For instance, kidney cells (hepatocytes) are completely different in structure and function from skin cells (fibroblast). Interestingly, all these different kind of cell types in an adult organism are actually originated from the same kind of cells, called pluripotent stem cells. These are the cells that have the potential to give rise to any kind of fetal or adult cell type. Stem cells have this potential to give rise to any kind of lineage at the embryonic developmental stage, but this plasticity is lost upon differentiation into a certain somatic cell type.

3.1   The Generation of iPSCs

Recent advancements in molecular biology have enabled researchers to obtain induced pluripotent stem cells (iPSCs) from the somatic cells by following a reliable cell conversion methodology – usually referred as cellular reprogramming. By following the established protocols of applying a particular recipe of TFs in the medium of in vitro culture of somatic cells, iPSCs can be grown in culture and will have the same plasticity potential as those of stem cells from embryos. The very first and a well-known example of cellular reprogramming is the conversion of mouse fibroblasts into iPSCs by introducing four TFs (*Oct4*, *Sox2*, *c-Myc*, and *Klf4*) [19]. iPSCs provide a new framework to obtain a renewable source of healthy cells which can help in treating a wide spectrum of diseases, such as the neurodegenerative and cardiovascular ones. Nevertheless, a bottleneck in cellular reprogramming is the identification of reprogramming determinants (TFs) that can trigger a transition between cellular phenotypes with high efficiency and fidelity.

3.2   Transdifferentiation

Similar to reprogramming, where we want the iPSCs to differentiate into a particular lineage and cell type, another approach to obtain the same cell type of interest without undergoing an intermediate pluripotent state is transdifferentiation. Transdifferentiation is the direct and irreversible conversion of one somatic cell type to another. Various examples of transdifferentiation were reported in the literature where a defined TF recipe or a combination of TFs and microRNA (miRNA) or small molecules was introduced in a somatic cell type culture and the desired mature cell type was obtained within days. For example, the TF *MyoD1* was used to transdifferentiate the mouse embryonic fibroblast into myoblast [20]. Since this first case reported in literature in 1990, there have been numerous examples of successful somatic cell conversions with defined factors and small molecules [21–23]. Moreover, various computational methods have been reported to systematically predict the candidate TFs that can help in converting one fully differentiated cell type to another, and their predictions have been also experimentally validated in the laboratory [24]. Transdifferentiation is emerging as a promising approach to directly transdifferentiate cells while avoiding the use of iPSCs to derive patient-specific cells. This remarkable potential of transdifferentiation is proving to be the most promising source of regenerative medicine for tissue regeneration and disease therapy. Nevertheless, an important roadblock to efficient transdifferentiation is the limited number of successful cellular conversions obtained so far, albeit with low to intermediate efficiency. Furthermore, the role of changes in the epigenetic landscape for achieving an efficient transdifferentiation is not yet systematically explored.

3.3 Modeling Cellular Phenotypes and Conversions

In some modeling approaches, a cellular phenotype is modeled as a network of genes with a particular gene expression pattern and a unique stable steady state (attractor). Phenotypic transitions in such models are introduced by identifying the genes in the network that can destabilize this attractor and lead the system into another attractor. This concept has been applied to model diseases as a transition from a healthy phenotype to a diseased state, caused by a mutation or a chemical compound [25]. Moreover, it has been also applied in modeling the cellular conversions [26] (reprogramming, differentiation, and transdifferentiation), where researcher first identified the attractors of two phenotypes (starting and destination cell types) and then detected the minimal set of genes in the network's elementary circuits whose perturbation (up- or downregulation) led from the attractor of the starting cell type to the attractor of the destination cell type [13, 14].

Modeling the cellular phenotype requires the inference of condition-specific GRNs. Literature suggests a number of different GRN inference methods, which rely on different underlying rationales, such as modeling formalism (Boolean and Bayesian) and different updating schemes (synchronous and asynchronous). Furthermore, there have been methods introducing the concept of contextualization, which is the removal of non-specific edges that are not compatible with the gene expression program of the cell type under consideration [13, 15]. Most of these methods rely only on gene expression data, but more recently, approaches using gene expression as well as epigenetic information have also been introduced [16]. Nevertheless, a bottleneck in the GRN inference problem is the benchmarking of inferred networks. Most of these methods rely on the interactions of a set of specific

TFs in a particular cell type diagnosed by experimental ChIP-seq to validate the networks. Unfortunately, this benchmarking approach can only validate a part of the network as the complete benchmarking information, ChIP-seq for all the TFs in one cell type, is not available for even a single cell type. Moreover, the ChIP-seq cannot be a perfect gold standard as some TF-DNA interactions might be incorrectly labeled as positives because TF binding does not necessarily indicate a functional interaction. Besides ChIP-seq, SNP data as well as random network inference has been used as a reference for the benchmarking of inferred networks [16], but none of these approaches offer a complete and systematic network inference validation.

4 Computational Disease Modeling

The advances in molecular biology have resulted in the establishment of fast and efficient protocols for generating iPSCs cells in vitro. This *cells in a petri dish* approach has immensely contributed in modeling the human diseases and uncovering the molecular basis of disease-related dysregulations. Moreover, the generation of patient-specific iPSC-derived cell types having disease-related mutations provides an extremely viable in vitro system for the investigation of disease-related perturbations and to apply drug screening. However, the complex nature of human diseases, which affect multiple genes, hinders our knowledge about disease-specific dysregulations [27]. These dysregulations initiate a cascade of failures, which causes malfunctioning at the systems level and result in specific disease phenotypes. Therefore, instead of investigating the individual genes in a system, we may rather focus on their interactions as a channel to propagate disease-related perturbations. In this context, healthy and ill states can be represented as cellular network phenotypes with stable steady states, where a disease-specific perturbation shifts the steady state of a healthy network into the steady states of a disease network. Thus, the construction of complex regulatory interaction networks offers a new method for gaining a system-level understanding of disease pathology. These network-based models have proved to be a promising framework for identifying disease-related genes based on network topology [28]. For example, disease-gene-drug associations have already been predicted based on the differential network analysis [15]. Furthermore, disease-gene relationships have also been reported based on the identification of disease-related subnetworks and prediction of network neighbors of disease-associated genes [29, 30].

4.1 Differential Network Analysis and Disease Models

There have been an increasing number of approaches exploring the associations between genes, drugs, and diseases. Some of them include the construction of data repositories where different compounds were tested experimentally to associate drugs with genes and diseases, for example, the connectivity map [31] and gene perturbation atlas [32]. These approaches have provided immense help in linking drugs to their target genes, which has also benefited in drug repositioning based on a particular gene expression signatures produced after drug perturbation. However, these approaches neglect the mechanisms underlying gene regulation and avoid the indirect targets of drugs. Moreover, only a limited set of drugs and cell types have been used to carry out these experiments, which implicitly means that these approaches cannot cover the entire spectrum of human diseases. In this regard, the approaches relying on network pharmacology have proved to be promising in identifying candidate genes whose perturbation might lead to a desired therapeutic phenotype. Recently, there have been few reports relying on unique and differential network topologies [15, 33, 34] to identify the differential regulatory mechanisms leading to a given pathology. These approaches allow the building of condition-specific networks by collecting gene-gene interaction information from literature-curated resources and to predict target genes and drugs that could maximize the reversion from a disease phenotype to a healthy one. For example, by using the differential network-based approach, cyclosporine was predicted as a candidate drug to treat systemic lupus and rheumatoid arthritis. Surprisingly, this blindfold prediction was in agreement with existing literature, as cyclosporine has been successfully applied to treat these diseases [35, 36].

These findings suggest that network-based approaches hold a great potential to identify new disease-related genes and biomarkers for complex diseases. These approaches can uncover the regulatory mechanisms underlying disease pathologies by analyzing the differences in gene regulatory interactions of condition-specific networks. Furthermore, in silico simulations to mimic the network response upon drug application can boost the quest of identifying a putative drug for therapeutic intervention. Nevertheless, a prominent limitation of cell reprogramming approaches is the availability of good-quality interactome maps. For only a limited number of human diseases, we are able to gather enough omics data to construct a reliable interactome, which can help in exploring the underlying disease mechanisms. In order to overcome this information gap, research teams throughout the world are profiling next-generation sequencing experiments to obtain high-quality interaction maps of specific human disorders [37–39], while other consortiums like Roadmap Epigenomics [40] and ENCODE [41] are striving to create the reference human epigenomes and large-scale ChIP-seq profiling for different TFs across different cell types, respectively. Nonetheless, this information is still far from being complete and will require extensive future efforts to develop complete, high-quality, and noise-free interaction maps for all well-studied human diseases. We strongly believe that bridging this information gap will play a crucial role in the future of biomedical research.

References

1. Teng S, Madej T, Panchenko A, Alexov E. Modeling effects of human single nucleotide polymorphisms on protein-protein interactions. Biophys J. 2009;96(6):2178–88.
2. Ay A, Arnosti DN. Mathematical modeling of gene expression: a guide for the perplexed biologist. Crit Rev Biochem Mol Biol. 2011;46(2):137–51.
3. Angelini C, Costa V. Understanding gene regulatory mechanisms by integrating ChIP-seq and RNA-seq data: statistical solutions to biological problems. Front Cell Dev Biol. 2014;2:51.
4. Stolyar S, Van Dien S, Hillesland KL, Pinel N, Lie TJ, Leigh JA, et al. Metabolic modeling of a mutualistic microbial community. Mol Syst Biol. 2007;3(1):92.
5. Ledyankina OA, Mikhailov SA. Composite model of a research flight simulator for a helicopter with the hingeless main rotor. Rus Aeronaut. 2016;59(4):495–9.
6. Ahluwalia U, Katyal N, Deep S. Models of protein folding. J Protein Proteomics. 2013;3(2):85–93.
7. Agatonovic-Kustrin S, Beresford R. Basic concepts of artificial neural network (ANN) modeling and its application in pharmaceutical research. J Pharm Biomed Anal. 2000;22(5):717–27.
8. Brodland GW. How computational models can help unlock biological systems. Semin Cell Dev Biol. 2015;47-8((Supplement C)):62–73.
9. Nikitin A, Egorov S, Daraselia N, Mazo I. Pathway studio the analysis and navigation of molecular networks. Bioinformatics. 2003;19(16):2155–7.
10. Milo R, Shen-Orr S, Itzkovitz S, Kashtan N, Chklovskii D, Alon U. Network motifs: simple building blocks of complex networks. Science. 2002;298(5594):824–7.
11. Soliman S. A stronger necessary condition for the multistationarity of chemical reaction networks. Bull Math Biol. 2013;75(11):2289–303.
12. Plahte E, Mestl T, Omholt SW. Feedback loops, stability and multistationarity in dynamical systems. J Biol Syst. 1995;03(02):409–13.
13. Crespo I, Perumal TM, Jurkowski W, del Sol A. Detecting cellular reprogramming determinants by differential stability analysis of gene regulatory networks. BMC Syst Biol. 2013;7:140.
14. Okawa S, Angarica VE, Lemischka I, Moore K, del Sol A. A differential network analysis approach for lineage specifier prediction in stem cell subpopulations. Syst Biol Appl. 2015;1:15012.
15. Zickenrott S, Angarica VE, Upadhyaya BB, del Sol A. Prediction of disease-gene-drug relationships following a differential network analysis. Cell Death Dis. 2016;7:e2040.
16. Marbach D, Lamparter D, Quon G, Kellis M, Kutalik Z, Bergmann S. Tissue-specific regulatory circuits reveal variable modular perturbations across complex diseases. Nat Meth. 2016;13(4):366–70.
17. Boulesteix A-L, Strimmer K. Predicting transcription factor activities from combined analysis of microarray and ChIP data: a partial least squares approach. Theor Biol Med Model. 2005;2:23.
18. Vickaryous MK, Hall BK. Human cell type diversity, evolution, development, and classification with special reference to cells derived from the neural crest. Biol Rev. 2006;81(3):425–55.
19. Takahashi K, Yamanaka S. Induction of pluripotent stem cells from mouse embryonic and adult fibroblast cultures by defined factors. Cell. 2006;126(4):663–76.
20. Choi J, Costa ML, Mermelstein CS, Chagas C, Holtzer S, Holtzer H. MyoD converts primary dermal fibroblasts, chondroblasts, smooth muscle, and retinal pigmented epithelial cells into striated mononucleated myoblasts and multinucleated myotubes. Proc Natl Acad Sci U S A. 1990;87(20):7988–92.
21. Caiazzo M, Giannelli S, Valente P, Lignani G, Carissimo A, Sessa A, et al. Direct conversion of fibroblasts into functional astrocytes by defined transcription factors. Stem Cell Reports. 2015;4(1):25–36.
22. Vierbuchen T, Ostermeier A, Pang ZP, Kokubu Y, Sudhof TC, Wernig M. Direct conversion of fibroblasts to functional neurons by defined factors. Nature. 2010;463(7284):1035–41.

23. Pang ZP, Yang N, Vierbuchen T, Ostermeier A, Fuentes DR, Yang TQ, et al. Induction of human neuronal cells by defined transcription factors. Nature. 2011;476(7359):220–3.
24. Kamaraj US, Gough J, Polo JM, Petretto E, Rackham OJL. Computational methods for direct cell conversion. Cell Cycle. 2016;15(24):3343–54.
25. del Sol A, Balling R, Hood L, Galas D. Diseases as network perturbations. Curr Opin Biotechnol. 2010;21(4):566–71.
26. Rackham OJL, Firas J, Fang H, Oates ME, Holmes ML, Knaupp AS, et al. A predictive computational framework for direct reprogramming between human cell types. Nat Genet. 2016;48(3):331–5.
27. The Cancer Genome Atlas Research Network. Comprehensive genomic characterization defines human glioblastoma genes and core pathways. Nature. 2008;455:1061.
28. Lage K, Karlberg EO, Størling ZM, Olason PI, Pedersen AG, Rigina O, et al. A human phenome-interactome network of protein complexes implicated in genetic disorders. Nat Biotechnol. 2007;25:309–16.
29. Franke L, van Bakel H, Fokkens L, de Jong ED, Egmont-Petersen M, Wijmenga C. Reconstruction of a functional human gene network, with an application for prioritizing positional candidate genes. Am J Hum Genet. 2006;78(6):1011–25.
30. Barabási A-L, Gulbahce N, Loscalzo J. Network medicine: a network-based approach to human disease. Nat Rev Genet. 2011;12(1):56–68.
31. Lamb J, Crawford ED, Peck D, Modell JW, Blat IC, Wrobel MJ, et al. The connectivity map: using gene-expression signatures to connect small molecules, genes, and disease. Science. 2006;313(5795):1929–35.
32. Xiao Y, Gong Y, Lv Y, Lan Y, Hu J, Li F, et al. Gene Perturbation Atlas (GPA): a single-gene perturbation repository for characterizing functional mechanisms of coding and non-coding genes. Sci Rep. 2015;5:10889.
33. Ideker T, Krogan NJ. Differential network biology. Mol Syst Biol. 2012;8(1):565.
34. Mitra K, Carvunis A-R, Ramesh SK, Ideker T. Integrative approaches for finding modular structure in biological networks. Nat Rev Genet. 2013;14(10):719–32.
35. Caccavo D, Lagan B, Mitterhofer AP, Ferri GM, Afeltra A, Amoroso A, et al. Long-term treatment of systemic lupus erythematosus with cyclosporin A. Arthritis Rheum. 1997;40(1):27–35.
36. Wells GA, Haguenauer D, Shea B, Suarez-Almazor ME, Welch V, Tugwell P, et al. Cyclosporine for treating rheumatoid arthritis. Cochrane Database Syst Rev. 1998;2(2):CD001083. https://doi.org/10.1002/14651858.CD001083.
37. Ewing RM, Chu P, Elisma F, Li H, Taylor P, Climie S, et al. Large-scale mapping of human protein-protein interactions by mass spectrometry. Mol Syst Biol. 2007;3(1):89.
38. Lim J, Hao T, Shaw C, Patel AJ, Szabó G, Rual J-F, et al. A protein-protein interaction network for human inherited ataxias and disorders of Purkinje cell degeneration. Cell. 2006;125(4):801–14.
39. Kaltenbach LS, Romero E, Becklin RR, Chettier R, Bell R, Phansalkar A, et al. Huntingtin interacting proteins are genetic modifiers of neurodegeneration. PLoS Genet. 2007;3(5):e82.
40. Bernstein BE, Stamatoyannopoulos JA, Costello JF, Ren B, Milosavljevic A, Meissner A, et al. The NIH roadmap epigenomics mapping consortium. Nat Biotech. 2010;28(10):1045–8.
41. ENCODE Project Consortium. The ENCODE (ENCyclopedia of DNA elements) project. Science. 2004;306(5696):636–40.

Using Thermodynamic Functions as an Organizing Principle in Cancer Biology

Edward Rietman and Jack A. Tuszynski

Abstract One of the most powerful concepts in physics, introduced by Boltzmann, is the idea of entropy. All closed physical systems tend to a state of maximum entropy, which is a very penetrating observation that is yet to be contradicted by experimental evidence. A close examination of entropy and information from the microscopic to the macroscopic level is bound to provide deep insights into how living cells evolve from normal to malignant phenotype respecting the laws of physics. In this short review, we elaborate on the hypothesis that concepts borrowed from statistical thermodynamics, such as entropy and Gibbs free energy, can provide very powerful quantitative measures when applied to cancer research. We discuss how, on all length scales of biological organization hierarchy from cell to tissue and organ representation, cancer progression can be correlated with these thermodynamic measures. We illustrate how this can inform us about grade and stage of cancer and suggest a possible choice of optimal combination therapy. Significant diagnostic, prognostic, and therapeutic implications of these new organizing principles are presented.

1 Entropy and Information

Physics has evolved over the past 400 years from an empirical science to a fundamental basis of human knowledge about the universe. This took place as a result of several revolutions in our understanding of nature ushered by the discovery of the laws of physics. Newton's laws of mechanics; Maxwell's theory of electro-

E. Rietman
Computer Science Department, University of Massachusetts, Amherst, MA, USA

J. A. Tuszynski (✉)
Department of Oncology, University of Alberta, Edmonton, AB, Canada

Department of Physics, University of Alberta, Edmonton, AB, Canada

Department of Mechanical and Aerospace Engineering, Politecnico di Torino, Turin, Italy
e-mail: jackt@ualberta.ca

© Springer International Publishing AG, part of Springer Nature 2018   139
F. A. B. da Silva et al. (eds.), *Theoretical and Applied Aspects of Systems Biology*,
Computational Biology 27, https://doi.org/10.1007/978-3-319-74974-7_8

magnetism; quantum mechanics as embodied by Schrödinger's, Heisenberg's, and Dirac's equations for the time dependence of states and operator representations; and, finally, Einstein's theories of special and general relativity form a major conceptual edifice. These mathematical representations of physical reality changed the way we understand, interpret, and shape the world around us. No less profound was the introduction of the laws of thermodynamics into the physics vocabulary by Ludwig Boltzmann [38]. As is typically the case with breakthrough discoveries, Boltzmann's concept of entropy, despite being one of the most powerful ideas in physics, was fiercely resisted by his contemporaries. Yet, his finding that all closed physical systems tend to a state of maximum entropy is a very powerful observation, which is yet to be contradicted by any experimental evidence. It has found numerous applications not only in physics but also in fields as diverse as sociology, financial markets, and drug discovery [51, 52].

Biology, especially molecular biology, is an explosively expanding field of science, which today resembles the state of physics at the turn of the nineteenth century where reams of data had been collected regarding physical systems, but there was a dire lack of organizing principles such as the postulates of quantum mechanics. Today, molecular biology and its sister fields such as genetics, cell biology, and others have collected masses of data that only computational methods are able to sift through, visualize, and organize. However, most of the research in the area of life sciences is advanced on the basis of ad hoc hypotheses and their empirical validation. The lack of organizing principles leaves researchers at the mercy of computational tools. In 1944 in his book entitled *What is Life*, Erwin Schrödinger, a Nobel Prize-winning physicist, exposed the main challenges for biology from a physics point of view [46]. He stated that the reduction of entropy in living systems is a paradox that seems to contradict the second law of thermodynamics. The answer to this conundrum lies not only in physics but also in information science because entropy is negatively correlated with information as defined by the great computer scientist Claude Shannon [47].

In thermodynamics, entropy, S, is a measure of the number of possible microscopic configurations that correspond to a thermodynamic state of a system specified by its macroscopic measurable variables. Entropy is a measure of molecular disorder within a macroscopic system. The second law of thermodynamics states that an isolated system's entropy never decreases. Thermodynamic systems evolve toward thermodynamic equilibrium, which is a state with maximum entropy hence minimum free energy. Open systems, i.e., those that interact with their environment, may reduce their entropy, provided their environment's entropy increases by at least the same amount. Entropy is an extensive thermodynamic property which means that it is additive, so that the entropy of a system composed of subsystems is the sum of their respective entropies. This is very useful in the context of biological systems, which by definition are heterogeneous and hierarchical. In statistical thermodynamics, the most general mathematical formula for the thermodynamic entropy S is the so-called Gibbs entropy given by

$$S = -k_B \Sigma p_i \ln p_i \tag{1}$$

where k_B is the Boltzmann constant and p_i is the probability of a particular microstate i. The connection between thermodynamics and information theory was first made by Boltzmann and expressed by his famous equation

$$S = k_B \ln(W) \tag{2}$$

where S is thermodynamic entropy and W the number of microstates corresponding to the given macrostate. Since it is assumed that each microstate is a priori equally likely, the probability of a given microstate is $p_i = 1/W$. In other terms, the information entropy of a system is the amount of "missing" information needed to determine a microstate, given the macrostate. The average amount of information, I, that is gained with every event is equal to the opposite of entropy (also called negentropy), i.e.:

$$I = \sum_i p_i \log \frac{1}{p_i}. \tag{3}$$

For interacting systems, achieving thermodynamic equilibrium means a tendency to reach a minimum value of an appropriate thermodynamic potential such as the Gibbs free energy. This provides a route to solutions for numerous applications of statistical mechanics, including immensely complex biological systems giving us a single organizing principle. Many such examples include complex networks, which are ubiquitous in nature. A network may be described by a directed or undirected graph with node and edge sets. An edge signifies an interaction between two nodes. An example of a complex network is a protein-protein interaction network where nodes represent proteins and edges their interactions. Network complexity can be quantified by Shannon entropy using the distribution of connections in protein interactomes represented as oriented graphs. The concept of mathematically analyzing complexity of networks has a long history since the study of topology applied to networks was introduced more than 60 years ago, when degree entropy was proposed as a network complexity measure [41]. The extension of information theory to thermodynamics in networks was more recently made by Dehmer and Mowshowitz [16], where various entropy measures in network analysis were discussed.

2 Motivation

Although there is an acute need for developing better anticancer drugs, the lengthy time and enormous costs associated with cancer drug development, together with their high failure rates and limited efficacy of even targeted drugs, justify alternative

Table 1 Network entropy values for specific cancer types and scale-free and random networks

Pathway	Entropy
Scale-free	0.888
Thyroid	1.48
Bladder	1.67
Melanoma	1.68
Renal	1.77
Colorectal	1.80
Endometrial	1.84
Basal cell carcinoma	1.88
Pancreatic	2.05
Acute myeloid leukemia	2.10
Chronic myeloid leukemia	2.16
Small-cell lung	2.21
Glioma	2.26
Non-small cell lung	2.36
Prostate	2.40
Random	2.45

approaches to cancer drug discovery. Within the context of this chapter, new drug optimization strategies are being developed that include the identification of specific protein targets based on interactome networks with bioinformatic algorithms. Their experimental validation offers the advantage of a rational approach that may be further refined in stages from cell-based to in vivo assays. Also, the use of such a molecular network approach promises to minimize the size, costs, and failure rates of potential cancer inhibitors in clinical trials. The primary motivation for this chapter is to analyze cancer at the network level in order to develop potential uses of systems biology applied to chemotherapy. The main idea is to find a way to quantify the robustness of cancer signaling pathways as well as their weaknesses as potential targets for chemotherapeutic inhibition. The definition of robustness involves the ability of the pathway to properly function in the face of random (or targeted) perturbation. Using this definition, perturbations of these pathways approximate the inhibition of genes or their product proteins by chemotherapeutics. The assumption here is that the most robust pathways are most malignant and hence resistant to drugs. In order to quantify the robustness of the pathways, we use thermodynamic measures such as network entropy and Gibbs free energy. The literature on network research indicates that scale-free networks are highly robust (according to the random perturbation definition), whereas random networks are very non-robust. Since the equation for network entropy only requires information on the probability distribution of the graph, we calculated the entropy of the scale-free and random networks using a Poisson distribution for the random and a negative exponential distribution for the scale-free ($k^{-2.5}$) networks. The distributions were chosen based on the literature on scale-free and random networks. Below is the table of the entropies calculated for the various graphs describing cancer signaling networks. Their detailed descriptions can be found in the KEGG database (Table 1).

It is interesting to note that these values seem to indicate that lower entropy is indicative of higher robustness, which is somewhat counterintuitive. In addition the cross entropy of each signaling pathway was calculated using the scale-free network as the reference in order to rank the pathways better. Although scale-free networks are resilient to random attack, they are also exceptionally vulnerable to targeted attack. This seems to imply the networks, which are most similar to scale-free, should be more susceptible to chemotherapy, assuming the targets of those treatments are integral to the pathways. Alternatively, more robust pathways may still be more malignant or perhaps have higher incidence rates because resilience to random disruption might lead to an evolutionary or functional advantage. In addition to calculating the effect of standard chemotherapy treatments on the pathways, it is informative to identify the nodes in each pathway with the highest betweenness centrality. The idea behind this is that the most important nodes in each pathway might make ideal targets for new chemotherapeutics.

3 Complexity of Cancer

Current cancer treatments consist of surgery, radiation, and chemotherapy. However, for many types of cancer, for example, ovarian cancer, mortality statistics have changed very little over the last 40 years. Therefore, improved strategies to overcome drug and radiation resistance are required. The critical question is the optimal selection of the molecular targets in order to control cancer cell proliferation. Recent progress in data mining and high-throughput data generation with respect to gene, protein, and metabolic networks presents an opportunity to identify proteins of marginal significance in normal cells, which are signaling hubs in cancer cells and hence represent ideal targets for inhibition. We propose to use novel bioinformatic methods and their experimental validation to assess the relevance of specific protein targets in a cancer signaling network.

One of the main areas of the life sciences where an organizing principle should and can be found is oncology. We believe that close examination of entropy and information from the microscopic to the macroscopic level is bound to provide deep insights into how malignant cells and their clusters operate within the physical laws to achieve a winning strategy against the host organism. Cancer research has so far been overwhelmingly directed toward the biochemistry, genomics, and cell biology of cancer [23, 24] with far less attention paid to the biophysics of the cancer state. While the field of cancer research is vast in terms of empirical knowledge, it appears to have a dearth of quantitative measures of the sort typical in the physical sciences. Despite vast amounts of genetic, molecular, cellular, histological, and epidemiological information, and despite intense efforts to identify predisposing factors, cancer remains as yet an unresolved enigma. Cells, the fundamental units of organization in living matter, can exist in two main physiological states: (a) normal cells, which are well differentiated, reproduce themselves faithfully, undergo apoptosis when damaged or stimulated by their internal clock, and adhere to each other to form regular tissues or organs, and (b) cancer cells, which are poorly differentiated, reproduce unfaithfully and sometimes without limit, evade apoptosis, colonize organs where they do not belong, and associate in relatively disordered assemblages (tumors) rather than forming well-defined tissues and organs. Malignant cells selectively evolve to proliferate at the expense of the normal host cells. They evade programmed cell death (apoptosis), invade surrounding tissues and distort normal tissue architecture, and stimulate inflammation and formation of the blood vessels (angiogenesis). The process leads to lack of cell maturation and causes dedifferentiation or even trans-differentiation in the form of epithelial to mesenchymal transformation [23]. In the past, it was commonly assumed that the abnormal distributions of chromosomes (called genomic instability) were a result, rather than a cause, of malignant transformation. Hence cancer was thought to originate from intrinsic mutational changes [26]. However, it now appears that genomic instability is a logical consequence of abnormal mitosis and that cancer resulting from randomly accumulated genetic mutations has become a widely accepted "dogma." Two types of cancer-related genes were identified: (a) tumor suppressor genes, which lead to cancer predisposition, and (b) oncogenes, which drive malignant transformation. Tumor suppressor genes can remove an inhibitor on a proliferative pathway (e.g., PTEN), create chromosomal instability (e.g., p53), or cause an abnormal DNA repair (RB). In each case a second mutation is needed for cancer to develop. Oncogenes, on the other hand, are genomic alterations resulting in overexpression or constitutional activation of genes that stimulate growth and cell division or inhibit apoptosis. However, the oncogene/tumor suppressor gene theory does not fully explain carcinogenesis or cancer progression since we are not able to identify a consistent set of gene mutations that correlate with cancer initiation, progression, or metastases. This is evident in the immense genetic variability occurring within a single tumor. Both the theory of sequential accumulation of random mutations leading to a proliferative cancer clone and the theory of imbalanced ("aneuploid") distribution of DNA leading to cancer can be combined since these theories must account for the physical pressures within the tumor microenvironment. The development of a heterogeneous tumor subclones may indeed be a direct consequence of thermodynamic evolutionary pressures due to the environment. We believe that mutations and in situ evolution play a significant role but are driven by molecular, cellular, tissue, and organ-specific thermodynamics. In a sentence, while thermodynamics dictates the possible, chemical kinetics dictates the probable.

If a normal cell undergoes a transition to avoid apoptosis as a result of the accumulating genetic mutations [34] or due to somatic damage (e.g., due to ionizing radiation or toxins), two types of changes commonly take place: (a) at the cell level cancer initiation and (b) at the population level cancer progression [35]. The former type includes changes in cell metabolism, specifically a shift from oxidative phosphorylation to glycolysis (the so-called Warburg effect) [54], the epithelial-to-mesenchymal transition (EMT) involving changes in cell morphology and motility, as well as activation of signaling and protein expression alterations. These physiological, morphological, and molecular changes are correlated with epigenetic transformations. At the cell population level, alterations result in the replacement of one group of cells (those that adhere to each other to form a differentiated tissue) by another group of cells (which form a heterogeneous and more motile assemblage), i.e., a tumor or neoplasm. These changes are reminiscent of phase transitions in physical systems, which are suggestive that thermodynamics may be a driving force for cancer initiation and progression [43] and it may act on several scales as we elaborate on below.

4   Molecular Scale

A cell is comprised of a large variety of molecules of different sizes, shapes, and physical properties interacting synchronously in a complex network. Whenever a molecular species loses the normal state of chemical equilibrium with its reaction partners, this may result in chemical potential differences within the network causing perturbations that can further lead to mutation, fusion genes, or aneuploidy. Adding evolutionary pressures to this picture can cause an entirely new set of mutational adaptations, which may eventually lead to the development of a new attractor state. In biology, a persistent change in intracellular environment leads to preservation/selection of a "protective" mutation. An approach called the "modified dogma" [35] states that random mutations, on average, affect only one gene per cell in a lifetime but other factors such as the presence of carcinogens, reactive oxidants, malfunctions in DNA duplication, and repair machinery tend to increase the incidence of random mutations leading to the emergence of the cancer phenotype. Recently, Tomasetti and Vogelstein [49] presented epidemiological evidence, which indicates that as many as 2/3 of all cancers are a result of random mutations since their incidence correlates with the frequency of cell divisions. The remaining cancer cases can be linked to environmental and genetic factors.

To quantify the processes involving protein-protein interactions (PPI) in cancer cells, one can use the degree entropy of established PPI networks to assess cancer risk and survival. Breitkreutz et al. [3] found that the degree entropy of cancer PPI networks included in the KEGG database inversely correlated with 5-year survival of cancer patients. Each cancer PPI network is characterized by a type of different entropy, but in all cases studied entropy of the network could be inversely correlated with 5-year survival. The observed degree entropy corresponded to the complexity of the molecular PPI network, and a mathematical elimination of proteins leading to decrease in network complexity could be correlated to improved survival rates. In continuing work along this line, Benzekry et al. [5] discussed a topological metric on the KEGG PPI network that also inversely correlates with cancer survival. Due to the importance of this methodology, we discuss it in greater detail below.

5   Gibbs Free Energy of Protein-Protein Interaction Networks

The dynamics of a cell is coordinated by proteins interacting with other proteins. The set of all protein-protein interactions (PPI) defines a complex network. The foremost database of PPI networks is at BioGrid (http://thebiogrid.org/), which is described by Breitkreutz et al. [4] and Stark et al. [45]. Since the proteome is not yet fully mapped from open reading frames to genes and proteins, calculations of the networks' properties such as entropy or the Gibbs free energy are only estimates reflecting the present state of knowledge about these networks. Investigations of PPI networks suggest that changes in PPI network architecture correlate not only with survival statistics but also with stage. Paliouras et al. [40] analyzed prostate cancer samples and showed that changes in the PPI network architecture relate to Gleason score and prostate-specific antigen (PSA). Freije et al. [18] demonstrated that gene expression in gliomas correlates with patient survival. However, it is important to note that there are several ways of measuring complexity of protein-protein interaction networks. Hinow et al. [25] and Benzekry et al. [5] describe various *topological metrics* of PPI cancer networks that correlate with 5-year cancer patient survival. Correlation between transcription data and survival was also shown by Liu et al. [33] who defined a measure called state-transition-based local network entropy (SNE). It is a Shannon information measure that is probabilistically dependent on the previous state of a local dynamical network; hence it involves a Markov process. They used RNA expression data at different stages of tumor development, overlayed it on protein-protein interaction (PPI) network data, and showed that SNE values change significantly with cancer progression. Gibbs free energy is also a thermodynamic measure that encompasses both network complexity and cell thermodynamics (as represented by transcriptome), and it has been shown to correlate with cancer stage and survival. We introduce it below.

All biological cells are maintained in a homeostatic state by a complex and dynamic network of interacting molecules. If the concentration of any one of these molecular species changes dramatically, it alters the chemical balance of other species that interact with it in the network. These changes then percolate through the network affecting the chemical potential of other species giving rise to changes in the energetic landscape of the cell. These energetic changes can be described as chemical potential on an energetic landscape [1]. The PPI networks we consider in this chapter are viewed as being time invariant and hence show no dynamics. The reactions involving interacting species can be schematically represented as

$$A + B \underset{k_r}{\overset{k_f}{\rightleftharpoons}} AB \tag{4}$$

where A and B are two proteins and their interaction product is AB and k_f and k_r are the forward and reverse reaction rate constants, respectively. Each reaction is associated with it a binding free energy. From standard physical chemistry, the Gibbs free energy of this reaction is $\Delta G = \Delta H - T\Delta S$, where the symbols represent the change in Gibbs free energy, G; the change in enthalpy, H; and the change in entropy, S [32]. Proteins do not interact with large numbers of neighbors simultaneously; hence we *assume* an ensemble of the proteins of interest is reasonably well described as an *ideal gas mixture*. Therefore, we calculate the Gibbs free energy from the transcriptome and the PPI level, which is an undirected network because there is no directionality assigned to the links. This is related to the entropy of mixing [36]. These nominal chemical potentials, represented with either concentration or expression, can be used to calculate a nominal Gibbs free energy for not only a single protein with its neighbors but also for the whole network and thus for the cell or the tumor represented by the transcriptome. Since we do not have information on the molar fractions, or molar concentrations, we substitute a normalized, (rescaled) [0,1] RNA transcription value in place of the concentrations. The general equation for Gibbs free energy can thus be written as

$$G_i = c_i \ln \frac{c_i}{\sum_j c_j} \tag{5}$$

where the sum is over all neighbors j to node i and the sum includes the concentration of node i. We can compute this quasi-Gibbs free energy for the tumor by summing over all the nodes in the network: $G = \sum_i G_i$.

6 Cellular Scale

At the cellular scale one can calculate the entropy of an individual cell from karyotype and draw a similar conclusion for the relevant molecular network interactions. Davies et al. [15] discussed thermodynamic entropy of self-organization of biological cells, Metze et al. [37] used similar concepts for pathophysiology of cancer and calculated the image entropy for cancer cells and tissues, and Castro et al. [10] described information entropy applying karyotypic analysis of 14 different epithelial tumor types. Computing Shannon information from the karyotype, they found a very high Spearman rho correlation with 5-year survival of cancer patients.

Carels et al. [8] applied entropy maximization in the context of different breast cancer cell lines and extended it to develop a strategy for the optimized selection of protein targets for drug development. By combining human interactome and transcriptome data from malignant and control cell lines, they identified the most highly connected proteins in the selected PPI networks. They assumed that proteins that are most upregulated in malignant cell lines are suitable targets for chemotherapy. In addition to traditional drug targets such as EGFR, MAPK13, or HSP90, they found several proteins, not generally targeted by drug treatments, which calls for an extension of existing therapeutic agents to include novel inhibitors designed against these newly found target proteins. Their study also showed that signaling mechanisms in the luminal A, B, and triple-negative subtypes are different. These results have significant clinical implications in the personalized treatment of cancer patients by pointing to a rational repurposing approach for the available drugs. In a follow-up study, Carels et al. [9] investigated breast cancer cell lines and found that the entropy of their PPI networks correlates negatively with their sensitivity to target-specific drugs. Conversely, they have found no correlation for drugs that are either of low potency or with no specific molecular targets. As a result of this study, all anticancer drugs can be divided into target specific and generally cytotoxic according to the GI_{50} they produce in malignant cell lines. Interestingly, these authors have predicted that the inhibition of the top-5 upregulated protein hubs by targeted drugs may reduce the protein network entropy by 2%, which is expected to provide major clinical benefit for patient survival anticipating complete remission over a 5-year period.

7 Tissue Scale

As described above, one can calculate the Gibbs free energy from the chemical potential. For computing the Gibbs energy of a cell, one should use mRNA expression or RNAseq counts as a surrogate for protein concentration. Greenbaum et al. [22], Kim et al. [30], Wilhelm et al. [55], Liu et al. [33], and Berretta et al. [2] all point to the use of RNA abundance as a measure of protein concentration. Carrying this out on several TCGA cancers, this can be first computed as indicated in the equation for the individual Gibbs free energy for a particular protein, but then it should be summed over all proteins to obtain the total Gibbs free energy of the network. This now represents an average of the Gibbs energy for that tissue sample from the biopsy. Rietman et al. [44] showed that the Gibbs free energy closely correlates with 5-year survival as well as with the cancer stage [42]. Their Gibbs free energy is a genuine thermodynamic measure computed from using the mRNA expression values for cancer patient tissues and overlaying that on the human PPI from BioGrid (http://thebiogrid.org/). Correlation between the Gibbs free energy and cancer stage was also found using TCGA cancer datasets and including two GEO datasets for prostate cancer GSE3933 [31] and GSE6099 [50] and a GEO dataset for liver cancer GSE6764 [56]. Importantly, several of these cancers analyzed show very significantly linear correlation of Gibbs free energy to cancer stage, which is highly indicative that the Gibbs free energy can be interpreted as a real thermodynamic measure of cancer stage.

8 Metabolic Entropy Increase in Cancer

Complex cellular mechanisms exist to maintain cell integrity and function. One of the most important ones involves the production of biochemical energy in the form of ATP through oxidative phosphorylation in the mitochondria. It is now generally accepted that tumors have an increased uptake of glucose. High demand for glucose, even in the presence of adequate oxygen supply, is called the Warburg effect. The causes and advantages of increased glucose consumption of tumors have been extensively discussed (Vander Heiden et al. [53]), and they include anti-apoptotic factors, acidic microenvironment, and rapid generation of the biomass. This increased consumption of glucose is associated with an anaerobic mechanism, which causes a significant energy burden to the cancer patient not previously integrated into resting energy expenditure (REE) estimates. Friesen et al. [19] developed a mathematical model incorporating tumor's energy metabolism in the calculation of an energetic burden that leads to cachexia in order to meet the tumor's energetic demands. As tumors grow, this energetic cost may eventually become prohibitive and combined with reduced caloric intake may lead to a catabolic state. A corresponding representation of cancer development and progression through the stages ending with cachexia and death can be viewed as a continuous increase in metabolic entropy produced as a result of highly inefficient glucose metabolism with an associated increase of the glycolytic rate as predicted by Warburg [54].

9   Organismal Scale

One of the most fundamental differences between animate and inanimate matter from the point of view of thermodynamics is that the former exists in states that are far from thermodynamic equilibrium. Living systems survive due to a continuous flux of matter and energy between them and their surroundings. This involves excess entropy transfer into the environment (heat and waste) to compensate for the creation and maintenance of structural order (entropy reduction) and functional organization. In the case of a transition from normal to cancer cells, the nature of the transformation occurring at both the molecular and cellular scales involves a drastic elimination of cell cycle checkpoints and a simplification of the cell's functional program. The cancer phenotype is one, which is aimed mainly at survival and proliferation. In terms of PPI networks, cells function not only as signaling networks but also as metabolic networks defined by a large ensemble of interacting enzymes within a substrate. Their activities are mediated by processes that are best described using chemical kinetics which transforms one metabolite into another. A qualitative description of the transformation from normal to cancer tissue using the concepts of phase transition that include order parameters, control parameters, Gibbs free energy, and entropy and susceptibility has been provided by Davies et al. [14], and it yields further support to the use of thermodynamic measures in the description of cancer.

10   Epidemiological Scale

Rietman et al. [44] described the concept of Gibbs free energy for cancer stage and applied a topological concept known in the area as "filtration" [57] to produce a persistent homology from the energy landscape that the PPI created with transcriptome data represent. At any given threshold, an energetic subnetwork is produced, which is characteristic of not only an individual patient but also a point item that can represent a different stage in cancer progression. Different patients have different energetic persistent homology networks. Furthermore, one can apply another topological concept known as the Betti number [5], which represents a count of the number of rings of four or more proteins in a PPI network. This allows one to find which node in the network when removed causes the greatest drop in the Betti number as described by Rietman et al. [44]. Using this method, one can identify a suitable target for protein inhibition in treating a particular cancer patient.

11   Data Sources and Methods

Data for several cancers from The Cancer Genome Atlas (TCGA) hosted by the National Institute of Health (http://cancergnome.nih.gov/) have been collected and can be used as a very important resource for investigations. We can also use the human protein-protein interaction network from BioGrid (http://thebiogrid. org/), discussed by Breitkreutz et al. [4] and Stark et al. [45]. Detailed statistical information about the 5-year survival rates of patients with cancer is readily available at the Surveillance, Epidemiology, and End Results (SEER) National Cancer Institute database. In addition we used the National Brain Tumor Society database. The data for specific cancer types can also be obtained from Gene Expression Omnibus (GEO) at ncbi.nlm.nih.giv. This dataset is well described in Wurmbach et al. [56].

The empirical equation for the linear fit of the Gibbs free energy with survival has been found as [42, 44]: $G = 8.112\sigma + 5753.9$. Using the data from Breitkreutz et al. [3], an empirical equation for the linear fit of entropy was obtained as $S = -0.0087\sigma + 2.2731$. Solving both these equation for 5-year survival probability, σ, one finds a relationship between Gibbs free energy and entropy as $G = 7873 - 932S$ which is consistent with the thermodynamic relation linking Gibbs free energy and entropy, namely, $G = H - TS$. In summary, the expression data and the PPI network analysis are both needed for the generation of a meaningful Gibbs free energy for a cancer cell. As shown above, the topology of the network provides a structure to the expression data.

The results discussed above provide conceptual support for the work reported by Zhang et al. [59] and Suva et al. [48]. Zhang et al. [59] described reprogramming sarcoma cells to transform them to a pluripotent-like state, which then differentiates into connective tissue or red blood cells. On the other hand, Suva et al.

[48] described reprogramming the tumor-propagating cells of glioblastoma. We represent cancer as a dynamical system capable of undergoing state changes on an energy landscape and discussed how this can be associated with a quantitative measure of the PPI network (either entropy of Gibbs free energy) and correlated with the malignancy level of the tumor. This conceptual framework may lead to a new way of designing cancer therapy, which does not rely on inhibiting a specific protein from a mutated gene (or two). Instead, it may be possible to treat cancer by reprogramming of the PPI network using an associated Gibbs free energy landscape. This new perspective on cancer considers not just the oncogenes and highly mutated genes but emphasizes the role of the PPI network and its thermodynamic profile.

12 Application of Group Theory to Systems Biology of Cancer

In the past, the overriding mathematical theme has been evolution processes that can be described by ordinary differential equations, partial differential equations, or difference equations, sometimes infused with elements of stochasticity. This is hardly a surprise since it is very natural to try to predict the development of a biological system given a known initial state. As it is well known, topology and group theory have had a long and successful history of applications to physics and chemistry, for example, in understanding the symmetries of elementary particles or in classifying isoforms of molecules. The fact that biological information is passed on in form of discrete DNA sequences suggests that discrete mathematics can play an important role in biology. Mathematically, such a network may be described by a directed or undirected graph $G = (V, E)$ with vertex and edge sets V and E, respectively. An automorphism is a permutation of the set V that preserves the adjacency relation and, if present, the orientation of arrows between vertices. With the operation of composition, the automorphisms form a group Aut(G). This is in contrast to large random graphs, such as Erdős-Rényi graphs, the majority of which are rigid, that is, they have only the trivial automorphism [6, 7]. The difference is not that surprising if one realizes that real networks display a modular structure, with vertices organized in communities tightly connected internally and loosely connected to each other [21]. This results in the presence of symmetric subgraphs such as trees and complete cliques. This also helps to classify the nodes of a network into a "backbone" (those that remain fixed under the automorphisms) and "appendages" (those that get mapped to other vertices). An edge appears in the graph if there is a known interaction between the two partners, either by direct binding or by enzymatic catalysis. The complexity of a graph can be measured in a variety of ways [16, 39]. One of them being the degree entropy, which is the sum over i vertices of the product of the likelihood of their connection number, i.e. the vertex degree – $d(v)$, by the logarithm of this likelihood:

$$H(G) = -\sum_{i=1}^{n} p\left(d(vi)\right) \log p\left(d(vi)\right), \text{ with } vi \in V \tag{6}$$

In addition, one can order the vertices of the network according to different measures of centrality. Selecting the three nodes of each network of highest betweenness centrality shows that certain proteins are highly central in several different cancer interaction networks. This suggests that such proteins may be better targets of anticancer drugs than others as we have briefly discussed above. As an example, the PPI network for pancreatic cancer was investigated for its specific symmetry that was found with the help of SAUCY [29].

13 The Maximum Entropy Principle

The maximum entropy principle is deeply rooted in thermodynamics. Jaynes' pioneering work [27, 28] launched the maximum entropy principle as a reasoning tool to process information with the least bias. When dealing with complicated biological systems, which involve either many-body interactions at a microscopic level, or complicating regulating PPI networks at a cell level, we often do not have sufficient knowledge to completely understand these systems. Both in silico and in vitro methods in drug discovery also employ a similar concept. Drug discovery starts with identifying molecular fragments that can bind to potential pockets in the identified molecular targets and then develop chemical bonds between those fragments to form a single molecule that is designed to have the highest binding affinity under specific molecular constraints. This approach significantly reduces the conformational space and increases the structural diversity of molecules. However, the drawback is the associated combinatorial explosion leading to a bottleneck in this type of drug design. However, maximum entropy can be introduced to drug discovery as a useful tool to solve such problems of relevance to drug discovery. It is important to ask the right question to solve these problems. Once the right question is posed, the use of maximum entropy becomes a unique and straightforward tool to answer it and solve the problem posed. A related issue involves systems biology. This concerns the choice of biological molecules that are relevant drug targets, and it requires proper interpretation and characterization of network data. Yang et al. [58] defined the activity of a molecular target based on the reaction rates within the corresponding PPI network. Alternatively, Fuhrman et al. [20] proposed to instead rely on Shannon entropy to quantify information contained in molecular activities specifically to analyze information changes in targets' gene expression patterns. Here, entropy is viewed as a measure of variation in a series of gene expression data that describe a biological processes. Chang et al. [13] suggested a better alternative, namely, MEDock, which utilizes the maximum entropy principle as a guide and ask the following question: "What is the probability of finding the deepest energy valley in the ligand-target interaction energy landscape?"

Caticha [11, 12] argued that since entropy has many faces including degree of randomness, information measure, and being a tool for inductive inference, there is no need to seek precise interpretation. The key aspect for solving problems within an entropy-based scheme is to ask the right question and present information relevant to the problem. Consequently, the principle of maximum entropy leads to the most honest inference available to solve the problem at hand. Maximum entropy is the central idea to govern the choice of right drug targets, structures of optimal modulators, and their best combinations [52].

Chang et al. [13] proposed an entropy-based scheme for complete genome comparison, which aims to define the probability distribution representing our current state of knowledge regarding the different combinations of the four bases in DNA sequences. They specified that k-mer nucleotides in the sequence may encode certain genetic information, where k is an arbitrary number. Based on this definition, information in sequences can be quantified using the definition of Shannon information. Subsequently, Chang et al. [13] introduced the concept of reduced Shannon information, which is defined as the ratio of the Shannon information of the genome to the Shannon information of random sequences, in order to quantify the extent to which the information contained in the genome is different from the information in a random DNA sequence. Hence, it indicates the degree of our belief that a genome is not a random sequence. This concept is similar to relative entropy.

The development of high-throughput screening techniques, such as microarray technology, has generated numerous PPI data and has revealed regulatory mechanisms of the biological species in the networks. Robustness of biological networks may be the key for identifying systems that can tolerate external perturbations and uncertainty triggered by external factors. Biological networks share a global feature as they represent "scale-free networks" with characteristic power-law distribution functions in these networks. Dover [17] suggested that the emergence of the power-law distributions is a consequence of the maximum entropy principle when the internal order of subnetworks of a complicated large network remained fixed. Within entropy-based inference, the power-law distributions of biological networks simply represent the most preferred choice that maintains the fixed internal order of the subnetworks.

14 Conclusions

While incremental progress in cancer treatments has been made including advances in immunotherapy, targeted chemotherapy, radiation tomotherapy, and more accurate diagnostic tools, we are still almost clueless regarding the molecular-level causes of cancer and methods of arresting cancer initiation and progression. Hence clinical outcomes across the board have been far from impressive. Many of the cancer chemotherapy drugs are very expensive, provide modest clinical improvements, and have significant negative side effects. We believe that to make serious progress in cancer therapy, we must strive to uncover an organizing principle in cancer cell transformations at a molecular and cellular levels. We know that cells become cancerous as a result of complex genetic and epigenetic reprogramming involving complicated regulatory networks leading to their immortality and uncontrolled division. Hundreds of oncotargets have been identified and some therapeutics developed aiming at their inhibition. However, it is clear that we cannot inhibit all oncotargets at once since multiple overlapping toxicities would first kill the patient. Moreover, we cannot possibly inhibit the oncotargets once and for all because cancer cells are plastic and evasive. When challenged, they start quiescence programs and "learn" how to survive developing drug resistance. Much of our selection of cancer targets and the development of therapies is very ad hoc and lacks a rationale. Based on the various investigations reviewed here, we hypothesize that cancer can be characterized by a trend toward thermodynamic stability defined by a corresponding thermodynamic function of state (entropy maximization or Gibbs free energy minimization). These trends are likely to persist at all levels of the hierarchical organization from DNA to tumor tissue. In all cases discussed, cancer initiation and progression follow a predictive trajectory in a thermodynamic phase space. This is in contrast to normal cells whose main dynamical objective is determined by homeostasis, i.e., thermodynamic stability around its equilibrium state. Normal cells also satisfy an orchestrated functioning in concert with other tissues to serve the organism as whole. As is now abundantly clear, cancer cells do not fulfill these functional objectives. Instead, they eliminate a number of cell cycle checkpoints and simplify their program to achieve two main goals: immortality and cell division. Analyzing cancer hallmarks from DNA mutations, to histone methylation, to DNA packaging, to aneuploidy, to cell metabolism (the Warburg effect), to cell morphology, to cell organization (epithelial to mesenchymal transformation), to (fractal) tumor morphology, and even to metastases, one can introduce an organizing principle at all levels of transformations in cancer, which involves a tendency to evolve its thermodynamic state function toward maximum stability, which in biological terms can be called robustness. In this chapter we have brought evidence to support the statement that cancer progression is associated with directional changes in both entropy and Gibbs free energy, both of which are anticorrelated.

We have advanced a hypothesis at all levels of biological organization and drawn practical conclusions in terms of both prognostic information for cancer patients and therapeutic interventions. The former can be used to better predict the life expectancy of individual patients, while the latter could be aimed at reversing or at least stalling the process in order to reduce entropy with the objective of slowing down or even halting the progression of cancer. This involves the issue of the optimal selection of the molecular therapeutic target and their inhibitors in order to control cancer cell proliferation. In essence, we have proposed a groundbreaking paradigm shift from a descriptive to a quantitative measure of cancer based on the application of a physical organizing principle that of entropy maximization or Gibbs free energy minimization as a specific representation of this principle under thermodynamic constraints. In a nutshell, as Erwin Schrödinger [46] famously pondered in his seminal book *What is Life?*", life is a tendency to reduce entropy of a biological system, which, if left to itself, would satisfy the second law of thermodynamics to increase entropy. We hypothesize that cancer, as a pathological, unsustainable state of a living organism, is characterized by entropy increase leading to disorder, disorganization, and ultimately death.

References

1. Atkins PW. Physical chemistry, 3rd. Oxford: Oxford University Press; 1986.
2. Berretta R, Moscato P. Cancer biomarker discovery: the entropic hallmark. PLoS One. 2010;5(8):e12262.
3. Breitkreutz D, Hlatky L, Rietman E, Tuszynski JA. Molecular signaling network complexity is correlated with cancer patient survivability. Proc Natl Acad Sci. 2012;109(23):9209–12.
4. Breitkreutz BJ, Stark C, Tyers M. The GRID: the general repository for interaction datasets. Genome Biol. 2003;4(3):R23.
5. Benzekry S, Tuszynski JA, Rietman EA, Klement GL. Design principles for cancer therapy guided by changes in complexity of protein-protein interaction networks. Biol Direct. 2015;10(1):32.
6. Bollobás B. Graduate texts in mathematics. Modern graph theory. Springer: Berlin; 1998.
7. Bollobás B, Fulton W, Katok A, Kirwan F, Sarnak P. Cambridge studies in advanced mathematics. In: Random graphs (vol. 73). G. Tourlakis (ed.), Cambridge: Cambridge University Press; 2001.
8. Carels N, Tilli T, Tuszynski JA. A computational strategy to select optimized protein targets for drug development toward the control of cancer diseases. PLoS One. 2015a;10(1):e0115054.
9. Carels N, Tilli TM, Tuszynski JA. Optimization of combination chemotherapy based on the calculation of network entropy for protein-protein interactions in breast cancer cell lines. EPJ Nonlinear Biomed Phys. 2015b;3(1):6.
10. Castro MA, Onsten TT, de Almeida RM, Moreira JC. Profiling cytogenetic diversity with entropy-based karyotypic analysis. J Theor Biol. 2005;234(4):487–95.
11. Caticha A. Information and entropy. In: AIP conference proceedings (vol. 954, no. 1). AIP; 2007. p. 11–22.
12. Caticha A. Relative entropy and inductive inference. In: AIP conference proceedings (707(1)). AIP; 2004. p. 75–96.
13. Chang DT, Oyang YJ, Lin JH. MEDock: a web server for efficient prediction of ligand binding sites based on a novel optimization algorithm. Nucleic Acids Res. 2005;33(suppl_2):W233–8.
14. Davies PC, Demetrius L, Tuszynski JA. Cancer as a dynamical phase transition. Theor Biol Med Model. 2011;8(1):30.
15. Davies PC, Rieper E, Tuszynski JA. Self-organization and entropy reduction in a living cell. Biosystems. 2013;111(1):1–10.
16. Dehmer M, Mowshowitz A. A history of graph entropy measures. Inf Sci. 2011;181(1):57–78.
17. Dover Y. A short account of a connection of power laws to the information entropy. Physica A: Stat Mech Appl. 2004;334(3):591–9.
18. Freije WA, Castro-Vargas FE, Fang Z, Horvath S, Cloughesy T, Liau LM, Mischel PS, Nelson SF. Gene expression profiling of gliomas strongly predicts survival. Cancer Res. 2004;64(18):6503–10.
19. Friesen DE, Baracos VE, Tuszynski JA. Modeling the energetic cost of cancer as a result of altered energy metabolism: implications for cachexia. Theor Biol Med Model. 2015;12(1):17.
20. Fuhrman S, Cunningham MJ, Wen X, Zweiger G, Seilhamer JJ, Somogyi R. The application of Shannon entropy in the identification of putative drug targets. Biosystems. 2000;55(1):5–14.

21. Garlaschelli D, Ruzzenenti F, Basosi R. Complex networks and symmetry I: a review. Symmetry. 2010;2(3):1683–709.
22. Greenbaum D, Colangelo C, Williams K, Gerstein M. Comparing protein abundance and mRNA expression levels on a genomic scale. Genome Biol. 2003;4(9):117.
23. Hanahan D, Weinberg RA. The hallmarks of cancer. Cell. 2000;100(1):57–70.
24. Hanahan D, Weinberg RA. Hallmarks of cancer: the next generation. Cell. 2011;144(5): 646–74.
25. Hinow P, Rietman EA, Omar SI, Tuszynski JA. Algebraic and topological indices of molecular pathway networks in human cancers. Math Biosci Eng. 2015;12(6):1289–1302.
26. Holland AJ, Cleveland DW. Boveri revisited: chromosomal instability, aneuploidy and tumorigenesis. Nat Rev Mol Cell Biol. 2009;10(7):478–87.
27. Jaynes ET. Information theory and statistical mechanics. Phys Rev. 1957a;106(4):620.
28. Jaynes ET. Information theory and statistical mechanics. II. Phys Rev. 1957b;108(2):171.
29. Katebi H, Sakallah KA, Markov IL. Graph symmetry detection and canonical labeling: differences and synergies. arXiv preprint arXiv:1208.6271. 2012.
30. Kim MS, Pinto SM, Getnet D, Nirujogi RS, Manda SS, Chaerkady R, Madugundu AK, Kelkar DS, Isserlin R, Jain S, Thomas JK. A draft map of the human proteome. Nature. 2014;509(7502):575–81.
31. Lapointe J, Li C, Higgins JP, Van De Rijn M, Bair E, Montgomery K, Ferrari M, Egevad L, Rayford W, Bergerheim U, Ekman P. Gene expression profiling identifies clinically relevant subtypes of prostate cancer. Proc Natl Acad Sci U S A. 2004;101(3):811–6.
32. Liberles A. Introduction to theoretical organic chemistry. New York: McMillan; 1968.
33. Liu R, Li M, Liu ZP, Wu J, Chen L, Aihara K. Identifying critical transitions and their leading biomolecular networks in complex diseases. Sci Rep. 2012;2:813.
34. Loeb LA, Springgate CF, Battula N. Errors in DNA replication as a basis of malignant changes. Cancer Res. 1974;34(9):2311–21.
35. Loeb LA, Loeb KR, Anderson JP. Multiple mutations and cancer. Proc Natl Acad Sci. 2003;100(3):776–81.
36. Maskill H. The physical basis of organic chemistry. Oxford: Oxford University Press; 1985.
37. Metze K, Adam RL, Kayser G, Kayser K. Pathophysiology of cancer and the entropy concept. Model Based Reason Sci Technol. 2010;114:199–206.
38. McQuarrie DA. Statistical thermodynamics. New York: Harper and Row; 1973.
39. Mowshowitz A, Dehmer M. Entropy and the complexity of graphs revisited. Entropy. 2012;14(3):559–70.
40. Paliouras M, Zaman N, Lumbroso R, Kapogeorgakis L, Beitel LK, Wang E, Trifiro M. Dynamic rewiring of the androgen receptor protein interaction network correlates with prostate cancer clinical outcomes. Integr Biol. 2011;3(10):1020–32.
41. Rashevsky N. Life, information theory, and topology. Bull Math Biol. 1955;17(3):229–35.
42. Rietman E, Bloemendal A, Platig J, Tuszynski J, Klement GL. Gibbs free energy of protein-protein interactions reflects tumor stage. BioRxiv. 2015, January 1:022491.
43. Rietman EA, Friesen DE, Hahnfeldt P, Gatenby R, Hlatky L, Tuszynski JA. An integrated multidisciplinary model describing initiation of cancer and the Warburg hypothesis. Theor Biol Med Model. 2013;10(1):39.
44. Rietman EA, Platig J, Tuszynski JA, Klement GL. Thermodynamic measures of cancer: Gibbs free energy and entropy of protein–protein interactions. J Biol Phys. 2016;42(3):339–50.
45. Stark C, Breitkreutz BJ, Reguly T, Boucher L, Breitkreutz A, Tyers M. BioGRID: a general repository for interaction datasets. Nucleic Acids Res. 2006;34(suppl_1):D535–9.
46. Schrodinger E. What is life? And, mind and matter, by Erwin Schrodinger. London: Cambridge University Press; 1967.
47. Shannon CE. A mathematical theory of communication. ACM SIGMOBILE Mob Comput Commun Rev. 2001;5(1):3–55.
48. Suvà ML, Rheinbay E, Gillespie SM, Patel AP, Wakimoto H, Rabkin SD, Riggi N, Chi AS, Cahill DP, Nahed BV, Curry WT. Reconstructing and reprogramming the tumor-propagating potential of glioblastoma stem-like cells. Cell. 2014;157(3):580–94.

49. Tomasetti C, Vogelstein B. Variation in cancer risk among tissues can be explained by the number of stem cell divisions. Science. 2015;347(6217):78–81.
50. Tomlins SA, Mehra R, Rhodes DR, Cao X, Wang L, Dhanasekaran SM, Kalyana-Sundaram S, Wei JT, Rubin MA, Pienta KJ, Shah RB. Integrative molecular concept modeling of prostate cancer progression. Nat Genet. 2007;39(1):41–51.
51. Tseng CY, Tuszynski JA. Using entropy leads to a better understanding of biological systems. Entropy. 2010;12(12):2450–69.
52. Tseng CY, Tuszynski J. A unified approach to computational drug discovery. Drug Discov Today. 2015;20(11):1328–36.
53. Vander Heiden MG, Locasale JW, Swanson KD, Sharfi H, Heffron GJ, Amador-Noguez D, Christofk HR, Wagner G, Rabinowitz JD, Asara JM, Cantley LC. Evidence for an alternative glycolytic pathway in rapidly proliferating cells. Science. 2010;329(5998):1492–9.
54. Warburg O. On the origin of cancer. Science. 1956;123(3191):309–14.
55. Wilhelm M, Schlegl J, Hahne H, Gholami AM, Lieberenz M, Savitski MM, Ziegler E, Butzmann L, Gessulat S, Marx H, Mathieson T. Mass-spectrometry-based draft of the human proteome. Nature. 2014;509(7502):582–7.
56. Wurmbach E, Chen YB, Khitrov G, Zhang W, Roayaie S, Schwartz M, Fiel I, Thung S, Mazzaferro V, Bruix J, Bottinger E. Genome-wide molecular profiles of HCV-induced dysplasia and hepatocellular carcinoma. Hepatology. 2007;45(4):938–47.
57. Weinan E, Lu J, Yao Y. The landscape of complex networks. arXiv preprint arXiv:1204.6376. 2012.
58. Yang K, Bai H, Ouyang Q, Lai L, Tang C. Finding multiple target optimal intervention in disease-related molecular network. Mol Syst Biol. 2008;4(1):228.
59. Zhang X, Cruz FD, Terry M, Remotti F, Matushansky I. Terminal differentiation and loss of tumorigenicity of human cancers via pluripotency-based reprogramming. Oncogene. 2013;32(18):2249–60.

Systems Immunology

Melissa Lever, Thiago C. Hirata, Pedro S. T. Russo, and Helder I. Nakaya

Abstract The advance of systems biology approaches now means that much of the immune response to pathogens and vaccines can be assessed. Modern immunologists have at their disposal an arsenal of high-throughput technologies and tools that generate data relating the quantities of genes, metabolites and proteins within immune cells. The challenge posed is how to interpret this abundance of data to accurately understand and predict the immune response. Systems immunology is the discipline that uses computational and mathematical approaches to integrate these measurements and explain the nonintuitive interactions between biological components. In this chapter we will provide an overview of this interdisciplinary approach, its challenges, and highlight some of the applications of systems biology to assess the complexity of our immune system.

1 A Brief Overview of the Immune System

The immune system is a complex and versatile defence system that provides protection against infection and disease. It is composed of a network of diverse cell types and molecules that collaborate to recognise and eliminate a variety of foreign intruders. Great progress has been made in understanding the components of the immune system, but there are still many basic immunological questions that remain unresolved. Broader issues must be tackled too, such as understanding how autoimmunity develops, how to harness the immune system to target cancer and how to develop new vaccines and understanding the interaction between the immune system and the microbiota.

When the body is invaded by a pathogen, the first line of defence is effected by innate immunity, which is fast acting and equipped to respond to a broad range of foreign intruders. Its components include the skin and mucus membranes

M. Lever · T. C. Hirata · P. S. T. Russo · H. I. Nakaya (✉)
Department of Clinical and Toxicological Analyses, School of Pharmaceutical Sciences,
University of São Paulo, São Paulo, Brazil
e-mail: hnakaya@usp.br

© Springer International Publishing AG, part of Springer Nature 2018   159
F. A. B. da Silva et al. (eds.), *Theoretical and Applied Aspects of Systems Biology*,
Computational Biology 27, https://doi.org/10.1007/978-3-319-74974-7_9

which form physical batters, an array of cells derived from white blood cells and systems of proteins that reside in the blood and bodily fluids. The complement system comprises a set of plasma proteins that form a biochemical cascade that can tag or destroy pathogens and alert other immune cells to their presence. Cells of innate immune system detect pathogens through germ line-encoded sensors called pattern recognition receptors (PRRs), which recognise invariant features of pathogenic molecules [1]. These cells include neutrophils, which can ingest foreign microorganisms and release toxic enzymes that can kill them. Natural killer (NK) cells can detect cells infected with viruses and induce their death through apoptosis. Microorganisms and infected cells can also be engulfed by cells such as neutrophils, dendritic cells and macrophages, where they are then destroyed by a process called phagocytosis [2]. Innate immune cells also secrete signalling molecules called cytokines that can alert other cells to the presence of a pathogen.

While the innate immune system reacts rapidly, the adaptive immune system provides a more specific response that can take days or weeks to develop. The adaptive immune system can be activated when short antigenic peptides are presented by antigen-presenting cells (APCs) to T lymphocytes (T cells). A dendritic cell is a type of APC that presents fragments of the microorganisms it has phagocytosed on its cell surface. T cells detect this antigen through their surface-bound T cell receptors (TCRs), which initiates a signal transduction cascade within the T cell. This drives transcriptional and metabolic reprogramming to enable T cell proliferation and differentiation into cells that can perform effector functions such as cell killing [3]. Some of these T cells will also differentiate into memory cells that can confer immediate and long-lasting protection against the antigen they are specific to [4]. The structure of the TCR expressed on each unactivated T cell is unique, since during T cell development, the DNA segments that code for the TCR undergo recombination. This development process, which takes place in the thymus, enables T cells to detect a large repertoire of antigen.

Immunity is maintained by immune cells that transit through the blood, the lymphatic system and the peripheral tissues. The bridging of innate immunity with adaptive immunity takes place in the lymph nodes, where APCs carrying antigen from the peripheral tissues will meet with naive T cells [5]. If a naive T cell is activated, it will then migrate to the peripheral tissues to perform its effector functions [6]. The other type of adaptive immunity response is mediated by the B lymphocyte (B cell), which also enters the lymph nodes. B cells produce large proteins called antibodies that can neutralise pathogens and alert other immune cells [7]. Antibodies can either be surface bound (B cell receptor) or secreted as soluble proteins that will circulate through the blood. Like the TCR, the B cell receptor is unique and randomly determined. B cells can be activated through the help of a class of T cell called a T helper cell, and once activated, B cells secrete antibodies and interact with T cells and innate cells to govern the outcome of an immune response [8]. Figure 1 shows the main outcomes to vaccination and infection which are used as endpoints in systems immunology studies.

A diverse variety of T cell and B cell subsets exist, whose function can be activatory, performing cytotoxic functions or immune cell help, or regulatory,

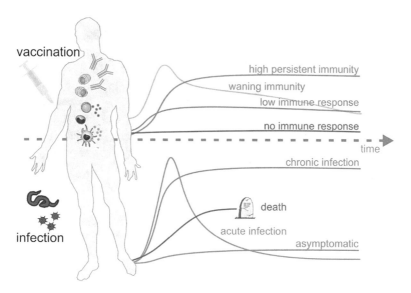

Fig. 1 Immune responses and health outcomes to vaccination and infection. Vaccination (top part) often induces a B cell response (i.e. antibodies) and/or a T cell-mediated response (i.e. cytotoxic CD8+ and helper CD4+ T cells) that can either persist for years or wane after few weeks. For some vaccinees, the response can be low or even absent. Infection (bottom part) can lead to the death of some individuals or cause no clinical symptoms to others. Pathogens may persist in the body for years (chronic infection) or be cleared by the immune system (acute infection) or drug treatment

dampening the immune response. Each cell type can be characterised by the cytokines it secretes, the cell surface receptors it expresses and its transcriptional programs [9, 10]. Characterising the regulatory mechanisms that control the release of cytokines or the signalling pathways induced by them during an immune response is imperative for understanding how immunity works. This knowledge may lead to better vaccine design and development, as well as improvement in immunotherapy and drug treatment.

2 High-Throughput Approaches to Immunology

Advances in experimental technologies now allow the immune system to be probed with unprecedented detail across biological scales. An immune cell can now be examined at the level of its DNA (genome), the expression of this DNA into RNA (transcriptome), its intracellular and surface-bound proteins (proteome) and the metabolites that are the products of biochemical processes (metabolome). What remains a challenge is understanding how the quantities of these components relate to the resultant cell functionality. This relationship is in no way linear or straightforward [11], because a variety of mechanisms influence biological

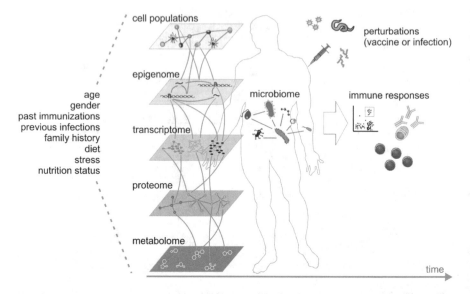

Fig. 2 Systems immunology integrates distinct layers of biological information. Each layer contains an intricate network of components that are associated with components from other layers. Factors on the left have direct influence on these networks and therefore play a major role on the immune response induced by vaccination or infection

phenotype (Fig. 2). This includes the epigenome, which comprises the chemical modifications that affect the accessibility of the transcriptional machinery to DNA [12], and resultant gene expression.

In addition, only a fraction of RNA is protein coding, and many non-coding RNAs influence gene expression on a transcriptional and post-transcriptional level [13]. What's more, the levels of protein within a cell are governed by the rate of protein degradation, as well as translation [14]. Nevertheless, probing various immune cell types, resident in different tissues and cytokine environments, has led to advances in our understanding of how the immune system operates. Headway has also been made in obtaining biomarkers that can define and predict certain immune states.

A reduction in the cost and speed of next-generation DNA sequencing has meant that whole-genome sequencing of humans has become more common. This has allowed researchers to identify genetic polymorphisms that are relevant to disease [15]. One approach is expression quantitative trait loci analysis (eQTL), which can identify the genetic locus that can affect the levels of expression of genes associated with a given disease. Sequencing of the BCR and TCR repertoires has provided insights into how the adaptive immune system defends itself from the diverse array of potential pathogens [16, 17]. DNA sequencing is also used to identify transcription factor binding sites and protein interactions using chromatin immunoprecipitation-seq (ChIP-seq). In this method, transcription factors bound to DNA are immunoprecipitated, and the DNA is then sequenced [18].

The most powerful tool to target the transcriptome is RNA sequencing, which uses deep-sequencing technologies. As well as detecting the expression of genes, RNA-seq can detect gene isoforms and novel transcribed regions [19]. Microarray technology is more dated but still routinely used because it offers cheap and high-throughput detection of the transcriptome using predetermined probes. These technologies have revealed the transcriptional programs of many immune cells including macrophages [20], T cells [21] and dendritic cells [22]. The role of non-coding RNAs (ncRNAs) has also been investigated, including their role in T cell fate [23], the regulation of cytotoxic CD8+ [24] and helper CD4+ [25] T cells, and their role in innate immunity [26].

High-resolution mass spectrometry has been harnessed to investigate the proteome and metabolome. A recent study characterised the proteome of 28 human haematopoietic cell populations in steady and activated states [27], revealing the social network architecture of their proteomes and the proteins that they secrete. Metabolomics is a field still in its infancy but offers the potential to provide better understanding of how the other "omics" fields relate to each other [28]. It has been used to detect biomarkers in diseases including asthma [29], inflammatory bowel disease [30] and type 2 diabetes [31]. Mass spectrometry technology is also used to detect cell-surface protein expression in time-of-flight mass spectrometry (CyTOF). This method improves on flow cytometry by tagging surface proteins with metal isotypes rather than with antibodies, thus allowing many proteins to be profiled simultaneously [32]. This allows immune cell types to be characterised with greater specificity, since the surface markers on its surface define the functional activity of the cell.

Alongside these deep profiling technologies, the Luminex assay is used to determine the quantities of cytokines secreted by cells. And although RNA-Seq and microarrays can profile the whole transcriptome, RT-PCR remains the gold standard for determining RNA expression for a small number of RNAs.

3   The Rise of Systems Immunology

Through the use of these experimental technologies, biologists have made great progress in characterising the molecular constituents of life. The challenge remains to understand how these components interact with each other to produce biological function [33]. "Systems biology" is the term given to the approach that attempts to do just this; it is the endeavour to explain the emergent phenomena of biological systems using mathematical and computational techniques that integrate the data on biological components. The immune system, owing to its complexity and intricacy, has always demanded a systems approach for its workings to be understood. Now the experimental and computational technology has arrived at a point where this is feasible. Systems immunology promises to explain how the diverse array of immune cells and their cell-bound and soluble mediators are coordinated to bring into effect an immune response [34].

Systems immunology studies begin by perturbing an immunological system, which can be through the administration of a vaccine, exposure of cells to certain cytokines or stimulation of surface receptors, amongst others. Experimental approaches can involve the traditional "reductionist" approaches as well as the high-throughput "omic" technologies reviewed in the last section [35]. The responses are then quantified and integrated in order to formulate predictions and generate new hypotheses. Through these methods, new insights have been gained in diverse immunological areas, including basic immunological studies on signal transduction within T cells [36–38], Toll-like receptor signalling [39] and transcriptional control of T helper 17 cell differentiation [40]. Broader themes have also been investigated, such as the effect of ageing on transcriptional programs in mouse haematopoietic stem cells [41]. Another study analysed the different regulatory mechanisms between mouse and human immune systems [42], which is important given the use of the mouse as a model for the human immune system. A systems approach also devised the molecular circuit that drives influenza infection, integrating transcriptome and flow cytometry data [43].

Vaccinology, a field that has developed alongside, but not always integrated with, immunology, has also benefited from systems analyses. A challenge facing the field is the lack of mechanistic understanding of vaccine-induced immunity, due to the often empirical nature of vaccine design. Systems approaches to vaccinology offer the potential to understand how the immune system is perturbed by the administration of a vaccine and identify biomarkers that can be designated as the immune correlates of protection [44]. A landmark study identified a new role for the gene GCN2/EIF2AK4 in response to the yellow fever vaccine [45] using transcriptome data and computational classification methods. It was later confirmed experimentally that this gene is in the stress-response pathway and promotes CD8+ T cell responses [46]. Following this initial study, systems vaccinology approaches have been applied to influenza [47, 48], HIV [49], malaria [50] and smallpox [51].

4 Mathematical Modelling of Immunity

Mathematical modelling plays an important role in systems immunology because it transforms informal associations between biological components into quantitative descriptions [52]. As our understanding of the complexity of the immune system increases, it has become evident that mathematical models are necessary to explain the non-linear relationships between biological components that are hard to deduce by intuition alone. What's more, mathematical models produce quantifiable predictions that can be tested experimentally [53].

A biological system can be modelled using a "bottom-up" approach, which means that mechanistic knowledge gathered from experiments is used to develop a set of equations (Fig. 3). A commonly used system of equations is called ordinary differential equations (ODEs) [54]. Variables are used to represent the quantities of biological components (such as proteins, cells and tissues), and parameters

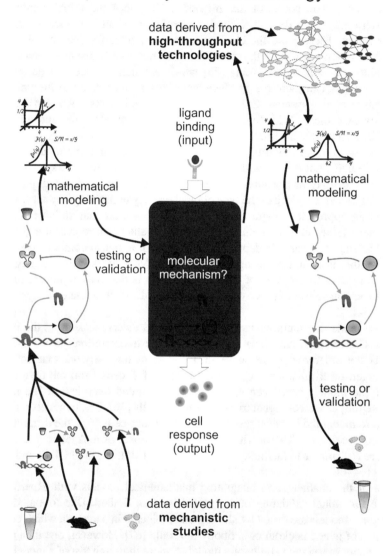

Top-Down Systems Immunology

data derived from **high-throughput technologies**

ligand binding (input)

mathematical modeling

mathematical modeling

testing or validation

molecular mechanism?

testing or validation

cell response (output)

data derived from **mechanistic studies**

Bottom-Up Systems Immunology

Fig. 3 The two approaches available for systems immunology. The "top-down" approach starts with the large amount of data derived from high-throughput technologies. Mathematical modelling is then used to reduce information to obtain a putative mechanism that can be later tested or validated in animal models or in vitro and in vivo experiments. The "bottom-up" approach integrates mechanistic data derived from previous experiments in order to construct a network or circuit that may explain the biological process studied. Mathematical modelling is then applied to predict the behaviour of the system, which can be later tested or validated in a real experiment

determine the rates of the reactions (such as catalysis, binding, degradation) between them. The power of this method is that once the initial conditions of the system have been described, the time evolution of all biological components can be found. Therein lies the challenge, however, since for even modestly sized biological systems, not all of the parameters are known and must be estimated. Such uncertainty can lead to overfitting [55] which is when a model has an excess of free parameters can therefore fit phenomena that is not related to the underlying mechanism of the system, such as experimental noise. As John von Neumann famously explained, "With four parameters I can fit an elephant, and with five I can make him wiggle his trunk".

One way to avoid this difficulty is to formulate a phenotypic model, which is based on ODEs, but only includes the coarse-grained features of a system that are essential to display its phenotype. While not mechanistically detailed, phenotypic modelling can give an idea of the type of biochemistry underlying a system. Another simplifying approach is logic modelling, which assumes that variables can take only binary values or a small number of discrete values that represent a qualitative state. The state of a variable depends on the input of a logical combination of other variables that is computed using OR, AND and NOT gates [56]. Although logic models are easy to use, one of the disadvantages is that they represent time and variable outputs in a qualitative way, which can be difficult to relate to experimental results [54].

These modelling methods have been applied to diverse aspects of the immune system. ODE models have been used to model the transcription factor NF-kB that controls the inflammatory response [57], the signalling network that controls T cell antigen discrimination [58] and the output of T cells from the thymus [59]. Phenotypic models have been applied to give a broad overview of the immune response [60] as well as signal transduction in T cells [36]. Logic models have been applied to model the immune response in Langerhans cells [61], the differentiation of T helper cells into Th1 or Th2 [62] and the survival of leukaemic T cells [63]. Software is available to facilitate the generation and simulation of ODEs and logical models [64–66].

One of the challenges to integrating mathematical models with experiment is that the functional validation of the model is often limited. Due to experimental constraints, models are tested by altering one molecule in a system, which could be by means of gene knockout or antibody blockade [67]. However, one data point is not sufficient to test the validity of a model. A more thorough test of a model can be done by titrating the abundance of a molecule over a broad range of concentrations to produce a dose-response curve. Another obstacle to accurate model generation is that modellers generally do not build on work generated by previous modellers. The utility of a model is limited if no one actually uses it, and a more collaborative approach would improve model reliability.

5 Predictive and Integrative Analyses

While mathematical modelling attempts have generally focused on quantifying biological information from one scale, such as gene expression or protein signalling, a grand ambition of the life sciences is to integrate biological information from different sources. With the advent of new high-throughput technologies, there has been an ever-increasing abundance of information, such as DNA-sequencing data, microarray or RNA-seq expression data, proteomics spectra, metabolomics data, protein-protein interactions and epigenetic interactions such as methylation and chromatin structure information. Much of this information is accessible and stored in online databases, but it remains a significant challenge to transform it into cohesive, meaningful knowledge.

This enormous amount of data has motivated "top-down" data analysis approaches, which look at the wider picture of biological interactions instead of focusing on individual genes or proteins. Graph theory, a subarea of mathematics, provides several tools which allow the analysis of a network of interacting elements, defining important concepts such as nodes, edges, interactions, clusters, paths, etc. Starting from this mathematical framework, systems biologists have applied and advanced on these methods to analyse biological data. These networks are constructed by using statistical techniques such as clustering approaches, Bayesian methods and information-theoretic processes [68].

Gene expression data has proved to be particularly suitable for this type of analysis, since they have been shown to be inherently modular in nature [69]. Network analysis methods attempt to construct networks of highly correlated genes that share expression patterns across a group of samples and then investigate these genes' shared behaviour across different conditions [11, 70]. One of the most popular tools for this kind of method is the weighted gene co-expression network analysis (WGCNA) framework [71]. WGCNA transforms a matrix of microarray/RNA-seq expression values into gene modules, which are clusters of highly correlated genes that tend to be co-regulated. It does this by calculating the correlations between gene expression values and soft thresholding the values, to penalise weak correlations and strengthen higher ones. The genes are then sorted into modules based on their similarity. These modules can then be used to test for significant biological functions using other tools such as gene set enrichment analysis [72], with the underlying premise that similarly expressed genes should act together in coordinating similar biological processes [11].

Many other methods have also been proposed for the construction of gene co-expression networks. DICER is a method which focuses on identifying modules differentially correlated between sample groups [73]; CoXpress aims to identify if genes co-expressed in one sample group are also co-expressed in other sample groups and provides different network visualisation methods [74]; and DINGO is a tool that is capable of grouping genes taking into account the behaviour of groups of genes in a specific subset of samples, such as the ones perturbed in a particular disease or condition [75]. Finally, CEMiTool (Russo 2017) is a new method that provides co-expression modular analyses in a user-friendly way, requiring little to no prior user experience in programming. It also returns several other analyses, including functional annotation of gene modules based on biological pathway databases such as Reactome [76] and integration of gene modules with protein-protein interaction data.

The advantages of data-driven modelling are that it provides an unbiased examination of the data and thus offers the potential to make predictions that could not be made with the existing prior knowledge [77]. WGCNA analysis has been applied to diverse studies including whole-blood transcriptional profiling of patients with psoriasis [78], leishmaniasis [79] and an investigation of gene networks in dendritic cell subsets [80]. Another tool that considers the perturbation of gene expression on a per-sample basis is called the molecular distance to health and has been used to assess the disease severity of patients infected with tuberculosis [81] and respiratory syncytial virus [82].

Despite the success in elucidating novel mechanisms and pathways with gene co-expression networks, methods for integrating these with different types of information, on different spatial and time scales, are still in their infancy. The challenge to creating a multi-scale model is that the complexity required to exhibit the totality of a biological system would be too complicated to compute, and therefore at present, only subparts can be modelled [53]. Packages have been developed that integrate information of transcription factor binding from ChIP-seq experiments with gene expression data in order to infer protein-DNA interactions and their regulation, such as the Binding and Expression Target Analysis (BETA) [83] and the Aggregation and Correlation Toolbox [84]. Progress has also been made using "bottom-up" modelling methods through the cardiac physiome project, which aims to integrate subcellular, cellular, tissue and organ models of the heart, using markup languages search as CellML [85].

6 Conclusion

Systems immunology approaches will soon become the norm in modern immunological studies as high-throughput screening methods are becoming affordable in most laboratories. It is imperative that the new generation of immunologists are not only technically equipped to run the computational tools necessary to analyse this data but also that they have an understanding of the principles behind these tools. When a judicious choice is made over the type of tool and the parameters to be used, there is greater chance that useful information will be extracted from the data. The immunologist of the future will therefore need to develop technical skills beyond the traditional wet lab, and at the same time, computational scientists and bioinformaticians need to ensure that the tools they develop are immunologist-friendly. Computational tools should no longer be the domain of those with advanced programming skills but rather be made accessible to all.

Systems biology technologies provide greater resolution than conventional immunological tools. Together with the appropriate computational tools, a modern immunologist alone will be able to assess the intricate networks of DNA modifications, genes, metabolites and proteins inside immune cells. This will lead to an unprecedented understanding of the full complexity of our own immune system.

References

1. Iwasaki A, Medzhitov R. Control of adaptive immunity by the innate immune system. Nat Immunol. 2015;16:343–53.
2. Chaplin DD. Overview of the immune response. J Allergy Clin Immunol. 2010;125:S3–23.
3. Smith-Garvin JE, Koretzky GA, Jordan MS. T cell activation. Annu Rev Immunol. 2009;27:591–619.
4. Metz PJ, Arsenio J, Kakaradov B, Kim SH, Remedios KA, Oakley K, et al. Regulation of asymmetric division and CD8+ T lymphocyte fate specification by protein kinase Cζ and protein kinase Cλ/ι. J Immunol. 2015;194:2249–59.
5. von Andrian UH, Mempel TR. Homing and cellular traffic in lymph nodes. Nat Rev Immunol. 2003;3:867–78.
6. Krummel MF, Bartumeus F, Gérard A. T cell migration, search strategies and mechanisms. Nat Rev Immunol. 2016;16:193–201.
7. Gonzalez SF, Degn SE, Pitcher LA, Woodruff M, Heesters BA, Carroll MC. Trafficking of B cell antigen in lymph nodes. Annu Rev Immunol. 2011;29:215–33.
8. Hoffman W, Lakkis FG, Chalasani G. B cells, antibodies, and more. Clin J Am Soc Nephrol. 2016;11:137–54.
9. Allman D, Pillai S. Peripheral B cell subsets. Curr Opin Immunol. 2008;20:149–57.
10. Golubovskaya V, Wu L. Different subsets of T cells, memory, effector functions, and CAR-T immunotherapy. Cancers (Basel). 2016;8:E36.
11. Vidal M, Cusick ME, Barabási A-L. Interactome networks and human disease. Cell. 2011;144:986–98.
12. Goldberg AD, Allis CD, Bernstein E. Epigenetics: a landscape takes shape. Cell. 2007;128:635–8.
13. Cech TR, Steitz JA. The noncoding RNA revolution-trashing old rules to forge new ones. Cell. 2014;157:77–94.
14. de Sousa AR, Penalva LO, Marcotte EM, Vogel C. Global signatures of protein and mRNA expression levels. Mol BioSyst. 2009;5:1512–26.
15. Soon WW, Hariharan M, Snyder MP. High-throughput sequencing for biology and medicine. Mol Syst Biol. 2013;9:640.
16. Elhanati Y, Murugan A, Callan CG, Mora T, Walczak AM. Quantifying selection in immune receptor repertoires. Proc Natl Acad Sci U S A. 2014;111:9875–80.
17. Roy B, Neumann RS, Snir O, Iversen R, Sandve GK, Lundin KEA, et al. High-throughput single-cell analysis of B cell receptor usage among autoantigen-specific plasma cells in celiac disease. J Immunol. 2017;199:782–91.
18. Raha D, Hong M, Snyder M. ChIP-Seq: a method for global identification of regulatory elements in the genome. Curr Protoc Mol Biol. 2010;Chapter 21:Unit 21.19.1-14. https://doi.org/10.1002/0471142727.mb2119s91
19. Wang Z, Gerstein M, Snyder M. RNA-Seq: a revolutionary tool for transcriptomics. Nat Rev Genet. 2009;10:57–63.
20. Martinez FO, Gordon S, Locati M, Mantovani A. Transcriptional profiling of the human monocyte-to-macrophage differentiation and polarization: new molecules and patterns of gene expression. J Immunol. 2006;177:7303–11.

21. Zhang JA, Mortazavi A, Williams BA, Wold BJ, Rothenberg EV. Dynamic transformations of genome-wide epigenetic marking and transcriptional control establish T cell identity. Cell. 2012;149:467–82.
22. Watchmaker PB, Lahl K, Lee M, Baumjohann D, Morton J, Kim SJ, et al. Comparative transcriptional and functional profiling defines conserved programs of intestinal DC differentiation in humans and mice. Nat Immunol. 2014;15:98–108.
23. Isoda T, Moore AJ, He Z, Chandra V, Aida M, Denholtz M, et al. Non-coding transcription instructs chromatin folding and compartmentalization to dictate enhancer-promoter communication and T cell fate. Cell. 2017;171:103–19. e18
24. Wu H, Neilson JR, Kumar P, Manocha M, Shankar P, Sharp PA, et al. miRNA profiling of naïve, effector and memory CD8 T cells. PLoS One. 2007;2:e1020.
25. Gutiérrez-Vázquez C, Rodríguez-Galán A, Fernández-Alfara M, Mittelbrunn M, Sánchez-Cabo F, Martínez-Herrera DJ, et al. miRNA profiling during antigen-dependent T cell activation: a role for miR-132-3p. Sci Rep. 2017;7:3508.
26. Lu Y, Liu X, Xie M, Liu M, Ye M, Li M, et al. The NF-κB-responsive long noncoding RNA FIRRE regulates posttranscriptional regulation of inflammatory gene expression through interacting with hnRNPU. J Immunol. 2017;199:3571–82.
27. Rieckmann JC, Geiger R, Hornburg D, Wolf T, Kveler K, Jarrossay D, et al. Social network architecture of human immune cells unveiled by quantitative proteomics. Nat Immunol. 2017;18:583–93.
28. Veenstra TD. Metabolomics: the final frontier? Genome Med. 2012;4:40.
29. Jung J, Kim SH, Lee HS, Choi GS, Jung YS, Ryu DH, et al. Serum metabolomics reveals pathways and biomarkers associated with asthma pathogenesis. Clin Exp Allergy. 2013;43:425–33.
30. Stephens NS, Siffledeen J, Su X, Murdoch TB, Fedorak RN, Slupsky CM. Urinary NMR metabolomic profiles discriminate inflammatory bowel disease from healthy. J Crohns Colitis. 2013;7:e42–8.
31. Menni C, Fauman E, Erte I, Perry JRB, Kastenmüller G, Shin S-Y, et al. Biomarkers for type 2 diabetes and impaired fasting glucose using a nontargeted metabolomics approach. Diabetes. 2013;62:4270–6.
32. Spitzer MH, Nolan GP. Mass cytometry: single cells, many features. Cell. 2016;165:780–91.
33. Bruggeman FJ, Westerhoff HV. The nature of systems biology. Trends Microbiol. 2007; 15:45–50.
34. Davis MM, Tato CM, Furman D. Systems immunology: just getting started. Nat Immunol. 2017;18:725–32.
35. Germain RN. Will systems biology deliver its promise and contribute to the development of new or improved vaccines? What really constitutes the study of "systems biology" and how might such an approach facilitate vaccine design. Cold Spring Harb Perspect Biol. 2017. https://doi.org/10.1101/cshperspect.a033308
36. Lever M, Lim H-S, Kruger P, Nguyen J, Trendel N, Abu-Shah E, et al. Architecture of a minimal signaling pathway explains the T-cell response to a 1 million-fold variation in antigen affinity and dose. Proc Natl Acad Sci U S A. 2016;113:E6630–8.
37. François P, Voisinne G, Siggia ED, Altan-Bonnet G, Vergassola M. Phenotypic model for early T-cell activation displaying sensitivity, specificity, and antagonism. Proc Natl Acad Sci U S A. 2013;110:E888–97.
38. Mukhopadhyay H, de Wet B, Clemens L, Maini PK, Allard J, van der Merwe PA, et al. Multisite phosphorylation modulates the T cell receptor ζ-chain potency but not the Switchlike response. Biophys J. 2016;110:1896–906.
39. Chevrier N, Mertins P, Artyomov MN, Shalek AK, Iannacone M, Ciaccio MF, et al. Systematic discovery of TLR signaling components delineates viral-sensing circuits. Cell. 2011;147:853–67.

40. Yosef N, Shalek AK, Gaublomme JT, Jin H, Lee Y, Awasthi A, et al. Dynamic regulatory network controlling TH17 cell differentiation. Nature. 2013;496:461–8.
41. Kowalczyk MS, Tirosh I, Heckl D, Rao TN, Dixit A, Haas BJ, et al. Single-cell RNA-seq reveals changes in cell cycle and differentiation programs upon aging of hematopoietic stem cells. Genome Res. 2015;25:1860–72.
42. Shay T, Jojic V, Zuk O, Rothamel K, Puyraimond-Zemmour D, Feng T, et al. Conservation and divergence in the transcriptional programs of the human and mouse immune systems. Proc Natl Acad Sci U S A. 2013;110:2946–51.
43. Brandes M, Klauschen F, Kuchen S, Germain RN. A systems analysis identifies a feedforward inflammatory circuit leading to lethal influenza infection. Cell. 2013;154:197–212.
44. Nakaya HI, Pulendran B. Vaccinology in the era of high-throughput biology. Philos Trans R Soc Lond Ser B Biol Sci. 2015;370:20140146.
45. Querec TD, Akondy RS, Lee EK, Cao W, Nakaya HI, Teuwen D, et al. Systems biology approach predicts immunogenicity of the yellow fever vaccine in humans. Nat Immunol. 2009;10:116–25.
46. Ravindran R, Khan N, Nakaya HI, Li S, Loebbermann J, Maddur MS, et al. Vaccine activation of the nutrient sensor GCN2 in dendritic cells enhances antigen presentation. Science. 2014;343:313–7.
47. Nakaya HI, Wrammert J, Lee EK, Racioppi L, Marie-Kunze S, Haining WN, et al. Systems biology of vaccination for seasonal influenza in humans. Nat Immunol. 2011;12:786–95.
48. Bucasas KL, Franco LM, Shaw CA, Bray MS, Wells JM, Niño D, et al. Early patterns of gene expression correlate with the humoral immune response to influenza vaccination in humans. J Infect Dis. 2011;203:921–9.
49. Zak DE, Andersen-Nissen E, Peterson ER, Sato A, Hamilton MK, Borgerding J, et al. Merck Ad5/HIV induces broad innate immune activation that predicts CD8+ T-cell responses but is attenuated by preexisting Ad5 immunity. Proc Natl Acad Sci U S A. 2012;109:E3503–12.
50. Vahey MT, Wang Z, Kester KE, Cummings J, Heppner DG, Nau ME, et al. Expression of genes associated with immunoproteasome processing of major histocompatibility complex peptides is indicative of protection with adjuvanted RTS,S malaria vaccine. J Infect Dis. 2010; 201:580–9.
51. Reif DM, Motsinger-Reif AA, McKinney BA, Rock MT, Crowe JE, Moore JH. Integrated analysis of genetic and proteomic data identifies biomarkers associated with adverse events following smallpox vaccination. Genes Immun. 2009;10:112–9.
52. Gunawardena J. Beware the tail that wags the dog: informal and formal models in biology. Mol Biol Cell. 2014;25:3441–4.
53. Motta S, Pappalardo F. Mathematical modeling of biological systems. Brief Bioinf. 2013;14:411–22.
54. Le Novère N. Quantitative and logic modelling of molecular and gene networks. Nat Rev Genet. 2015;16:146–58.
55. Hawkins DM. The problem of overfitting. J Chem Inf Comput Sci. 2004;44:1–12.
56. Morris MK, Saez-Rodriguez J, Sorger PK, Lauffenburger DA. Logic-based models for the analysis of cell signaling networks. Biochemistry. 2010;49:3216–24.
57. Hoffmann A, Levchenko A, Scott ML, Baltimore D. The IkappaB-NF-kappaB signaling module: temporal control and selective gene activation. Science. 2002;298:1241–5.
58. Altan-Bonnet G, Germain RN. Modeling T cell antigen discrimination based on feedback control of digital ERK responses. PLoS Biol. 2005;3:e356.
59. Bains I, Thiébaut R, Yates AJ, Callard R. Quantifying thymic export: combining models of naive T cell proliferation and TCR excision circle dynamics gives an explicit measure of thymic output. J Immunol. 2009;183:4329–36.
60. Mayer H, Zaenker KS, An Der Heiden U. A basic mathematical model of the immune response. Chaos. 1995;5:155–61.

61. Polak ME, Ung CY, Masapust J, Freeman TC, Ardern-Jones MR. Petri Net computational modelling of Langerhans cell interferon regulatory factor network predicts their role in T cell activation. Sci Rep. 2017;7:668.
62. Mendoza L. A network model for the control of the differentiation process in Th cells. Biosystems. 2006;84:101–14.
63. Zhang R, Shah MV, Yang J, Nyland SB, Liu X, Yun JK, et al. Network model of survival signaling in large granular lymphocyte leukemia. Proc Natl Acad Sci U S A. 2008; 105:16308–13.
64. Terfve C, Cokelaer T, Henriques D, MacNamara A, Goncalves E, Morris MK, et al. CellNOptR: a flexible toolkit to train protein signaling networks to data using multiple logic formalisms. BMC Syst Biol. 2012;6:133.
65. Batt G, Besson B, Ciron P-E, de Jong H, Dumas E, Geiselmann J, et al. Genetic network analyzer: a tool for the qualitative modeling and simulation of bacterial regulatory networks. Methods Mol Biol. 2012;804:439–62.
66. Müssel C, Hopfensitz M, Kestler HA. BoolNet – an R package for generation, reconstruction and analysis of Boolean networks. Bioinformatics. 2010;26:1378–80.
67. Benoist C, Germain RN, Mathis D. A plaidoyer for "systems immunology". Immunol Rev. 2006;210:229–34.
68. Bansal M, Belcastro V, Ambesi-Impiombato A, di Bernardo D. How to infer gene networks from expression profiles. Mol Syst Biol. 2007;3:78.
69. Barabási A-L, Oltvai ZN. Network biology: understanding the cell's functional organization. Nat Rev Genet. 2004;5:101–13.
70. Carter SL, Brechbühler CM, Griffin M, Bond AT. Gene co-expression network topology provides a framework for molecular characterization of cellular state. Bioinformatics. 2004;20:2242–50.
71. Zhang B, Horvath S. A general framework for weighted gene co-expression network analysis. Stat Appl Genet Mol Biol. 2005;4:17.
72. Subramanian A, Tamayo P, Mootha VK, Mukherjee S, Ebert BL, Gillette MA, et al. Gene set enrichment analysis: a knowledge-based approach for interpreting genome-wide expression profiles. Proc Natl Acad Sci U S A. 2005;102:15545–50.
73. Amar D, Safer H, Shamir R. Dissection of regulatory networks that are altered in disease via differential co-expression. PLoS Comput Biol. 2013;9:e1002955.
74. Watson M. CoXpress: differential co-expression in gene expression data. BMC Bioinform. 2006;7:509.
75. Ha MJ, Baladandayuthapani V, Do K-A. DINGO: differential network analysis in genomics. Bioinformatics. 2015;31:3413–20.
76. Croft D, O'Kelly G, Wu G, Haw R, Gillespie M, Matthews L, et al. Reactome: a database of reactions, pathways and biological processes. Nucleic Acids Res. 2011;39:D691–7.
77. Dolinski K, Troyanskaya OG. Implications of Big Data for cell biology. Mol Biol Cell. 2015;26:2575–8.
78. Li B, Tsoi LC, Swindell WR, Gudjonsson JE, Tejasvi T, Johnston A, et al. Transcriptome analysis of psoriasis in a large case-control sample: RNA-seq provides insights into disease mechanisms. J Invest Dermatol. 2014;134:1828–38.
79. Gardinassi LG, Garcia GR, Costa CHN, Costa Silva V, de Miranda Santos IKF. Blood transcriptional profiling reveals immunological signatures of distinct states of infection of humans with Leishmania infantum. PLoS Negl Trop Dis. 2016;10:e0005123.
80. Pandey G, Cohain A, Miller J, Merad M. Decoding dendritic cell function through module and network analysis. J Immunol Methods. 2013;387:71–80.
81. Berry MPR, Graham CM, McNab FW, Xu Z, Bloch SAA, Oni T, et al. An interferon-inducible neutrophil-driven blood transcriptional signature in human tuberculosis. Nature. 2010;466:973–7.
82. Mejias A, Dimo B, Suarez NM, Garcia C, Suarez-Arrabal MC, Jartti T, et al. Whole blood gene expression profiles to assess pathogenesis and disease severity in infants with respiratory syncytial virus infection. PLoS Med. 2013;10:e1001549.

83. Wang S, Sun H, Ma J, Zang C, Wang C, Wang J, et al. Target analysis by integration of transcriptome and ChIP-seq data with BETA. Nat Protoc. 2013;8:2502–15.
84. Jee J, Rozowsky J, Yip KY, Lochovsky L, Bjornson R, Zhong G, et al. ACT: aggregation and correlation toolbox for analyses of genome tracks. Bioinformatics. 2011;27:1152–4.
85. Hunter PJ, Crampin EJ, Nielsen PMF. Bioinformatics, multiscale modeling and the IUPS Physiome project. Brief Bioinf. 2008;9:333–43.

The Challenge of Translating System Biology into Targeted Therapy of Cancer

Alessandra Jordano Conforte, Milena Magalhães, Tatiana Martins Tilli, Fabricio Alves Barbosa da Silva, and Nicolas Carels

Abstract Translational medicine has been leveraging new technologies and tools for data analysis to promote the development of new treatments. Integration of translational medicine with system biology allows the study of diseases from a holistic perspective. Cancer is a disease of cell regulation that affects genome integrity and ultimately disrupts cell homeostasis. The inter-patient heterogeneity is well characterized, and the scientific community has been seeking for more precise diagnoses in personalized medicine. The use of precision diagnosis would maximize therapeutic efficiency and minimize noxious collateral effects of treatments to patients. System biology addresses such challenge by its ability to identify key genes from dysregulated processes in malignant cells. Currently, the integration of science and technology makes possible to develop new methodologies to analyze a disease as a system. Consequently, a rational approach can be taken in the selection of the most promising treatment for a patient given the multidimensional nature of the cancer system. In this chapter, we describe this integrative journey from system biology investigation toward patient treatment, focusing on molecular diagnosis. We view tumors as unique evolving dynamical systems, and their evaluation at molecular level is important to determine the best treatment options for patients.

1 Introduction

The number of cancer cases increased significantly with the improvement of economic conditions and the raise of the populations' average age [1], despite the significant improvements in diagnosis and treatment obtained in recent years.

A. J. Conforte · M. Magalhães · T. M. Tilli · N. Carels (✉)
Laboratório de Modelagem de Sistemas Biológicos, Centro de Desenvolvimento Tecnológico em Saúde, Fundação Oswaldo Cruz, Rio de Janeiro, Brazil
e-mail: nicolas.carels@cdts.fiocruz.br

F. A. B. da Silva
Laboratório de Modelagem Computacional de Sistemas Biológicos, Programa de Computação Científica, Fundação Oswaldo Cruz, Rio de Janeiro, Brazil
e-mail: fabricio.silva@fiocruz.br

© Springer International Publishing AG, part of Springer Nature 2018 175
F. A. B. da Silva et al. (eds.), *Theoretical and Applied Aspects of Systems Biology*, Computational Biology 27, https://doi.org/10.1007/978-3-319-74974-7_10

Estimates indicate 13.2 million cancer deaths by 2030 worldwide [2]. In Brazil, the National Cancer Institute (INCA) expected about 600,000 new cases of cancer in 2016–2017, which appears as the second main cause of death in the country.

Oncogenesis occurs as a result of genetic and epigenetic reprogramming processes involving complex regulatory circuits resulting in cell immortalization and uncontrolled cell division [3]. Both processes of immortalization and uncontrolled cell proliferation lead to the increase in tumor mass [4], which results in localized physiological disturbances often accompanied by loss of organ function. The evolution of this disease reaches metastasis formation, which compromises other tissues and organs, and ultimately leads to death.

Hanahan and Weinberg [5] grouped the features acquired by malignant cells into ten classes of alterations that interfere with the normal physiology of cells and tissues. They may be cited as (i) growth factor self-sufficiency, (ii) insensitivity to the inhibition of proliferation factors, (iii) evasion of apoptosis or programmed cell death, (iv) unlimited replicative potential, (v) angiogenesis, (vi) tissue invasion and metastasis, (vii) inflammation, (viii) reprogramming of energy metabolism, (ix) evading destruction by the immune system, and (x) genomic instability and mutation. It is worth noting here that metastasis constitutes the greatest challenge for cancer clinical management and is the main cause of patient death.

Three distinct tumor phases have been traditionally recognized: initiation, promotion, and progression [6]. *Initiation* is the transformation process from a normal cell into a malignant one. Within this phase, the incubation step is long and typically takes years or decades of mutation accumulation to finally reach a critical stage where the control of genome integrity by molecular check points is bypassed. This incubation step is followed by the constitutive induction of proliferation due to mutations in genes related to cell division. These genes can be classified in suppressor and oncogenes. Suppressor genes are involved in the repression of processes promoting the malignant stage, such as uncontrolled division, for instance, i.e., to keep processes related to cancer hallmarks under control. By contrast, oncogenes are those that may stimulate tumor formation, activating signaling pathways under abnormal regulation. The defects of cell homeostasis induced by mutation accumulation may cause abnormalities in methylation pattern, gene expression, DNA repair activity, cell replication, chromosome number, and telomere size.

The *promotion* phase involves the contribution of inflammatory processes in tumor microenvironment, which is called *stroma*, a tissue that is modulated by factors synthesized in the tumor. This phase also involves the spread of immature malignant cells.

The *progression* phase is reached when a tumor evolves toward angiogenesis and metastasis. The growth of new blood vessels formed by angiogenesis in the tumor is necessary to feed it, supporting its growth and discard its wastes. These blood vessels are erratic and constitute a way for malignant cells to enter the blood circulation. Once these cells reach suitable locations, they may adapt to the new environment of other tissues and establish a secondary malignant site.

The drugs usually administrated in traditional treatments are considered beneficial for a whole group with similar tumor characteristics, and consequently, these treatments are also called *one-size-fits-all*. Unfortunately, these drugs often present reduced molecular precision and may affect both cancer and healthy cells, causing several noxious effects to patients. Furthermore, many drugs currently available are costly and provide few improvements in the overall survival rate.

Many advances were obtained in cancer therapy from a diagnostic and therapeutic standpoint. From 2011 ahead, new drugs related to target therapies, such as imatinib mesylate, rituximab, and trastuzumab, used against leukemia, lymphomas, and breast cancer, respectively, were incorporated to the SUS (Brazilian public unique healthy system) list of cancer treatments. According to standard protocols, chemotherapy, radiotherapy, and surgery can be used to treat cancer. These protocols are adapted according to tumor tissue or molecular characteristics, and physicians may adopt different therapeutic combinations in line with patients' response.

Nonetheless, the inactivation of one therapeutic target may be followed by signaling network reshuffling, which promotes the use of alternative pathways with the consequence of preserving the malignant state [7].

For this reason, an urgent issue in drug development is to precisely identify the molecular target whose inactivation would not lead to signaling network reshuffling. If a target protein plays a central role in normal cell signaling or metabolism, it may be a questionable target as its inhibition would cause significant adverse side effects for patients' health [8]. The identification of suitable therapeutic targets for treatment with cocktails of drugs is not straightforward because malignant and healthy cells lack evident molecular differences. Actually, these differences lie in the signaling pathways regulation [9].

In the last three decades, hundreds of molecular targets related to cancer were identified, and new drugs were developed. Several approaches were described in the literature for the development of drugs against essential molecular targets, such as US20110287953 and US20070038385 patents. The first patent is related to the identification of a treatment strategy based on protein-protein interaction (PPI) network reconstruction for a specific cancer case, prioritizing important gene targets and available drugs. The US20070038385 patent is about a metabolic system reconstruction that integrates sequence, clinical and experimental data for the identification of biomarkers. This patent uses lists of targets recently described as seeds to feed the metabolic model.

As stated above, these methodologies seek to identify treatment strategies in important pathways, functional modules, or driver genes. Consequently, these methodologies focus on a limited number of routes already described as important pathways for oncogenesis processes.

In contrast to *one-size-fits-all* treatment, personalized medicine uses patients' genetic profiles to determine the appropriate treatment aiming at successful outcome and minimization of noxious side effects.

The development of personalized medicine concept is directly related to the huge technological advances reached in the last years. The large-scale molecular approaches currently available, such as DNA and RNA sequencing, proteomics, metabolomics, epigenomics, are important tools for the study of gene expression patterns in tumors and healthy cells. These high-throughput techniques have already revealed a series of tumor-specific somatic mutations related to the constitutive activation of signaling circuits in malignant cells [5]. These technologies allowed a better understanding of tumor biology and led to the understanding that each tumor should be considered as unique, which is the aim of *precision medicine*.

Personalized medicine can be defined as a concept based on individual clinical, genetic, and environmental information. Due to the tumor intrinsic heterogeneity [10], it is also necessary to identify common targets between all malignant cells to avoid the development of resistance against the administered drugs that often results into highly aggressive forms of cancer. These common targets are biological molecules called biomarkers and act as indicators for a specific biological condition, such as the carcinogenic state.

Biomarkers identified from heterogeneous tumors are being used in diagnostic, prognostic, and cancer treatments based on personalized medicine concept, and several commercial diagnostic kits, such as MammaPrint (Agendia BV, Amsterdam), Oncotype DX (Genomic Health, California), MapQuant DX (Ipsogen, France), and Theros (BioTheranostics, California), were proposed. However, their gene lists show limited or absent overlap. The reason for this disparity has been attributed to molecular and clinical differences between the patients' groups analyzed, sample preparation, microarray platforms, and the statistical methods used to establish their significance. The lack of standardization in the methodologies for the implementation of these tests resulted in poor prognostic reproducibility, which has affected drastically their reliability [11–13].

The technologies that enabled the first sequencing of human genome in the year 2000, at a very high cost, evolved a lot, and it is currently possible to perform individualized human sequencing at an affordable price. The great question is how to transform molecular biology knowledge into clinical practice regarding the prevention and treatment of cancer. In this chapter we describe different approaches that can be, or have been, used to approximate translational data from personalized medicine.

2   Epigenetic Influence

Epigenetic changes such as DNA methylation and chromatin modifications have a determinant role on gene expression and can activate or inhibit signaling pathways that lead to cancer.

DNA methylation is an epigenetic mechanism that consists in the addition of a methyl group on the 5-cytosine of CpG dinucleotides. CpG sites occur throughout the genome, yet they may be grouped in clusters called CpG islands, mainly found at gene promoters.

This mechanism is controlled by different DNA methyltransferases (DNMTs), and the alteration of their activity may accelerate the propagation of noxious genomic remodeling. Blockade of DNMT1 expression, for example, resulted in the loss of promoter methylation, the reactivation of the tumor suppressor gene p16^{ink4A}, and the inhibition of cell proliferation in a bladder cancer cell line [14]. However, in colon cancer cell line, the deletion of DNMT1 was not sufficient to cause promoter demethylation and gene reactivation [15]. Furthermore, gene silencing of both DNMT1 and DNMT3b reduced genomic DNA methylation by greater than 95% in a colon cancer cell line, showing that the two enzymes cooperatively maintain DNA methylation and gene silencing in human cancer cells [16]. According to these evidences, it is possible that DNMT genes act together to maintain or establish DNA methylation patterns in individual types of human cancers whose methylation is essential for optimal neoplastic proliferation.

Global DNA hypomethylation is normally acknowledged as a sign of malignant cells [17–19]. One of the mechanisms of demethylation is caused by suppressing the pattern of methylation managed by DNMTs during replication. If DNMT1 is inhibited or absent when the cell divides, the newly synthesized strand of DNA will not be methylated, and successive rounds of cell division will result in passive demethylation by dilution of DNA methylation [20]. Another way of demethylation is by the 10–11 translocation enzymes (TETs), which is a DNA demethylase that specifically removes 5-methylcytosine (5mC) [21]. Imbalance in TET-mediated DNA demethylation may participate in oncogenesis. Up- or downregulation of TET has also been observed in several solid cancers, such as breast, liver, colon, prostate, and gastric cancers [22–24].

Increased DNA methylation in a gene promoter is generally associated with downregulation of the downstream gene. However, a hypomethylated promoter does not necessarily indicate an active gene. The DNA methylation may impact gene expression through three different ways: it may (i) recruit proteins that have affinity for methylated DNA (methyl-CpG-binding proteins), which then mediate downstream biological effects; (ii) overlap with the binding site of a transcription factor; and (iii) directly alter the chromatin structure by acting synergistically with posttranslational modifications of histone tails. Therefore, mutations in histone methyltransferases (HMT), histone acetyltransferases (HAT), and histone deacety-lases (HDACs) may also have a role in oncogenesis [25, 26].

Defects in the homeostasis of DNA methylation also affect miRNAs, which regulates about 60% of transcriptional activity [27]. In breast cancer, several miRNAs were found to be dysregulated by DNA methylation [28]. Also aberrant promoter methylation or mutant p53 resulted in the downregulation of miRNA-145 expression due to the lack of p53-miRNA-145 binding in prostate cancer and several cancerous cell lines [29].

The DNA methylation status can also be influenced by environmental factors. Exposure to pollution and tobacco smoke induces oxidative stress, which is thought to favor the demethylation process. Chronic exposure to cigarette smoke, either through passive secondhand smoking or active smoking, leads to an increased risk of lung cancer [30]. It should be noted that exposure to radiation and toxic compounds has also been implicated in the modulation of DNA methylation. In this context, chemotherapy and radiotherapy can cause epigenetic alterations, but the clinical influence of this observation is unclear.

The DNA methylation status has been used as a biomarker. Some types of cancer such as breast, colon, and lung cancers are regularly diagnosed at a late stage. Robust diagnostic biomarkers with not only predictive but also prognostic value that allow the diagnosis of cancer patients at early stage are of great importance. In a breast cancer study, DNA methylation was successfully used to predict the outcome among different populations of breast cancer patients under chemotherapy, showing the role of DNA methylation as a biomarker [31]. Aberrant DNA methylation of *CDKN2A* and *MGMT* promoter regions was identified in lung cancer and chronic obstructive pulmonary disease (COPD) compared to healthy controls. Interestingly, promoter methylation at these genes was also strongly associated with active smoking [32]. The *MLH1* promoter, a mismatch-repair gene, was found hypermethylated in normal colonic epithelium of some colorectal cancer patients, suggesting that epigenetic abnormalities may precede classical genetic alterations such as deletions and mutations in the course of oncogenesis [33]. As a result, the methylation status of *MLH1* could be a biomarker for the early diagnosis of malignant tumors.

DNA methylation also has an important role in genomic stability. The demethylation, as consequence of DNA methyltransferase inactivation, increases chromosome instability and eventually promotes this improper recombination and breakage, increasing the rate of aneuploidy and karyotype aberrations [34]. At centromeres, DNA methylation is important to maintain chromosomal stability. A connection between hypomethylation and the stability of whole chromosome arms is also found in hypomethylation-induced T cell lymphomas in mice and several cancers [35]. Genetic instability results in mutations, which fuel oncogenesis, but mutations may also be caused by different processes such as (i) direct action of physicochemical agents on DNA, (ii) genetic predisposition to mistakes in nucleotide incorporation by polymerase (hypermutator phenotype), (iii) transposon activity with mutagenic effects, and (iv) defects in DNA repair machinery.

Epigenetic treatment has been approved by regulatory agencies for the therapy of some cancers [36]. For example, DNMT inhibitors (DNMTi) that prevent the repair of radiotherapy-induced DNA damage inhibit cell proliferation, block cell repopulation during radiotherapy [37], and may induce apoptosis [38]. The combination of inhibitors of histone deacetylases (HDACi) with immunotherapeutics is also useful, because of their capacity to moderate different factors and pathways involved in the interaction between tumor cells and the immune system [39]. Despite promising results, the toxicity to normal cells remains a challenge, as most epigenetic modulators have systemic effects.

3   Mathematical and Computational Modeling

Computer models can represent the biological world using mathematical, physical, and engineering concepts to reveal new aspects and deepen our knowledge about biological processes. The in silico network inference can facilitate the generation and testing of valuable hypothesis to be validated by in vitro and/or in vivo experimentation.

Cancer modeling is useful to obtain more information about signaling, regulatory, and metabolic networks of malignant cells. Modeling has been improving the knowledge of typical cancer dysregulated processes, such as cellular proliferation, differentiation, tumor morphology and angiogenesis [40]. Modeling can simulate tumor cell dynamics, drug pharmacokinetics, and potential new therapies. Consequently, new potential biomarkers and drug targets have already been identified through cancer modeling [41].

Signaling and regulatory networks of cancer are very complex and nonlinear. They involve many steps and different regulation mechanisms, including positive and negative feedback loops as well as cross-talks between different signals. For these reasons, genome-scale modeling of signaling and regulation network still has some hindrances. Nevertheless, mathematical modeling has been used to explore these complex networks and to elucidate many processes through the analysis of the relationship between gene functions and their network organization. For instance, we can cite the linear relationship between proliferation and amount of EGF receptor-ligand complex at steady state [42], different strategies to control cell proliferation [43], and the dynamics between IL-2 and IL-4 receptors with T cell proliferation [44].

The investigation of cancer metabolic networks, on the other hand, has been favored due to the availability of a large body of data because metabolism reprogramming has been described as one of the cancer hallmarks [5]. Thus, many generic human *genome-scale metabolic models* (GEM) have been reconstructed [45–47]. These models describe cancer genetic traits and have enabled Folger et al. to identify 52 drug targets, from which 31 are new and 21 present anticancer drugs in experimental or approved situation, and to predict combinations of lethal drug targets, whose synergy was validated with available data [48].

The *flux balance analysis* (FBA) is commonly applied to metabolic models. It can predict metabolic flux states and the consequences of genetic and environmental perturbations on metabolic phenotypes [49]. Even if this methodology is based on the assumption of system's steady state and only few parameters, it enables meaningful genome-wide predictions. Zielinski et al. have applied this strategy for cell lines and discovered that resistance to chemotherapeutic drugs broadly correlates with the amount of glucose uptake [50]. Gatto et al. also used FBA in cell lines and identified five essential genes for *clear cell renal cell carcinoma* (CCRCC) growth, which are not essential in normal cell metabolism as validated in vitro [51].

The reconstruction of cancer metabolic models is based on repositories of biochemical data including enzymatic reactions catalyzed by gene products regardless their state (active or inactive). Neglecting the expression state of a gene has the consequence of promoting the occurrence of false negatives in simulations of gene essentiality or genetic interaction [52]. In other words, predictions based on genome-scale modeling may fail for not considering the effects of gene regulation. The integration between different sources of omics data is crucial to produce accurate models allowing personalized genome-scale metabolic modeling. Data integration allows the construction of a functional model that is a mathematical representation capable of simulating molecular and cellular phenotypes. There are some available approaches to integrate the different omics data with metabolic network reconstruction. The main approaches used are the comparison between model simulations and omics data and the use of omics data to create cell- or tissue-specific models [53].

To improve the predictive capabilities of genome-scale models, one can implement probabilistic inferences, but there are some significant computational hurdles that need to be overcome [54]. The transcriptional regulation can be analyzed in a stoichiometric model, where a known chemical pathway structure is used to understand the state of the system, and the genes' states receive a Boolean representation (activated or inactivated) [55]. Boolean models were widely used to elucidate cancer networks [56]. They are based on discrete logic and provide a qualitative approximation of a biological system. Their limitation is related to the transfer function, which is not related to the physiological one, but the use of variable timescale and the execution of asynchronous updates [57] is sufficient to mimic satisfactorily the biological reality. This methodology is an important predictive tool in the absence of more information about the parameters of transfer functions. Lastly, ordinary differential equations also are commonly used for cancer modeling [58]. They describe biophysical relationships and can be used if the parameters of the interactions in a network are known [59].

In order to build predictive models of cancer considering both metabolism and signaling networks, a hybrid model should be used. This type of model is widely used in engineering and computational sciences; it combines discrete and continuous timescales. For instance, a tumor growth model can consider cells as discrete entities, while intracellular and extracellular signals are modeled as continuous phenomena [60, 61]. Singhania et al. [62] linked a continuous cell cycle model with a Boolean gene network that regulated essential substrates involved in the cell cycle process.

Currently, several tools are available to improve the search for more accurate models. The tINIT (task-driven integrative network inference for tissues) algorithm is one example and has been to automatically reconstruct functional GEM and integrate them with available knowledge about tissue-specific enzymatic reactions [63, 64]. GEMs allow the calculation of a metabolic phenotype from genomic data. The availability of databases such as those from the Human Metabolic Reaction [46]; the Human Protein Atlas [65], which provides cell type and tissue specific profiles; the Recon2 [66], together with human genome scale metabolic reconstruction; as well as the databases that provide expression profile for different malignant cell lines and tissues, such as NCI-60 panel, TCGA (*The Cancer Genome Atlas*), and ICGC (*International Cancer Genome Consortium*), are few examples that can be explored for personalized cancer research.

Personalized cancer GEMs were reconstructed and used to capture common and specific metabolic shifts across cancer patients, tissues, and cells and to identify selective anticancer drugs [63]. It is worth noting that antimetabolites were used as anticancer drugs. They inhibit the catalysis of one or more endogenous metabolites identified as beneficial for cancer development processes, such as those involved in the robustness and growth of malignant cells [67, 68].

The need to consider personalized cancer models is justified by the largely recognized heterogeneity of malignant cell lines and tissues; it indicates the inadequacy of reconstructing a generic molecular model of cancer and stands out for the improvement of integrated personalized models. The lack of multiple replicates from a same sample has been considered an obstacle to achieve robust personalized models [51] and has motivated some scientists to still adhere to generic models and, as consequence, to one-size-fits-all treatments [50, 63].

Besides the need for personalized cancer models, the combinatorial nature of treatment has also been discussed. Genome-scale biological networks have been widely used to identify single target genes, but they are also suitable to identify subnetworks. These subnetworks can be used to choose a combination of targets able of disarticulating several regulatory processes and leading to a synergistic effect against cancer proliferation [48, 69].

The potential contribution of mathematical and computational modeling to cancer research involves the identification of new gene targets, essential metabolites, the test of treatment combinations, the prediction of treatment outcome, the reduction of bench experimentation, as well as the indication of the most promising strategies in terms of time and cost saving.

4   Measures of Network Topology

A signaling network is built from the interactions between its proteins. Until recently, connections between proteins were considered as fixed, and there are several databases that follow this concept, such as psimitab, Biogrid, and STRING. However, a real interactome is dynamic. The development of nucleic acid programmable protein array (NAPPA, [70]) enables to map until 12,000 real-time interactions at once [71].

The topological network of protein interactions can be inferred using RNA-seq and interactome data. Quantifying signaling proteins can now be assessed by large-scale *precision proteomics* using mass spectrometry [72]; however an approximation can also be assessed at lower costs by mRNA sequencing and tag count, also called RNA-seq. The normalized tag count may act as an indication of protein concentration, since the translation rates across tissues correlate with protein level in about 80% [73].

Some characteristics should be considered when inferring the network topology:

(i) If a node is not expressed, it is not present.
(ii) A node with high expression level is more important to the network than a node with a low expression level.
(iii) Similarly, a node with a large number of connections is more important to the network than one with a low connection number.

Based on these node features, we can calculate the influence of each node in a network through the calculation of its degree-entropy.

The notion of entropy (S) has been introduced by Boltzmann; it measures the probability of possible microscopic configurations of a macroscopic system [74]. Independently, Shannon derived a similar equation, where entropy is a measure of information content [75]. The degree-entropy is the application of Shannon's formula to signaling network, where the connection number is indicative of information content. The higher the connection number, the higher the information content at a given node; in other words, the largest is its influence in a network. The degree-entropy of a node is calculated as the probability of the observed connection number of a node, i.e., the node connection number over the sum of all connections of the whole network, multiplied by the natural logarithm of that probability. The degree-entropy of a network is the sum of local degree-entropies across the network's nodes.

According to the three basic characteristics listed above, the expression level must also be considered to calculate the information content. This can be done with the appropriate weighting of the degree-entropy at each node with its corresponding expression level. This solution has been proposed by Banerji et al. [76] and correlates with the *t-test-based pluripotency score* (TPSC), which is a score of pluripotency based on the expression level of 19 genes traditionally used as fingerprint of totipotency.

Pluripotent cells are able to generate a large diversity of cellular phenotypes. As a result, they maintain many pathways in their activated network. By contrast, differentiated cells are more homogeneous and maintain only few activated pathways, the ones related to their function. Due to the number of activated nodes in each network, pluripotent cells are expected to have larger network entropy than the differentiated cells. For instance, the differentiation of stem cell follows an evolutionary pattern that matches their trajectories on Waddington's landscape, i.e., their trajectories go from a higher entropy level to a lower one according to the differentiation state [76].

Entropy is a valuable measure to rank cellular complexity; it has the benefit of (i) being a self-calibrating measure and (ii) being insensitive to the normalization method used.

The betweenness-centrality measure is also important in network science. It can be calculated as the summing of betweenness-centrality across all possible node pairs in a given network. Each node betweenness-centrality is calculated as the sum of shortest paths between two nodes that passes through the node under analysis over the number of all shortest paths between these two nodes. It has been described that connectivity and betweenness-centrality measures are highly correlated ($r = 0,91$) [77]. The node centrality and, as a consequence, the protein connectivity are important features to be considered in therapeutic design because they indicate the likelihood of a node to disarticulate the signaling network if this node is inactivated.

Therapeutic design also involves the understanding of the benefit that the patient might obtain from precision medicine. The patient benefit is related to its survival rate after treatment. Statistics are available for patients' survival associated with different types of cancer [78]. The patient outcome has a negative correlation with the degree-entropy of the subnetwork formed by the upregulated genes of its tumor. This indicates that the higher the tumor entropy, the higher its aggressiveness [7]. The association of the signaling network entropy of a tumor with its aggressiveness has allowed the inference of the benefit that a patient may draw from the composition of a cocktail [79]. This strategy is based on the following facts: (i) protein connectivity follows a power law [80] and (ii) the signaling networks of biological cells are scale-free [81]. According to the therapeutic design based on maximizing the patient benefit, a scale-free network is more effectively disarticulated when few nodes are rationally selected for inactivation than when the target selection is done at random. The nodes rationally selected are expected to be upregulated hubs in order to minimize noxious collateral effects to the patient [82].

5   A Molecular Approach: Paving the Way for Personalized Medicine – Breast Cancer Case Study

A strategy has been delineated to identify protein targets of breast cancer in order to implement a personalized treatment [83]. As outlined above, the protein targets identified following that strategy should allow the development of therapeutic agents minimizing the deleterious side effects to patients. The genes found to be upregulated in malignant cell lines by statistical comparison to the RNA-seq of a non-tumoral cell line of reference were considered potential targets for drug development because the transient inhibition of their expression did not affect the living condition of the reference cells. Among the 150–300 upregulated genes in malignant cells, some have a larger likelihood of being suitable targets for drug development than the others because they have a larger protein connectivity rate in the cell-line-specific subnetworks induced by signaling rewiring during the oncogenesis process [77].

Proteins acting as connectivity hubs in the signaling network of malignant cell lines were found by comparison with interactome data. The local degree-entropy associated with each expressed protein can be calculated from the interactome data and used to rank the relative connectivity rate according to the total degree-entropy associated with the whole network as well as to rank the comparative benefits of drug cocktails to patients according to the profile of their upregulated top connectivity hubs [79, 84]. The combinations of top-5 connection hubs effectively found to be upregulated were found to be specific to each cell line [77]. This result highlights the concept that each patient needs different inhibitor composition in his therapeutic cocktail, making clear the need for rational therapeutic design in the context of precision medicine.

The concept outlined above has been validated by RNA interference with a cocktail of interfering RNA (siRNA) designed to inactivate the top-5 upregulated connection *hubs* of the protein interactome of MDA-MB-231, a triple-negative malignant cell line of breast cancer (TNBC) [7, 77, 79, 84].

The top-5 genes identified for MDA-MB-231 have well-known contribution as individual components to cancer development and progression. TK1 promotes cell proliferation, decreases DNA repair efficiency, and induces cell death [85–87]. HSP90AB1 promotes angiogenesis not only as a protein chaperone but also as an mRNA stabilizer for pro-angiogenic genes, such as *BAZF* [88]. Vimentin is a marker of epithelial-mesenchymal transition (EMT) and promotes invasion and metastasis [89, 90]. YWHAB, a member of the 14-3-3 family of proteins, is involved in the activation of tumor/metastasis pathways and inhibition of apoptosis [91–94]. Similarly, CK-2β promotes EMT and metastasis while inhibiting apoptosis [95–98, 103].

The inactivation of these five targets in MDA-MB-231 cells has been shown to significantly decrease cell proliferation, colony formation, anchorage-independent cell growth, cell migration, and cell invasion [83]. This also validates that the prediction strategy based on bioinformatics inferences in the analysis of PPI network identifies potentially suitable targets for cancer treatment [77, 79]. This proof-of-concept study can serve as a preliminary step in the process of target discovery toward development of precision therapies.

Functional assays demonstrated that the expression of these five targets significantly contributed to the proliferative and invasive phenotype of MDA-MB-231. Consistent with the bioinformatics inferences [77], the knockdown of these network targets had little or no effects on growth, migration, and invasion of the noninvasive MCF-7 or reference MCF-10A cells. Furthermore, the strong differences in the response of MDA-MB-231 and MCF-7 cells to the inhibition of the top-5 targets reinforce the concept of precision medicine, which is considered as a shift away from the *one-size-fits-all* approach.

Several studies have reported the individual expression pattern and function of each selected hub in different tumors and tumor models in vitro [84, 85, 87, 88, 90–102]. However, single siRNA transfections had no detectable effect on the proliferative and invasive properties of the cell lines under study. These results are not necessarily in contradiction with other studies where inactivation of these hubs had measurable effects and may be due to different cell lines and culture/assay conditions.

The effect of combination knockdown is larger than the one expected from the sum of each knockdown. This is most likely a consequence of cell signaling pathway redundancy in malignant cells. Consequently, simultaneous inactivation of several hubs may be necessary for successful elimination of potential alternative/cross talking signaling pathways that malignant cell need for the maintenance of their phenotype.

In agreement with the reported inhibition of cell proliferation and metastasis, a continued cell death beyond siRNA action time was observed. This observation suggests that despite the short half-life of the top-5 gene mRNA, the transient knockdown was sufficient to significantly affect the stability of the malignant signaling network [104, 105]. As a confirmation, Fumiã and Martins [56] showed, using a modeling approach, that while monotherapies were ineffective, drug cocktails were highly effective and necessary for efficient reversal of all hallmarks of cancer.

The entire set of predicted drug targets identified has been experimentally validated by available drugs or siRNAs [101, 106–110]. Thus, conceivably, the described approach can be implemented for a substantial number of currently used chemotherapeutic drugs with well-described mechanisms of action as reported in various studies [56].

Compared to conventional cytotoxic drugs that affect both normal and tumoral cells, *synthetic lethality* [111] can address anticancer therapy by optimal hub targeting according to cancer type while sparing normal cells. However, despite the advances in siRNA targeting and compound screening, synthetic lethal interactions between genes and drugs have remained extremely difficult to predict on a global scale. Thus, the integration of interactome and transcriptome data allows the effective selection and prioritization of suitable protein targets for drug development and experimental testing or eventual clinical translation among the massive number of possible target combinations. Network-based methods provide a convenient platform to find functional interactions enabling the identification of targets and drug combinations for effective and personalized cancer therapies.

The induction of 40% cell death by transient inactivation of top-5 hub proteins in a TNBC cell line suggests that increasing the number of deactivated hubs or combining siRNA therapy with drugs may provide synergistic antitumor effects with the current protocol of radiotherapy and adjuvant chemotherapy. According to this rationale, cells sensitized by top-5 target-specific drugs (or more than five targets) should allow the reduction of concentrations of individual cytotoxic drugs and hence lead to reduced adverse side effects.

Based on the analysis of gene expression and PPI in signaling networks using bioinformatics, a combinatorial therapeutic approach has been tested and validated using cell-based assays. The results clearly demonstrate the effectiveness of this approach to significantly decrease malignant cell proliferation, migration, and invasion without any noticeable deleterious side effects to the reference cell line. While these results should be validated in vivo using animal models, they clearly support the power of bioinformatics inferences for identification and selection of hub targets as well as formulating synergistic combination of drug therapies for the treatment of breast cancer.

6 Remaining Challenges

Although many advances toward cancer precision therapy were achieved in the last decade, there are still many challenges to overcome. One problem is the use of generic models to study specific conditions. Although many steps were taken toward models of precision medicine, the low availability of signaling data limits the robustness of the inferences obtained and induces scientists to be more careful about their assumptions. Generally, when a scientist develop a model for a specific cell line or tissue, he refers to targets common to other models previously described in the literature because of the lack of replicates or little material availability when dealing with individual data in the context of personalized medicine [50, 51].

Different samples of control and tumor tissues from a same patient are also necessary for the precise identification of key differentially expressed genes for rational treatment design. The availability of biopsy from biorepositories is also important for the development of new methods capable of identifying efficient target combinations. In this context, *The Cancer Genome Atlas* (TCGA) provides free online access to several RNA-seq from patients together with their clinical data. Moreover, some of them, the minority, have both healthy and cancer tissue data from the same patient. This kind of initiative is essential to overcome the lack of sample data availability.

Another challenge is the need of replicates from a same sample or even a follow-up of a same patient to achieve robust prediction inferences. This is a very challenging issue because it would involve multiple biopsy surgeries, which is invasive for patients. Until now, this challenge can be assessed through mathematical and computational inferences to preserve the patient's life quality, but future developments in diagnosis of blood biopsy are expected to help in dealing with that issue.

The lack of standard protocols in sample preparation and mathematical analysis of the available data is also a limiting factor. Many studies have few or none overlap in results even though using the same data sources. This happens as a result of the lack of consensus regarding experimental and statistical methods to be applied [11–13].

Lastly, some recently discovered targets may already have related approved drugs or drugs in developmental process. Indeed, many drugs, or bioagents, were already approved, which makes possible to use rational approaches and drug repositioning for cocktail design and immediate therapeutic applications. However, if considering precision medicine and immediate care, we must invest in new forms of treatments that do not require the discovery and/or approval of new drugs, a process that typically takes at least 10 years of development.

An alternative to drug treatment is the use of new active pharmaceutical ingredients (APIs), such as RNAi, aptamers, and antibodies, which are theoretically capable of inactivating any gene. Due to its negative charge, siRNAs need a delivery system to overcome the electrostatic repulsion from cell membranes, also negatively charged [112]. In this context, nanoparticles are promising, which also solves other hurdles related to RNAi treatment, such as low cell absorption, intracellular instability, and fast renal absorption [113].

7 Conclusions

The identification of target proteins from interactome networks with algorithms of bioinformatics and experimental validation offers a rational approach to the development of cancer therapies. Inference of target proteins, based on molecular networks to identify individual specificities within a patient, is the heart of the personalized medicine concept. In addition, the use of this approach in the design of molecular diagnosis and treatment is expected to minimize costs, time, and failure rates of potential cancer inhibitors in clinical trials. Personalized therapy should be addressed through the quick diagnosis of key targets to be inhibited according to disease course and to the minimization of noxious side effects. In addition, intrinsic and acquired resistance to treatment, tumor heterogeneity, adaptation, and genetic instability of tumor cells must also be considered in the framework of the proposed precision therapy. Since the information encoding all these cell properties is included in the gene expression profile of patients, personalized treatment based on patients' RNA-seq profiling should increase the treatment efficacy and the patients' life quality.

Acknowledgment This study was supported by fellowships from the Oswaldo Cruz Institute (https://pgbcs.ioc.fiocruz.br/) to A.C., from *Instituto Nacional de Ciência e Tecnologia de Inovação em Doenças de Populações Negligenciadas* (#573642/2008-7) to M.M., and from *Convenio CAPES/Fiocruz* (cooperation term 001/2012 CAPESFiocruz) to T.M.T.

References

1. de Magalhães JP. How ageing processes influence cancer. Nat Rev Cancer. 2013;13(5): 357–65.
2. Ferlay J, Shin H-R, Bray F, Forman D, Mathers C, Parkin DM. Estimates of worldwide burden of cancer in 2008: GLOBOCAN 2008. Int J Cancer. 2010;127(12):2893–917.
3. Hanahan D. Rethinking the war on cancer. Lancet (London, England). 2014;383(9916): 558–63.
4. Wallace DI, Guo X. Properties of tumor spheroid growth exhibited by simple mathematical models. Front Oncol. 2013;3:51.
5. Hanahan D, Weinberg RA. Hallmarks of cancer: the next generation. Cell. 2011;144(5): 646–74.
6. Barcellos-Hoff MH, Lyden D, Wang TC. The evolution of the cancer niche during multistage carcinogenesis. Nat Rev Cancer. 2013;13(7):511–8.
7. Breitkreutz D, Hlatky L, Rietman E, Tuszynski JA. Molecular signaling network complexity is correlated with cancer patient survivability. Proc Natl Acad Sci. 2012;109(23):9209–12.
8. Wachi S, Yoneda K, Wu R. Interactome-transcriptome analysis reveals the high centrality of genes differentially expressed in lung cancer tissues. Bioinformatics. 2005;21(23):4205–8.
9. Lim DHK, Maher ER. Genomic imprinting syndromes and cancer. Adv Genet. 2010;70: 145–75.
10. de Bruin EC, Taylor TB, Swanton C. Intra-tumor heterogeneity: lessons from microbial evolution and clinical implications. Genome Med. 2013;5(11):101.
11. EGAPP. Recommendations from the EGAPP Working Group: can tumor gene expression profiling improve outcomes in patients with breast cancer? Genet Med. 2009;11(1):66–73.
12. Abba MC, Lacunza E, Butti M, Aldaz CM. Breast cancer biomarker discovery in the functional genomic age: a systematic review of 42 gene expression signatures. Biomark Insights. 2010;2010(5):103–18.
13. Gabrovska PN, Smith R a, Haupt LM, Griffiths LR. Gene expression profiling in human breast cancer – toward personalised therapeutics? Open Breast Cancer J. 2010;2:46–59.

14. Fournel M, Sapieha P, Beaulieu N, Besterman JM, Macleod AR. Down-regulation of human DNA- (cytosine-5) methyltransferase induces cell cycle regulators p16 (ink4A) and p21 (WAF/Cip1) by distinct mechanisms. J Biol Chem. 1999;274(34):24250–6.

15. Rhee I, Jair K-W, Yen R-WC, Lengauer C, Herman JG, Kinzler KW, et al. CpG methylation is maintained in human cancer cells lacking DNMT1. Nature. 2000;404(1998):1003–7.

16. Rhee I, Bachman KE, Park BH, Jair K-W, Yen R-WC, Schuebel KE, et al. DNMT1 and DNMT3b cooperate to silence genes in human cancer cells. Nature. 2002;416(6880):552–6.

17. Wasson GR, McGlynn AP, McNulty H, O'Reilly SL, McKelvey-Martin VJ, McKerr G, et al. Global DNA and p53 region-specific hypomethylation in human colonic cells is induced by folate depletion and reversed by folate supplementation. J Nutr. 2006;136(11):2748–53.

18. Wilson AS, Power BE, Molloy PL. DNA hypomethylation and human diseases. Biochim Biophys Acta. 2007;1775(1):138–62.

19. Vandiver AR, Idrizi A, Rizzardi L, Feinberg AP, Hansen KD. DNA methylation is stable during replication and cell cycle arrest. Sci Rep. 2015;5:17911.

20. Guo F, Li X, Liang D, Li T, Zhu P, Guo H, et al. Active and passive demethylation of male and female pronuclear DNA in the mammalian zygote. Cell Stem Cell. 2014;15(4):447–58.

21. Tahiliani M, Koh KP, Shen Y, Pastor WA, Bandukwala H, Brudno Y, et al. Conversion of 5-methylcytosine to 5-hydroxymethylcytosine in mammalian DNA by MLL partner TET1. Science. 2009;324(5929):930–5.

22. Wu MZ, Chen SF, Nieh S, Benner C, Ger LP, Jan CI, et al. Hypoxia drives breast tumor malignancy through a TET-TNFα-p38-MAPK signaling axis. Cancer Res. 2015;75(18):3912–24.

23. Yang H, Liu Y, Bai F, Zhang J-Y, Ma S-H, Liu J, et al. Tumor development is associated with decrease of TET gene expression and 5-methylcytosine hydroxylation. Oncogene. 2013;32(5):663–9.

24. Kudo Y, Tateishi K, Yamamoto K, Yamamoto S, Asaoka Y, Ijichi H, et al. Loss of 5-hydroxymethylcytosine is accompanied with malignant cellular transformation. Cancer Sci. 2012;103(4):670–6.

25. Hassler MR, Egger G. Epigenomics of cancer – emerging new concepts. Biochimie. 2012;94(11):2219–30.

26. Yoo CB, Jones PA. Epigenetic therapy of cancer: past, present and future. Nat Rev Drug Discov. 2006;5(1):37–50.

27. Friedman RC, Farh KKH, Burge CB, Bartel DP. Most mammalian mRNAs are conserved targets of microRNAs. Genome Res. 2009;19(1):92–105.

28. Qin W, Zhang K, Clarke K, Weiland T, Sauter ER. Methylation and miRNA effects of resveratrol on mammary tumors vs. normal tissue. Nutr Cancer. 2014;66(2):270–7.

29. Suh SO, Chen Y, Zaman MS, Hirata H, Yamamura S, Shahryari V, et al. MicroRNA-145 is regulated by DNA methylation and p53 gene mutation in prostate cancer. Carcinogenesis. 2011;32(5):772–8.

30. Philibert RA, Gunter TD, Beach SRH, Brody GH, Madan A. Rapid publication: MAOA methylation is associated with nicotine and alcohol dependence in women. Am J Med Genet Part B Neuropsychiatr Genet. 2008;147(5):565–70.

31. Hartmann O, Spyratos F, Harbeck N, Dietrich D, Fassbender A, Schmitt M, et al. DNA methylation markers predict outcome in node-positive, estrogen receptor-positive breast cancer with adjuvant anthracycline-based chemotherapy. Clin Cancer Res. 2009; 15(1):315–23.

32. Guzmán L, Depix M, Salinas A, Roldán R, Aguayo F, Silva A, et al. Analysis of aberrant methylation on promoter sequences of tumor suppressor genes and total DNA in sputum samples: a promising tool for early detection of COPD and lung cancer in smokers. Diagn Pathol. 2012;7:87.

33. Hitchins MP, Rapkins RW, Kwok CT, Srivastava S, Wong JJL, Khachigian LM, et al. Dominantly inherited constitutional epigenetic silencing of MLH1 in a cancer-affected family is linked to a single nucleotide variant within the 5′UTR. Cancer Cell. 2011;20(2):200–13.

34. Duesberg P, Li R, Fabarius A, Hehlmann R. The chromosomal basis of cancer. Cell Oncol. 2005;27(5–6):293–318.

35. Eden A, Gaudet F, Waghmare A, Jaenisch R. Chromosomal instability and tumors promoted by DNA hypomethylation. Science. 2003;300(April):2003.
36. Nervi C, De Marinis E, Codacci-Pisanelli G. Epigenetic treatment of solid tumours: a review of clinical trials. Clin Epigenetics. 2015;7(1):127.
37. Ikehata M, Ogawa M, Yamada Y, Tanaka S, Ueda K, Iwakawa S. Different effects of epigenetic modifiers on the cytotoxicity induced by 5-fluorouracil, irinotecan or oxaliplatin in colon cancer cells. Biol Pharm Bull. 2014;37(1):67–73.
38. Das DS, Ray A, Das A, Song Y, Tian Z, Oronsky B, et al. A novel hypoxia-selective epigenetic agent RRx-001 triggers apoptosis and overcomes drug resistance in multiple myeloma cells. Leukemia. 2016;30(11):2187–97.
39. Khan ANH, Gregorie CJ, Tomasi TB. Histone deacetylase inhibitors induce TAP, LMP, Tapasin genes and MHC class I antigen presentation by melanoma cells. Cancer Immunol Immunother. 2008;57(5):647–54.
40. Marcu LG, Harriss-Phillips WM. In silico modelling of treatment-induced tumour cell kill: developments and advances. Comput Math Methods Med. 2012;2012(i):1–16.
41. Mardinoglu A, Gatto F, Nielsen J. Genome-scale modeling of human metabolism – a systems biology approach. Biotechnol J. 2013;8(9):985–96.
42. Knauer DJ, Wiley HS, Cunningham DD. Relationship between epidermal growth factor receptor occupancy and mitogenic response. Quantitative analysis using a steady state model system. J Biol Chem. 1984;259(9):5623–31.
43. Starbuck C, Lauffenburger DA. Mathematical model for the effects of epidermal growth factor receptor trafficking dynamics on fibroblast proliferation responses. Biotechnol Prog. 1992;8(2):132–43.
44. Fallon EM, Lauffenburger DA. Computational model for effects of ligand/receptor binding properties on interleukin-2 trafficking dynamics and T cell proliferation response. Biotechnol Prog. 2000;16(5):905–16.
45. Mardinoglu A, Agren R, Kampf C, Asplund A, Nookaew I, Jacobson P, et al. Integration of clinical data with a genome-scale metabolic model of the human adipocyte. Mol Syst Biol. 2013;9:649.
46. Mardinoglu A, Agren R, Kampf C, Asplund A, Uhlen M, Nielsen J. Genome-scale metabolic modelling of hepatocytes reveals serine deficiency in patients with non-alcoholic fatty liver disease. Nat Commun. 2014;14:5.
47. Thiele I, Swainston N, Fleming RMT, Hoppe A, Sahoo S, Aurich MK, et al. A community-driven global reconstruction of human metabolism. Nat Biotechnol. 2013;31(5):419–25.
48. Folger O, Jerby L, Frezza C, Gottlieb E, Ruppin E, Shlomi T. Predicting selective drug targets in cancer through metabolic networks. Mol Syst Biol. 2011;7(1):517.
49. Joyce AR, Palsson BØ. Predicting gene essentiality using genome-scale in silico models. Methods Mol Biol. 2008;416:433–57.
50. Zielinski DC, Jamshidi N, Corbett AJ, Bordbar A, Thomas A, Palsson BO. Systems biology analysis of drivers underlying hallmarks of cancer cell metabolism. Sci Rep. 2017;7:41241.
51. Gatto F, Miess H, Schulze A, Nielsen J. Flux balance analysis predicts essential genes in clear cell renal cell carcinoma metabolism. Sci Rep. 2015;5(1):10738.
52. Szappanos B, Kovács K, Szamecz B, Honti F, Costanzo M, Baryshnikova A, et al. An integrated approach to characterize genetic interaction networks in yeast metabolism. Nat Genet. 2011;43(7):656–62.
53. Hyduke DR, Lewis NE, Palsson BØ. Analysis of omics data with genome-scale models of metabolism. Mol BioSyst. 2013;9(2):167–74.
54. Karr JR, Sanghvi JC, Macklin DN, Gutschow MV, Jacobs JM, Bolival B, et al. A whole-cell computational model predicts phenotype from genotype. Cell. 2012;150(2):389–401.
55. Simeonidis E, Price ND. Genome-scale modeling for metabolic engineering. J Ind Microbiol Biotechnol. 2015;42(3):327–38.
56. Fumiã HF, Martins ML. Boolean network model for cancer pathways: predicting carcinogenesis and targeted therapy outcomes. PLoS One. 2013;8(7):e69008.

57. Albert I, Thakar J, Li S, Zhang R, Albert R. Boolean network simulations for life scientists. Source Code Biol Med. 2008;3:16.
58. Chapman MP, Tomlin CJ. Member I. Ordinary differential equations in cancer biology. bioRxiv. 2016;1:2–4.
59. Turner TE, Schnell S, Burrage K. Stochastic approaches for modelling in vivo reactions. Comput Biol Chem. 2004;28(3):165–78.
60. Anderson ARA, Quaranta V. Integrative mathematical oncology. Nat Rev Cancer. 2008;8(3):227–34.
61. Alarcón T, Byrne HM, Maini PK. A multiple scale model for tumor growth. Multiscale Model Simul. 2005;3(2):440–75.
62. Singhania R, Sramkoski RM, Jacobberger JW, Tyson JJ. A hybrid model of mammalian cell cycle regulation. PLoS Comput Biol. 2011;7(2):e1001077.
63. Agren R, Mardinoglu A, Asplund A, Kampf C, Uhlen M, Nielsen J. Identification of anticancer drugs for hepatocellular carcinoma through personalized genome-scale metabolic modeling. Mol Syst Biol. 2014;10(3):1–13.
64. Agren R, Bordel S, Mardinoglu A, Pornputtapong N, Nookaew I, Nielsen J. Reconstruction of genome-scale active metabolic networks for 69 human cell types and 16 cancer types using INIT. PLoS Comput Biol. 2012;8(5):e1002518.
65. Uhlén M, Fagerberg L, Hallström BM, Lindskog C, Oksvold P, Mardinoglu A, et al. Proteomics. Tissue-based map of the human proteome. Science. 2015;347(6220):1260419.
66. Swainston N, Smallbone K, Hefzi H, Dobson PD, Brewer J, Hanscho M, et al. Recon 2.2: from reconstruction to model of human metabolism. Metabolomics. 2016;12(7):109.
67. Garg D, Henrich S, Salo-Ahen OMH, Myllykallio H, Costi MP, Wade RC. Novel approaches for targeting thymidylate synthase to overcome the resistance and toxicity of anticancer drugs. J Med Chem. 2010;53(18):6539–49.
68. Hebar A, Valent P, Selzer E. The impact of molecular targets in cancer drug development: major hurdles and future strategies. Expert Rev Clin Pharmacol. 2013;6(1):23–34.
69. Ghaffari P, Mardinoglu A, Asplund A, Shoaie S, Kampf C, Uhlen M, et al. Identifying anti-growth factors for human cancer cell lines through genome-scale metabolic modeling. Sci Rep. 2015;5(1):8183.
70. Ramachandran N, Hainsworth E, Bhullar B, Eisenstein S, Rosen B, Lau AY, et al. Self-assembling protein microarrays. Science. 2004;305(5680):86–90.
71. Yazaki J, Galli M, Kim AY, Nito K, Aleman F, Chang KN, et al. Mapping transcription factor interactome networks using HaloTag protein arrays. Proc Natl Acad Sci U S A. 2016;113(29):E4238–47.
72. Choudhary C, Mann M. Decoding signalling networks by mass spectrometry-based proteomics. Nat Rev Mol Cell Biol. 2010;11:427.
73. Wilhelm M, Schlegl J, Hahne H, Moghaddas Gholami A, Lieberenz M, Savitski MM, et al. Mass-spectrometry-based draft of the human proteome. Nature. 2014;509(7502):582–7.
74. Chakrabarti CG, De K. Boltzmann entropy: generalization and applications. J Biol Phys. 1997;23(3):163–70.
75. Schneider TD. A brief review of molecular information theory. Nano Commun Netw. 2010;1(3):173–80.
76. Banerji CRS, Miranda-Saavedra D, Severini S, Widschwendter M, Enver T, Zhou JX, et al. Cellular network entropy as the energy potential in Waddington's differentiation landscape. Sci Rep. 2013;3(1):3039.
77. Carels N, Tilli T, Tuszynski JA. A computational strategy to select optimized protein targets for drug development toward the control of cancer diseases. PLoS One. 2015;10(1):e0115054.
78. Parise CA, Caggiano V. Breast cancer survival defined by the ER/PR/HER2 subtypes and a surrogate classification according to tumor grade and immunohistochemical biomarkers. J Cancer Epidemiol. 2014;2014:1–11.

79. Carels N, Tilli TM, Tuszynski JA. Optimization of combination chemotherapy based on the calculation of network entropy for protein-protein interactions in breast cancer cell lines. EPJ Nonlinear Biomed Phys. 2015;3(1):6.
80. Álvarez-Silva MC, Yepes S, Torres MM, González Barrios AF. Proteins interaction network and modeling of IGVH mutational status in chronic lymphocytic leukemia. Theor Biol Med Model. 2015;12(1):12.
81. Barabasi AL, Oltvai ZN. Network biology: understanding the cell's functional organization [review]. Nat Rev Genet. 2004;5(2):101–NIL.
82. Albert R, Jeong H, Barabási A-L. Error and attack tolerance of complex networks. Nature. 2000;406(6794):378–82.
83. Tilli TM, Carels N, Tuszynski JA, Pasdar M. Validation of a network-based strategy for the optimization of combinatorial target selection in breast cancer therapy: siRNA knockdown of network targets in MDA-MB-231 cells as an in vitro model for inhibition of tumor development. Oncotarget. 2016;7(39):63189–203.
84. Watts JK, Corey DR. Silencing disease genes in the laboratory and the clinic. J Pathol. 2012;226(2):365–79.
85. Alegre MM, Robison RA, O'Neill KL. Thymidine kinase 1 upregulation is an early event in breast tumor formation. J Oncol. 2012;2012:1–5.
86. Chen Y-L, Eriksson S, Chang Z-F. Regulation and functional contribution of thymidine kinase 1 in repair of DNA damage. J Biol Chem. 2010;285(35):27327–35.
87. Di Cresce C, Figueredo R, Ferguson PJ, Vincent MD, Koropatnick J. Combining small interfering RNAs targeting thymidylate synthase and thymidine kinase 1 or 2 sensitizes human tumor cells to 5-fluorodeoxyuridine and pemetrexed. J Pharmacol Exp Ther. 2011;338(3):952–63.
88. Cheng Q, Chang JT, Geradts J, Neckers LM, Haystead T, Spector NL, et al. Amplification and high-level expression of heat shock protein 90 marks aggressive phenotypes of human epidermal growth factor receptor 2 negative breast cancer. Breast Cancer Res. 2012;14(2):R62.
89. Korsching E, Packeisen J, Liedtke C, Hungermann D, Wülfing P, van Diest PJ, et al. The origin of vimentin expression in invasive breast cancer: epithelial-mesenchymal transition, myoepithelial histogenesis or histogenesis from progenitor cells with bilinear differentiation potential? J Pathol. 2005;206(4):451–7.
90. Liu C-Y, Lin H-H, Tang M-J, Wang Y-K. Vimentin contributes to epithelial-mesenchymal transition cancer cell mechanics by mediating cytoskeletal organization and focal adhesion maturation. Oncotarget. 2015;6(18):15966–83.
91. Hodgkinson VC, Agarwal V, ELFadl D, Fox JN, McManus PL, Mahapatra TK, et al. Pilot and feasibility study: comparative proteomic analysis by 2-DE MALDI TOF/TOF MS reveals 14-3-3 proteins as putative biomarkers of response to neoadjuvant chemotherapy in ER-positive breast cancer. J Proteome. 2012;75(9):2745–52.
92. Kim Y, Kim H, Jang S-W, Ko J. The role of 14-3-3β in transcriptional activation of estrogen receptor α and its involvement in proliferation of breast cancer cells. Biochem Biophys Res Commun. 2011;414(1):199–204.
93. Akekawatchai C, Roytrakul S, Kittisenachai S, Isarankura-Na-Ayudhya P, Jitrapakdee S. Protein profiles associated with anoikis resistance of metastatic MDA-MB-231 breast cancer cells. Asian Pac J Cancer Prev. 2016;17(2):581–90.
94. Wilker E, Yaffe MB. 14-3-3 proteins – a focus on cancer and human disease. J Mol Cell Cardiol. 2004;37(3):633–42.
95. Ortega CE, Seidner Y, Dominguez I. Mining CK2 in cancer. Calogero RA, editor. PLoS One. 2014;9(12):e115609.
96. Filhol O, Giacosa S, Wallez Y, Cochet C. Protein kinase CK2 in breast cancer: the CK2β regulatory subunit takes center stage in epithelial plasticity. Cell Mol Life Sci. 2015;72(17):3305–22.

97. Deshiere A, Duchemin-Pelletier E, Spreux E, Ciais D, Forcet C, Cochet C, et al. Regulation of epithelial to mesenchymal transition: CK2β on stage. Mol Cell Biochem. 2011;356(1–2): 11–20.
98. Golden D, Cantley LG. Casein kinase 2 prevents mesenchymal transformation by maintaining Foxc2 in the cytoplasm. Oncogene. 2015;34(36):4702–12.
99. Phan L, Chou P-C, Velazquez-Torres G, Samudio I, Parreno K, Huang Y, et al. The cell cycle regulator 14-3-3σ opposes and reverses cancer metabolic reprogramming. Nat Commun. 2015;6:7530.
100. Boudreau A, Tanner K, Wang D, Geyer FC, Reis-Filho JS, Bissell MJ. 14-3-3σ stabilizes a complex of soluble actin and intermediate filament to enable breast tumor invasion. Proc Natl Acad Sci U S A. 2013;110(41):E3937–44.
101. Kren BT, Unger GM, Abedin MJ, Vogel RI, Henzler CM, Ahmed K, et al. Preclinical evaluation of cyclin dependent kinase 11 and casein kinase 2 survival kinases as RNA interference targets for triple negative breast cancer therapy. Breast Cancer Res. 2015;17:19.
102. Miwa D, Sakaue T, Inoue H, Takemori N, Kurokawa M, Fukuda S, et al. Protein kinase D2 and heat shock protein 90 beta are required for BCL6-associated zinc finger protein mRNA stabilization induced by vascular endothelial growth factor-A. Angiogenesis. 2013;16(3):675–88.
103. Pallares J, Llobet D, Santacana M, Eritja N, Velasco A, Cuevas D, et al. CK2β is expressed in endometrial carcinoma and has a role in apoptosis resistance and cell proliferation. Am J Pathol. 2009;174(1):287–96.
104. Kitano H. Biological robustness. Nat Rev Genet. 2004;5(11):826–37.
105. Kitano H. Towards a theory of biological robustness. Mol Syst Biol. 2007;18:3.
106. Atkinson DM, Clarke MJ, Mladek AC, Carlson BL, Trump DP, Jacobson MS, et al. Using fluorodeoxythymidine to monitor anti-EGFR inhibitor therapy in squamous cell carcinoma xenografts. Head Neck. 2008;30(6):790–9.
107. Didelot C, Lanneau D, Brunet M, Bouchot A, Cartier J, Jacquel A, et al. Interaction of heat-shock protein 90β isoform (HSP90β) with cellular inhibitor of apoptosis 1 (c-IAP1) is required for cell differentiation. Cell Death Differ. 2008;15(5):859–66.
108. Lahat G, Zhu Q-S, Huang K-L, Wang S, Bolshakov S, Liu J, et al. Vimentin is a novel anti-cancer therapeutic target; insights from in vitro and in vivo mice xenograft studies. Bauer JA, editor. PLoS One. 2010;5(4):e10105.
109. Cao W, Yang X, Zhou J, Teng Z, Cao L, Zhang X, et al. Targeting 14-3-3 protein, difopein induces apoptosis of human glioma cells and suppresses tumor growth in mice. Apoptosis. 2010;15(2):230–41.
110. Dong S, Kang S, Lonial S, Khoury HJ, Viallet J, Chen J. Targeting 14-3-3 sensitizes native and mutant BCR-ABL to inhibition with U0126, rapamycin and Bcl-2 inhibitor GX15-070. Leukemia. 2008;22(3):572–7.
111. Thompson JM, Nguyen QH, Singh M, Razarenova OV. Approaches to identifying synthetic lethal interactions in cancer. Yale J Biol Med. 2015;88(2):145–55.
112. Stegh AH. Toward personalized cancer nanomedicine – past, present, and future. Integr Biol. 2013 [cited 2016 Jan 11];5(1):48–65.
113. Reischl D, Zimmer A. Drug delivery of siRNA therapeutics: potentials and limits of nanosystems. Nanomed Nanotechnol Biol Med. 2009;5(1):8–20.

Computational Modeling
of Multidrug-Resistant Bacteria

Fabricio Alves Barbosa da Silva, Fernando Medeiros Filho,
Thiago Castanheira Merigueti, Thiago Giannini, Rafaela Brum,
Laura Machado de Faria, Ana Paula Barbosa do Nascimento,
Kele Teixeira Belloze, Floriano Paes Silva Jr., Rodolpho Mattos Albano,
Marcelo Trindade dos Santos, Maria Clicia Stelling de Castro,
Marcio Argollo de Menezes, and Ana Paula D'A. Carvalho-Assef

Abstract Understanding how complex phenotypes arise from individual molecules and their interactions is a primary challenge in biology, and computational approaches have been increasingly employed to tackle this task. In this chapter, we describe current efforts by FIOCRUZ and partners to develop integrated computational models of multidrug-resistant bacteria. The bacterium chosen as the main focus of this effort is *Pseudomonas aeruginosa*, an opportunistic pathogen associated with a broad spectrum of infections in humans. Nowadays, *P. aeruginosa* is one of the main problems of healthcare-associated infections (HAI) in the world, because of its great capacity of survival in hospital environments and its intrinsic resistance to many antibiotics. Our overall research objective is to use integrated computational models to accurately predict a wide range of observable cellular

F. A. B. da Silva (✉) · F. M. Filho · T. C. Merigueti · T. Giannini · L. M. de Faria ·
A. P. B. do Nascimento · F. P. Silva Jr. · A. P. D'A. Carvalho-Assef
Fundação Oswaldo Cruz – FIOCRUZ, Rio de Janeiro, Brazil
e-mail: fabricio.silva@fiocruz.br; floriano@ioc.fiocruz.br; anapdca@ioc.fiocruz.br

M. Trindade dos Santos
Laboratório Nacional de Computação Científica – LNCC/MCTI, Petrópolis, Brazil
e-mail: msantos@lncc.br

K. Belloze
Centro Federal de Educação Tecnológica Celso Suckow da Fonseca – CEFET/RJ, Rio de Janeiro, Brazil
e-mail: kele.belloze@cefet-rj.br

R. Brum · R. M. Albano · M. C. S. de Castro
Universidade do Estado do Rio de Janeiro – UERJ, Rio de Janeiro, Brazil
e-mail: albano@uerj.br; clicia@ime.uerj.br

M. A. de Menezes
Instituto de Física, Universidade Federal Fluminense, Niterói, Brazil

Instituto Nacional de Ciência e Tecnologia de Sistemas Complexos, INCT-SC, Rio de Janeiro, Brazil
e-mail: marcio@mail.if.uff.br

© Springer International Publishing AG, part of Springer Nature 2018   195
F. A. B. da Silva et al. (eds.), *Theoretical and Applied Aspects of Systems Biology*,
Computational Biology 27, https://doi.org/10.1007/978-3-319-74974-7_11

behaviors of multidrug-resistant *P. aeruginosa* CCBH4851, which is a strain belonging to the clone ST277, endemic in Brazil. In this chapter, after a brief introduction to *P. aeruginosa* biology, we discuss the construction of metabolic and gene regulatory networks of *P. aeruginosa* CCBH 4851 from its genome. We also illustrate how these networks can be integrated into a single model, and we discuss methods for identifying potential therapeutic targets through integrated models.

1 Introduction

Healthcare-associated infections (HAI) are a serious public health problem. Among the pathogens related to HAI, the group of bacteria is the one that stands out. More than 2 million of HAIs occur each year in the USA, with 50–60% being caused by antimicrobial-resistant bacteria. According to the Health Services Quality and Patient Safety Bulletin published in 2016 by Brazilian Health Surveillance Agency (ANVISA), 33,481 cases of primary bloodstream infections in adult patients hospitalized at Brazilian intensive care units were reported in 2015, with high resistance rates among the isolated microorganisms [1].

The rapid evolution and spread of bacterial resistance, as well as the slow development of new drugs, has dramatically affected the treatment of infections. In 2014, the World Health Organization (WHO) published the report *Antimicrobial Resistance: Global Report on Surveillance*, warning of the growing increase in antimicrobial resistance in the world. Antimicrobial resistance among hospital pathogens has increased at alarming levels, both in developed and developing countries. It is estimated that there will be a worldwide spread of untreatable infections both inside and outside hospitals. Based on the report titled "Tackling Drug-Resistant Infections Globally," presented in 2016 by Jim O'Neill, and the "Global Action Plan on Antimicrobial Resistance" published in 2017 by the WHO, it is estimated that, by 2050, about 10 million people per year and a cumulative $ 100 trillion in economic output are at risk due to increased antimicrobial resistance if proactive solutions are not taken seriously. According to the bulletin published in 2017 by WHO, there are 12 major antibiotic-resistant bacteria that deserve attention and urgently need more research and development (R&D) of new and effective antibiotic treatments, with the Gram-negative bacteria most involved in HAI (carbapenem-resistant *Acinetobacter baumannii*, *Pseudomonas aeruginosa*, and *Enterobacteriaceae*) being those at the top of the list considered to be of critical priority [2].

The computational modeling of biological systems can be understood as a way of constructing models to represent existing or hypothetical biological systems in order to, for example, explain the integrated function of each gene in a cell and thus to describe the behavior of the organism in several environments. Covert and colleagues [3] state that the modeling of biological processes has increased in scope and complexity and elapses a series of challenges, parameters, and functions that can be considered and optimized. An accurately modeled biological process supports the construction of theories and hypotheses of cellular behavior and the prediction of observable variables of the biological system.

In this manuscript, we discuss how to build integrated computational models to accurately predict a wide range of observable cellular behaviors of a *P. aeruginosa* strain (CCBH4851) belonging a clone endemic in Brazil (ST277) [4]. In an integrated model, several cellular processes are modeled using the most appropriate mathematical representation and then simulated concurrently in a common computational architecture. As examples of integrated models described in the literature, we have the modeling of the metabolic, gene regulatory, and signaling networks of *Escherichia coli* [5] and the whole-cell model of *Mycoplasma genitalium* [6]. Our analysis includes those behaviors related to multidrug resistance, susceptibility to new drugs, and identification of new therapeutic targets.

In this chapter, we first present a brief introduction to *P. aeruginosa* biology in section "*Pseudomonas aeruginosa*." In section "Multidrug Resistance in *P. aeruginosa*," we discuss aspects related to multidrug resistance in *P. aeruginosa*. Section "Network Models of Bacteria" presents networks models of *P. aeruginosa* and methods for generating those models. In section "Network Models of Bacteria," we emphasize metabolic and gene regulatory network models. Particularly, we discuss the construction of metabolic and gene regulatory networks of *P. aeruginosa* CCBH 4851 from its genome. Integrated models of bacteria are discussed in section "Integrated Models of Bacteria." Since there are no integrated models of *P. aeruginosa* described in the literature, we focus on models of *E. coli* and *M. genitalium*. Computational aspects of integrated models design and execution are discussed in section "Computational Aspects of Integrated Modeling." In section "Searching for Therapeutic Targets," we discuss the search of potential therapeutic targets through the analysis of integrated models. Section "Conclusion" contains our concluding remarks.

2   *Pseudomonas aeruginosa*

P. aeruginosa is a non-fermenting Gram-negative rod, with great nutritional versatility, able to use various organic compounds and to grow in culture media containing only acetate as a source of carbon and ammonium sulfate as a source of nitrogen. It is an aerobic microorganism, but it can grow under anaerobic conditions using nitrate or arginine as the final electron acceptor. It exhibits motility due to a single polar flagellum and produces water-soluble pigments such as pyoverdine (fluorescent pigment) and pyocyanin (blue phenazine pigment). The combination of these two pigments is responsible for the bright green color characteristic of *P. aeruginosa* colonies. It is a ubiquitous bacterium with a predilection for moist environments, being found in soil, water, and plants. In the hospital environment, it can be found in a variety of solutions (disinfectants, dialysis fluids, and eye drops) and equipment such as mechanical ventilation and dialysis [7].

It is an opportunistic pathogen associated with a broad spectrum of human infections, ranging from superficial infections to fulminant sepsis. In immuno-compromised patients, *P. aeruginosa* causes a significant level of morbidity and mortality. The severity of *P. aeruginosa* infections is due to the predisposing factors of the host but also to the great variety of virulence factors and the marked resistance to the majority of antimicrobials used in clinical use [8].

P. aeruginosa is able to live both in planktonic (free-floating) state and in biofilm, which is a three-dimensional multicellular community, where bacterial cells are involved in a matrix composed of polysaccharides, proteins, and nucleic acids. Binding of a single bacterial cell to the surface of the extracellular matrix promotes bacterial proliferation, leading to the formation of microcolonies that are grouped into several layers forming a bacterial "agglomerate" The ability to form biofilm confers greater resistance to biocides and antimicrobials compared to nonproducing ones, added to the fact that the biofilm is able to inhibit phagocytosis by cells of the immune system, thus conferring an advantage in bacterial protection [9]. Besides that, *P. aeruginosa* produce an arsenal of virulence attributes, including cell-associated determinants (e.g., lipopolysaccharide, adhesins, and flagellum) and soluble secreted factors (e.g., extracellular polysaccharides, exotoxins, proteases, and pyocyanin) which, after colonization, facilitate the rupture of epithelial integrity, invasion, and dissemination. However, the contribution of each of these factors varies with the type of infection [10].

The secretion of these virulence factors occurs through different export systems. Of particular importance is a type III secretion system which allows the bacteria to inject cytotoxins (ExoS, ExoU, ExoY, and ExoT) into the cytoplasm of host cells, evading the immune response of the host, preventing phagocytosis by modulating the actin cytoskeletal dynamics in host cells, and inhibiting the synthesis of DNA from mammals, leading to cell death [11]. *P. aeruginosa* is able to modulate its gene expression in response to environmental conditions, controlling the secretion of virulence factors. The regulation of the secretion of several virulence factors and the production of biofilm occur via the quorum sensing (QS) bacterial signaling system [10].

In recent years, *P. aeruginosa* served as a paradigm for the study of gene expression, metabolism, and pathogenesis [13]. The genome of *P. aeruginosa* (5–7 Mb) is one of the largest among prokaryotes, with 0.3% of these genes encoding proteins involved in antibiotic resistance. One of the reasons for the large size of the *P. aeruginosa* genome is that, unlike classical human pathogens, which are obligate intracellular and dependent on host cells to provide many of their nutritional requirements, *P. aeruginosa* is a free-floating bacterium and an opportunistic pathogen that must be nutritionally independent. In this way, it has an abundance of regulators facilitating its adaptation to a wide variety of environments [14]. During its evolution, the competition with other prokaryotes and the acquisition of defense mechanisms have allowed the maintenance of antibiotic resistance markers, degradation enzymes, and secretion systems that have an impact on human infection [13]. The large size and complexity of its genome are probably the bases for its ability to survive in diverse environments, to cause a variety of infections, and to resist a large number of antimicrobial agents [14].

3 Multidrug Resistance in *P. aeruginosa*

Nowadays, *P. aeruginosa* represents one of the main challenges of treatment and eradication in health institutions, both for its capacity to survive in the hospital environment and for resistance to treatment. Antimicrobial therapy of *P. aeruginosa* infections is very difficult because it is intrinsically resistant to many antibiotics, limiting therapeutic options [15]. In addition, treatment is being increasingly compromised due to acquired or mutational resistance. It is known that during antimicrobial treatment, *P. aeruginosa* is able to develop mutations in certain genes as a form of adaptation and become resistant to treatment. Among the mechanisms of resistance mediated by mutations, the ones particularly notable are those that lead to the repression or inactivation of the OprD porin expression, the overproduction of the cephalosporinase AmpC, or the overexpression of one of the various efflux pumps encoded by the genome of *P. aeruginosa*. In addition, this bacterium may also acquire new mechanisms of resistance through the acquisition of resistance genes in mobile genetic elements, like genes of beta-lactamases. The accumulation of these resistance determinants drastically limits the therapeutic choices available for the treatment of *P. aeruginosa* infections [4, 16].

The therapeutic options for the treatment of *P. aeruginosa* infections include three major groups of antimicrobials based on their mechanisms of action: β-lactams (inhibitors of cell wall synthesis) such as piperacillin, ticarcillin, third- and fourth-generation cephalosporin (ceftazidime and cefepime), and carbapenems (imipenem and meropenem); aminoglycosides (interfere with protein synthesis) such as gentamicin, amikacin, and tobramycin; or fluoroquinolones (interfere with replication), like ciprofloxacin [13].

The acquired resistance to aminoglycosides is related to changes in the outer membrane permeability, overexpression of efflux systems (MexXY-OprM), activity of rRNA methylases (e.g., RmtD), and production of aminoglycoside-modifying enzymes through acetylation, phosphorylation, and/or adenylation. The most common resistance mechanism in *P. aeruginosa* is the combination of enzymatic mechanisms and changes in permeability [17].

The resistance to high concentrations of fluoroquinolones, in *P. aeruginosa*, is usually mediated by site-specific mutations in DNA gyrase genes (mainly *gyr*A) and topoisomerase IV (especially *par*C), although the expression of efflux systems also has considerable importance, often occurring together [17].

Acquired resistance to beta-lactams is usually mediated by overexpression of chromosomal cephalosporins, which results in moderate resistance to all β-lactams, including those resistant to β-lactamases (except for carbapenems), but the acquisition of extended-spectrum beta-lactamases (ESBL) has also been described in *P. aeruginosa* [17].

In recent decades, there has been an increase of hospital outbreaks caused by multidrug-resistant (MDR) *P. aeruginosa* strains. The antimicrobials agents most commonly used for the treatment of *P. aeruginosa* MDR infections are carbapenems. However, high rates of resistance to these drugs have been observed. *P. aeruginosa* becomes resistant to carbapenems by different mechanisms, which may occur concomitantly or separately. The main mechanism of resistance is the decrease in the permeability of the outer membrane due to the loss of the OprD porin, which occurs in 50% of the infections treated for more than 7 days with imipenem. However, the production of carbapenemases (beta-lactamases with activity on carbapenems) is considered of major epidemiological importance, because they generally cause resistance to all beta-lactams and are disseminated through mobile genetic elements such as plasmids and transposons. Carbapenemases are divided into three classes: class A subgroup 2f, class B subgroup 3a, and class D subgroup 2df. Among these, the most important and most widespread in *P. aeruginosa* isolates are class B carbapenemases also called metallo-beta-lactamases (MBL) [18].

São Paulo metallo-beta-lactamase (SPM-1) is the main carbapenemase identified in *P. aeruginosa* in Brazil [19, 20]. SPM-1 confers resistance to virtually all antimicrobials of the beta-lactam class, except to aztreonam [21]. The first description of SPM-1 was in 2001 from a pediatric patient in São Paulo, Brazil [87]. Since then, SPM-1-producing *P. aeruginosa* has been associated with hospital outbreaks in several Brazilian states [21]. Most of SPM-1-producing *P. aeruginosa* isolates belong to sequence type ST277 through the MLST (multilocus sequence typing) [22]. In addition to SPM-1, isolates belonging to clone ST277 present resistance mechanisms to other important classes of antimicrobials, such as aminoglycosides and fluoroquinolones, making them multidrug-resistant. The presence of SPM-1 seems to be restricted to *P. aeruginosa* isolates from Brazil. To date, only two cases were reported outside Brazil, one in Switzerland and another in the UK, and both belonged to ST277 [23, 24].

The clone ST277 presents several genomic islands with remarkable genetic mobile elements inside and a type I-C CRISPR-Cas system, which constitute a bacterial adaptive genetic immune system controlling horizontal transfer of genetic elements such as phages and plasmids. Thus, this CRISPR-Cas system could be in part responsible for the genomic stability of this MDR clone [19].

For this study, we selected a representative MDR *P. aeruginosa* strain belonging to the clone ST277, endemic in Brazil (CCBH4851 strain). This strain was isolated in 2008 from the catheter of a patient hospitalized in a hospital in Goiás (Midwestern Brazil). It is resistant to all antimicrobials of clinical importance with the exception of polymyxin B and has several mechanisms of resistance and mobile genetic elements. This strain is deposited in the Culture Collection of Hospital-Acquired Bacteria (CCBH) of the Oswaldo Cruz Institute (WDCM947; CGEN022/2010), and its genome has already been sequenced and deposited in GenBank (accession number JPSS00000000) [4, 16].

4   Network Models of Bacteria

In this section, we illustrate the building of genome-scale network models of bacteria, with focus on metabolic and gene regulatory network models of *P. aeruginosa*. For the reconstruction of genome-scale models, access to a high-quality, annotated genome of the strain under consideration is mandatory. Therefore, the first step is to have an annotated version of the genome. We discuss genome annotation in the following subsection. Then we describe methods for the reconstruction of metabolic and gene regulatory networks of *P. aeruginosa* and discuss network models available in the literature.

4.1   Genome Annotation

The construction of network models relies on a precise description of the chemical reactions catalyzed by gene products of the organism being studied. This involves various reactions or interactions occurring in an organism. To assess the presence of these processes, we take as basis the genome annotation. For instance, if a gene responsible for the coding of an enzyme driving a reaction or a protein binding a regulatory DNA region is annotated, we assume that the reaction or interaction is present [25, 26].

Over the past years, next-generation sequencing technologies are becoming increasingly affordable and common. The resulting data are the input for genome assembly and annotation processes and should represent the sequenced organism as accurately as possible [27, 28].

The annotation follows genome assembly and is based on several protocols in order to predict genes and their functions. Different approaches are used to achieve this objective and are summarized in three main steps: ORF prediction, RNA prediction, and gene annotation. These approaches are described in the following paragraphs [27, 28].

4.1.1   ORF Prediction

The open reading frame (ORF) in prokaryotic organisms is a region delimited by a start and a stop codon. Prediction methods can be empiric or ab initio. The empiric method is based on similarity searches on databases to identify gene sequences. It is more reliable but requires the previous annotation of a phylogenetically related organism. The ab initio prediction is based in mathematical models to identify genes. These models take into account several traits to avoid false-positive and false-negative results. In addition to simply identify an ORF, algorithms are constructed to assess the presence of certain traits, e.g., the presence of upstream RNA polymerase binding site or the presence of ribosome binding site. The prediction programs can be trained using a reference set of sequences to generate a model, and then this model is used to predict ORFs. They can also be self-trained based on canonical traits frequently found in the genome. The algorithms used in these prediction programs are often based on hidden Markov models, Fourier transform, or neural networks [27, 29, 30]. The most common programs used to ORF prediction in prokaryotes are GLIMMER, GeneMarkS, and FGENESB [29, 31, 32]. An exclusive version of GeneMarkS was developed to compose the Prokaryotic Genome Annotation Pipeline (PGAP) offered by National Center for Biotechnology Information (NCBI) [33].

4.1.2 RNA Prediction

The identification of noncoding RNAs is required for a complete annotation process. Structural ribosomal RNAs in prokaryotes (23S, 16S, and 5S) are highly conserved in closely related species. The prediction of rRNA is based on sequence similarity. The RNA databases represent the families of RNAs by multiple alignments, secondary structure consensus, and covariance models; thus sequence comparison allows the identification of genes responsible for RNAs of interest [29, 33]. The program most commonly used for rRNA detection is RNAmmer. It is based on hidden Markov models trained on data of RNA databases [34].

The methods for prediction of transporter RNAs are mainly based on algorithms that assess the formation of secondary structure of a typical tRNA. They can also include other criteria like search for homology with RNA databases. The most common program used to tRNA prediction is tRNAscan-SE, but others like ARAGORN and tRNAfinder can also be useful [29, 35–37].

4.1.3 Gene Annotation

The previous steps produce a number of genes that require annotation. The gene annotation is basically the assignment of its function. The function is assigned by a comparative process with protein databases, e.g., NCBI and UniProt, or protein domain databases, e.g., Pfam, NCBI CDD, and InterPro [29, 35–37]. The comparison can generate four distinct groups of annotations as described below:

(a) A match with a direct orthologous will result in an annotation based on the name of the orthologous.
(b) A match with a conserved protein domain will result in the annotation of the corresponding domain.
(c) No match with a direct orthologous or protein domain will result in the annotation of a hypothetical protein.
(d) The match with a direct orthologous annotated as a hypothetical protein will result in the annotation of a conserved hypothetical protein.

Additional features can be annotated, e.g., pseudogenes, control regions, direct and inverted repeats, insertion sequences, transposons, and other mobile elements [33]. The minimum requirements for submission of complete genomes to NCBI database are:

(a) Structural RNAs (23S, 16S, and 5S) – at least one copy of each with appropriate length.
(b) tRNA – at least one copy for each amino acid.
(c) The number of protein-coding genes/genome length ratio is close to 1.
(d) No genes completely contained in another gene on the same or opposite strand.
(e) No partial features.

Due to the complexity and the large number of genomes, automatic annotation tools that are publicly available are largely employed. The automatic annotation mainly consists in a pipeline composed by available programs and/or original scripts performed in order to detect and annotate genes [28]. The NCBI PGAP is one of those tools and combines alignment-based methods with methods of predicting protein-coding and RNA genes and other functional elements directly from sequence [33]. Other tools like Rapid Annotation using Subsystem Technology (RAST) [38], Prokka, and MAKER2 [39] are also available. However, automatic pipelines can generate inaccurate genome annotation, and a manual curation is still very important to avoid submission of misinformation to public repositories [28].

The *P. aeruginosa* CCBH4851 genome was assembled and annotated following the procedures cited above. First, the sequencing was performed using both the Illumina and PacBio technologies. The resulting reads were used as input to the Maryland Super-Read Celera Assembler (MaSuRCA), which allows the combination of Illumina reads of distinct lengths with longer reads from other technologies to perform a de novo assembly [40]. A single chromosome was obtained with 6868867 bp size and 66.07 G+C content percent. The rapid annotation transfer tool (RATT) [41] was used to transfer annotation from *P. aeruginosa* PAO1 to CCBH4851 since these strains are very closely related. A difference of approximately 605 kb among these strains indicates the acquisition of new regions which genes should be annotated using the ab initio method. To identify genes that could not be transferred by RATT in consensus or new acquired regions, we used both GLIMMER [31] and GeneMarkS [32] gene prediction programs. Also, despite the RATT usage, rRNA and tRNA detection was performed to ensure their presence using RNAmmer [34] and tRNAscan-SE [35], respectively. After these steps, the remaining genes that still required annotation were compared to NCBI, UniProt, and InterPro databases to search for orthology and to have a function assigned. Additionally, KEGG and BRENDA databases were used to assign an enzyme commission (EC) number in order to facilitate the further metabolic network reconstruction. This curated annotation will be submitted to NCBI database under the accession number CP021380.2.

4.2 Metabolic Network Models

Genome-scale metabolic models (GEMs) are valuable sources of information that allow the prediction of physiological states from genomic data. Their construction demands a huge multidisciplinary effort [42], from genome assembly and annotation to the association of genes to enzymatic reactions and transport processes, ultimately complemented by gap-filling reactions. We are currently developing the metabolic model of CCBH4851, the strain of *P. aeruginosa* found in a Brazilian public hospital with data provided by LAPIH/IOC/FIOCRUZ (4). Next, we describe the protocol adopted and the current stage of development.

In an attempt to establish standard procedures for the generation of genome-scale metabolic networks, Thiele and Palsson [42] proposed in 2010 a protocol describing essential steps to the production of high-quality genome-scale metabolic reconstructions. The entire process is divided into four main phases:

(1) Creation of a draft model. Usually with algorithmic association of nucleotide/amino acid gene sequences from the target organism with known enzymatic proteins or EC numbers available from databases developed by diverse consortiums like UniProt [43], a collaborative effort generating freely accessible, manually curated data connecting high-quality protein sequences (556,196 to date) to functional information with links to biochemical reaction databases like KEGG [45] and Brenda [44].

(2) Manual curation. When reactions inferred in the previous step are validated by specialists, false-positives are eliminated and gaps on known reaction pathways are closed with the introduction of putative reactions. Organism-specific information, like biomass composition and growth/nongrowth-related ATP drains for the maintenance of cellular processes, like replication, are included in the reconstruction as pseudo-reactions (see more detailed information at [42]). This is a crucial step of the reconstruction, for which a consistent information platform is still lacking [46].

(3) Translation of the reconstruction into a mathematical/computational object.

(4) Assessment of the network. Basic capabilities of the organism, like non-zero synthesis of biomass precursors, are tested in the model. If necessary, gap-filling reactions are added to ensure non-zero biomass production.

To generate the metabolic network of CCBH4851, we recycle information from a high-quality reconstruction of a close relative, *P. aeruginosa* strain PAO1. The first metabolic model for PAO1, iMO1056 [47], accounted for 1056 genes encoding 1030 proteins catalyzing 883 reactions. More recent PAO1 reconstructions are distinct refinements of the original model: iMO1063 [48] is a reconstruction that includes reactions related to biofilm formation for the prediction of drug targets against *P. aeruginosa* biofilms by the investigation of system-wide differences on metabolic fluxes from planktonic and biofilm phenotypes. iPAE1146 [49], on the other hand, focuses on virulence mechanisms, adding reactions from virulence-linked pathways to assess the essentiality of virulence-linked targets and the contribution of virulence-related genes to metabolism.

To generate the metabolic network of *P. aeruginosa* CCBH4851, we rely on curated data from iMO1063. The more recent PAO1 reconstruction, iPAE1146, provides new name for all metabolites and reactions and no translation table. Furthermore, maximum biomass production rate is unrealistically high for *P. aeruginosa*, making this reconstruction less profitable for our purposes.

4.2.1 First Phase: Computational Draft Reconstruction

To benefit from the manual curation of reactions from iMO1063 and the GPRs defined for them, we seek which PAO1 genes included in the reconstruction have homologues in CCBH4851. We perform similarity analysis on all CCBH4851-PAO1 gene pairs with BLAST and assign homology to all reciprocal pairs with e-value smaller than 1e-4. We add to CCBH4851 reconstruction all reactions whose GPR relation is satisfied, given the subset of orthologue genes.

We also fetch from BRENDA database [44] the list of EC numbers associated to CCBH4851 genes. Reactions from iMO1063 linked to EC numbers from this list were also added to the reconstruction. The same applies to Transporter Classification (TC) numbers, defining transport processes. We downloaded the list of TC numbers and their associated genes from tcdb.org and, after performing similarity analysis with CCBH4851 genes, add to the reconstruction all iMO1063 transport reactions with matching TC numbers also associated to CCBH4851 genes.

Due to the high similarity between strains, only 12 reactions from iMO1063 are not included in CCBH4851 reconstruction. We also import the biomass production reaction inferred from PAO1 cellular composition.

4.2.2 Second Phase: Manual Curation of the Metabolic Network

In this phase, yet to be implemented, the goal is to correct errors and add information that could not be obtained in the first step. In order to promote the refinement of the network by specialists, a Web system is proposed. It is a *system of challenges* that brings with it a combination of computational techniques, called CurSystem (see Fig. 1). The CurSystem was developed to intermediate the work done by the specialists, so that they can propose and make possible adjustments in the metabolic network of *P. aeruginosa* CCBH4851 in an automated way. This step is very important because it will validate the data extracted automatically in the previous phase and correct any inconsistencies and gaps of the network. The proposal is to implement software that poses challenges for groups of specialists. These challenges will address issues related to network characteristics, and each approved resolution will be stored in the *P. aeruginosa* CCBH4851 knowledge base. The knowledge base is the reference database for model construction, analysis, and improvement. Therefore, the system has an export functionality in Systems Biology Markup

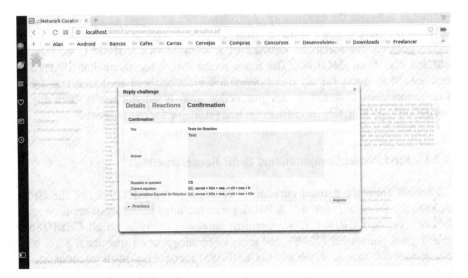

Fig. 1 CurSystem screenshot

Language (SBML) format, which is the standard format for representing models in systems biology.

With the execution and completion of these two steps, we expect to generate consistent information, allowing for the construction of a mathematical/computational model of *P. aeruginosa* CCBH4851 metabolic network.

4.3 Gene Regulatory Network Models

The methodology for the reconstruction of the gene regulatory network consists of four stages or steps. In the first step, we have the propagation, by orthology, of the already established knowledge about the regulatory interactions of evolutionarily close organisms. We will consider as references the networks of *P. aeruginosa* PAO1 and *Escherichia coli*. In the second step, we will collect data on the binding sites of the transcription factors identified in the previous step and the subsequent scanning of the transcription unit (TU) promoter regions in search of new site predictions. The third step consists of the manual refinement of the obtained network, based on the literature. Finally, we have the identification of the functional modules, which will be done through the direct comparison with the functional modules described in the *P. aeruginosa* PAO1 regulatory network and the analysis of the topological characteristics of the network. The steps outlined in this paragraph are described in more detail below. We illustrate the process by focusing on the reconstruction of the gene regulatory network of *P. aeruginosa* CCBH 4851.

4.3.1 Identification of Orthologs Through Reciprocal Best Hits (RBH)

In order to determine the regulatory interactions for *P. aeruginosa* CCBH 4851, we must first identify the corresponding regulatory interactions in *P. aeruginosa* PAO1 [25]. Given a transcription factor (TF) and a target gene (TG) belonging to the *P. aeruginosa* PAO1 network, this interaction will be propagated if CCBH 4851 has an orthologous gene to the transcription factor TF and another ortholog to the TG target gene. The criterion for establishing the relationship of orthology will be the existence of RBHs between the two genomes [50]. A pair of genes (**a**, **a**′) of genomes **A** and **A**′, respectively, will be considered orthologous if it is also an RBH, that is, if by aligning the sequence of **a**, against the list of genes of **A**′, we get **a**′ as the best alignment and if we aligned the sequence of **a**′ against the list of genes of genome **A**, we get **a** as the best hit. Once the complete set of RBHs between genomes **A** and **A**′ is obtained, a regulatory interaction between a TF (say gene **a**) and a TG (say gene **b**) is propagated from the reference network (PAO1 or *E. coli* as **A**) to CCBH 4851, if both of these genes have their respective RBHs in the CCBH4851 genome. That is, the propagation of a regulatory interaction **a**–**b** does occur if there exists a pair of genes **a**′–**b**′ in CCBH 4851 such that we have (**a**, **a**′) and (**b**, **b**′) as RBH pairs. This test for the propagation of regulatory interactions is performed for all interactions known in PAO1. For the alignments, in this project, we use the BLASTP program.

4.3.2 Identification of Transcription Factor Binding Sites (TFBS)

The first step here is to collect, from public databases such as RegPrecise [51], the set of binding sites for each of the transcription factors identified in *P. aeruginosa* CCBH 4851. Each set of sites, associated with one TF, will be represented statistically by a position weight matrix (PWM), which defines the probabilities of finding, at each position of the site, the bases A, C, T, or G. With this statistical representation, we can sweep a promoter region of orthologous transcription units (TU), evaluating iteratively for each window in the region its affinity with the PWM. This affinity is a probabilistic measure of the sequence in the evaluated window to be a TF binding site. In this step, we can use current tools such as MEME [52]. The search for the binding sites may reveal potential new interactions (or predictions), since the scanning with the matrix of each TF will be done in all the promoter regions of orthologous TUs. Two TUs from a reference genome and CCBH 4851 are considered orthologous if they have at least one RBH pair in common, that is, if there is a gene **a** in one TU from the reference genome, that is an RBH of **a**′ in the CCBH4851 genome.

Manual Refinement Manual refinement consists in searching the literature for confirmations of possible new interactions. This generic search for regulatory interactions will be done via the SwissProt/UniProt database also to identify new regulatory interactions in *P. aeruginosa* CCBH 4851 which have not been described in the networks considered as references.

4.3.3 Analysis of the Reconstructed Regulatory Network for *P. aeruginosa* CCBH 4851

The identification of the functional modules in the reconstructed network will be done through direct comparison with the functional modules described in the reference network of *P. aeruginosa* PAO1. However, our focus will be on the genes associated with virulence and resistance to antibiotics. Genes considered as potential therapeutic targets are usually hubs in the gene network, i.e., those with more neighbors connected [53], and these nodes are extremely important in several iterations [54]. Therefore, we will also perform the analysis of the topological properties of the network, such as calculating the distribution of probabilities of finding a node with degree k, P(k), to identify potential therapeutic targets. Both analyses of functional modules, graphic and topological, will be facilitated by the use of the Cytoscape program [55].

This methodology results in a comprehensive descriptive model for the regulatory network of *P. aeruginosa* CCBH 4851. We need further to translate it into a computational predictive model, to study its dynamics [56].

5 Integrated Models of Bacteria

In an integrated model of bacteria, several cellular processes are modeled using the most appropriate mathematical representation and then simulated concurrently in a common computational architecture. As examples of integrated models described in the literature, we have the joint modeling of the metabolic, gene regulatory, and signaling networks of *E. coli* [3, 57] and the whole-cell model of *M. genitalium* [6]. These models have shown to be very promising when confronted with data available in the literature and obtained in the laboratory.

It is worth noting that several methods have been proposed through the transcriptional, proteomics, metabolomics, and signal transduction layers [57]. Indeed, whole-cell (WC) models are developed combining multiple mathematically distinct submodels into a single multi-algorithm model [58], being the most appropriate mathematical representation to model each module dependent on how the biological process is experimentally characterized [6]. A crucial tool for this integrative modeling was the development of network inference algorithms, which could be used to generate topological models and consensus data networks [57].

A brief retrospective of integrative modeling starts approximately two decades ago. The E-Cell [12] emerged as a modular software widely used in 1999 and represented an initial work in integrative modeling. Its modular software environment for WC simulation included organelle cellular submodels. In 2002, an in silico methodology was published for genome-scale modeling of the metabolism of *Escherichia coli*, which considered also gene transcription regulation [59]. This regulatory/metabolic integrated model based on flux balance analysis (FBA) was named regulatory flux balance analysis (rFBA). This integrated model was used to simulate the dynamic behavior of *E. coli* under a variety of environmental conditions and genomic perturbations and provides more accurate results than FBA-only analysis of metabolic networks.

The development of a new methodology for the generation of an integrated *E. coli* model, which combined the existing rFBA methodology with Ordinary Differential Equations (ODE) models of *E. coli* central metabolism, was reported by Covert and colleagues [3]. This approach, named integrated FBA (iFBA), proposed an integrated model of *E. coli* which combines a FBA-based metabolic model and a transcriptional regulatory model with an ODE-based, detailed model of carbohydrate uptake control. The iFBA approach consists on an integration of a FBA metabolic network with a Boolean transcriptional regulatory network, as well as with a set of ordinary differential equations. The authors concluded that iFBA simulation results were more accurate than previous rFBA and ODE models.

In 2012, Karr and colleagues published the first whole-cell (WC) model representing each individual gene function in a bacterium [6]. The model represented the life cycle of a single bacterial cell of *M. genitalium* and predicted the dynamics of each molecular species. The model was composed of cellular submodels, independently modeled, representing distinct metabolic pathways, which were represented using multiple mathematical formalisms, including stochastic simulation, ordinary differential equations, flux balance analysis, and Boolean rules, all implemented in MATLAB. The cell was divided into 28 functional processes representing major cellular processes, such as DNA repair and replication, RNA synthesis and maturation, metabolism, protein synthesis, cytokinesis, and host interaction. The complete configuration of the modeled cell was represented with these submodels being structurally integrated by connecting their common inputs and outputs through 16 state cell variables. The cellular processes/variables can be grouped into five physiological categories: DNA, RNA, protein, metabolite, and others. The common inputs to the submodels were computationally determined at the beginning of each simulation time step. Each process was modeled independently considering a short simulation time step of 1 s but dependent on the values of the variables determined by other submodels in the previous time step. Therefore, for each 1 s time step, the submodels retrieved the current values of the cellular variables, calculated their contributions to the temporal evolution of the cell variables, and then updated the values of the cellular variables. This procedure was repeated thousands of times during the course of each simulation. Finally, simulations were terminated upon cell division or when the simulation time reaches a predefined maximum value.

Although the model has been extensively validated by independent experimental data, it did not model several cellular functions and did not predict certain phenotypes [6]. The Karr group has since then improved and extensively documented the model by publishing the source code of the simulator and tools that provide user-friendly computational interfaces of these models [60–62]. Among recent publications of the group, there is the work of Waltemath and colleagues, where the authors published the results of the "2015 Whole-Cell Modeling School" [63] in which ad hoc algorithms and rate laws, used by the original model, were replaced by the Gillespie algorithm and mass action kinetics. Frequent discussions about the state of the art of WC modeling underscore its tendency to overcome the challenges of building increasingly complex models with open-source modeling software, emphasizing the importance of reproducibility through the utilization of standards such as SBML, SED-ML (Simulation Experiment Description Markup Language), and SBGN (Systems Biology Graphical Notation).

In 2014, Carrera and colleagues published an integrative modeling methodology that unified, under a common framework, various biological processes and their interactions across multiple layers [57]. The authors used this methodology to generate a genome-scale model of *E. coli* integrating gene expression data for genetic and environmental perturbations, transcriptional regulation, signal transduction, metabolic pathways, and growth data. To allow for genetic and environmental perturbations, the authors developed a quadratic programming method named "Expression Balance Analysis" (EBA) that takes into account genetic, capacity, phenomenological, and environmental constraints to predict gene expression in *E.coli* and extended the current models for flux boundary calculations by developing a new method called "TRAnscription-based Metabolic flux Enrichment" (TRAME) that accounts for both metabolic and transcriptional interactions. This model was used for growth predictions after simulation of dozens of random environments where the cells grew in minimal medium. Statistical tests and subsequent experimental validation demonstrate the capacity of this integrative model to predict environmental and genetic perturbations when compared to stand-alone metabolic and gene expression models.

6   Computational Aspects of Integrated Modeling

In this section, we address the computational aspects involved in the simulation of integrated models of bacteria. Simulation is the process of elaborating a model of a real system and conducting experiments with this model, in order to understand the behavior of the system. Simulations allow the recreation of biological phenomena and the precise replication of the experiments. It is possible to test different alternatives for the system. Simulation can result in resource savings (time and material). In general, it is more economical than using the real system.

The simulation of a whole-cell (WC) model of bacteria has several objectives: to describe its behavior and to construct theories and hypotheses considering the observations of morphology, metabolism, growth, and reproduction. The model can be used to predict future behavior and the effects produced by changes in the environment. It also has several advantages. For example, we can control simulation time, allowing us to reproduce the biological phenomena in a slow or accelerated way, so that we can better study them. It also allows to simulate long periods in a reduced time.

Biological systems, in general, exhibit characteristics of complex systems. These systems have a large number of individual components interacting with each other, whose behavior may change over time. Analysis of the individual behaviors of components may not exhibit the collective behavior of the system. This characteristic is not easy to reproduce, because relating individual actions to collective behavior is not simple [64].

Aiming the development of a faster, scalable, WC modeling process, Goldberg and colleagues [58] have proposed a parallel, multi-algorithm, WC simulator that can be implemented as an application running on an optimistic parallel discrete event simulation (PEDS) system, such as ROSS [65]. The simulation consists of both reaction modules and species modules. Each module is executed inside a separate PDES logical process. The processes communicate via PDES event messages. Each reaction module is executed in a PDES logic process and uses a particular biochemical modeling method. Subsequently, pairs of reaction and species modules are co-located to minimize the network traffic. This simulator supports the most common modeling algorithms, including Stochastic Simulation Algorithm (SSA), ODEs, and FBA. The proposal is to reuse the existing simulation libraries that support these modeling algorithms.

In terms of the language to describe models, a recent proposal allows the creation of scalable methods of WC modeling of multicellular organisms [58]. The proposed language support multi-algorithmic modeling, allowing the modeler to specify the modeling algorithm of the reaction sets. In order to describe the combinatorial complexity of biological systems, the language will also support *data-based modeling*, i.e., the definition of species and reaction patterns in terms of patterns based on biochemistry, genomics, and other experimental data. It is noteworthy to mention that data-based modeling generalizes rule-based modeling and enables WC models to explicitly combine genomics with large-scale dynamic modeling – thereby avoiding a combinatorial explosion of reaction descriptions that could lead to non-feasible models [58].

Complex simulations involving several individual subsystems may require a large computational power [58, 66] and therefore need to be distributed in multiple cores or processing nodes through the use of, for example, a cluster. Distributed simulations can reduce the execution time of experiments by dividing the simulation model among nodes or through the simultaneous execution of multiple instances of the complete model, especially when considering stochastic models. To illustrate the need for distributed execution of integrated model simulations, the execution of 64 instances of the whole-cell model of *M. genitalium* may require up to 47.6 h of processing in 600 cores distributed in 10 processing nodes [67].

One important requirement is to develop distributed simulators using open-source software. This requirement would make possible to more research groups work with WC simulators. One open-source package that is compatible with MATLAB is GNU Octave [68]. In order to fulfil this requirement, our research group is porting the whole-cell *M. genitalium* simulator, originally written in

MATLAB language, from MATLAB to GNU Octave. GNU Octave treats any incompatibility with MATLAB as a bug; therefore each new released version has more compatible functions than the previous one. In the following paragraphs, we describe the current status, difficulties, and perspectives of this porting.

GNU Octave reads the file extension .m, the same one used in MATLAB. So, the first attempt when facing a new .m file is to execute it directly in GNU Octave. If there is some error, the corresponding file should be ported to Octave. We also observe if the obtained results are the same in GNU Octave and MATLAB. The main steps considered in the porting effort are shown in Fig. 2.

The porting starts with the non-specific libraries, called here as external libraries, such as CPLEX, GLPK, and LP solve. All of them are related to linear programming. Only the open-source libraries were executed in the Octave version of the WC simulator.

After this step, our research group analyzed the simulator code, looking first for general functions, such as plotting functions and functions that call external libraries. Following this step, more specific functions, such as FBA, have been tested on Octave.

The next step would be verifying each one of the 28 cellular processes of the *M. genitalium* simulator. Each method can execute and give results apart from the others. Our group verified if the results are the same in MATLAB and in GNU Octave. The final step would be WC simulation execution to verify if all methods can communicate with each other using the 16 cellular states and to compare the execution time in both software packages.

There are 26 external libraries in the simulator. From these, half of them were created by MATLAB users and are available in the MathWorks site. For the other half, only two do not have a license allowing modifications. Four others have the General Public License, version 2 or above, and two have other copyright licenses which allow modifications. For the last five, we could not find any license.

The result, after testing and fixing the external libraries used in the simulator, shows that six of them were not called inside the simulator. This happens because they were libraries used to control the process of code writing, like a subversion client or a database client. Thirteen libraries were executed with no or little fixing,

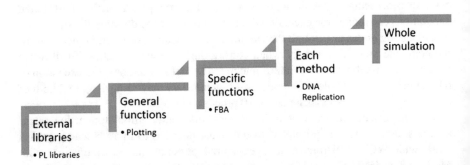

Fig. 2 Steps of porting the *M. genitalium* WC simulator from MATLAB to Octave

but six of them showed different results. These results may not interfere in the final simulation result, because they are related to random number generation or graphic handles. One external library has only .jar files and it could not yet be tested.

The six libraries initially presented execution errors, but four of the libraries were fixed in the porting. Three libraries had only unimplemented functions. These were ported by installing a patch that had the implementation of the most important functions and by encapsulating the other ones in a try/catch statement. The other ported library executed a driver that was calling an obsolete OS library. The missing OS library was fixed by recompiling the driver in the external library. Only two external libraries could not be ported, because they used a protected function file, called P-file. This file is precompiled in MATLAB, and it cannot be reversed to a .m source file. One of these libraries is for maximizing and minimizing windows, which is not crucial for simulation execution and results.

One major difference between GNU Octave and MATLAB is that GNU Octave does not support the object-oriented package scheme. While in MATLAB it is possible to consider folders to call the functions by their canonical name, GNU Octave outputs an error every time it executes those calls. The solution proposed was adding those functions' path in Octave, so that Octave would be able to find the functions when called by their short names. As this solution enlarges Octave's path, its impact on the execution time will be observed when the porting is finished.

The experience obtained with the porting of the *M. genitalium* WC simulator exposed the difficulties and intricacies of developing a similar simulator for a more complex microorganism like *P. aeruginosa*. Indeed, our research group intend to develop a distributed WC simulator, based on GNU Octave, to accurately predict a wide range of observable cellular behaviors of *P. aeruginosa* CCBH4851.

7 Searching for Therapeutic Targets

The dawn of the genomic era was accompanied with revitalized hope to find new antibacterial targets. However, such promise was not fulfilled at the expected level since very few lead compounds have entered drug discovery programs for new anti-infectives. Many creative and intricate bioinformatics pipelines were developed to analyze genomic data for both bacteria and the human host in the search for the ideal target (i.e., essential for infectious organism survival, specific for the microorganism and drugabble) [69].

It is becoming increasingly clear that modulating a single target, even with a very efficient drug, is unlikely to achieve the desired therapeutic result, i.e., rapid cure with no side effects. Thus, a growing perception is that we should increase the level of complexity of our proposed therapies by shifting the way we think about complex diseases from a gene-centered to a network-centered view [70].

For example, realizing the promise of molecularly targeted inhibitors for infectious diseases or cancer therapy will require a new level of knowledge about how the drug target is connected to the control circuit of a complex cellular control network.

The discovery of network-based drugs aims to tame this knowledge to identify fragility points in a biochemical network as a whole that specifically characterize an altered phenotype [71–73]. Therefore, the drug targeting process needs to also consider the positioning of these targets in the network, preferring these enclaves that are essential for moving the cellular trafficking and, at the same time, avoiding back-up circuits that could neutralize the effect of the drug.

In this section, we briefly review a few recent examples of the successful application of computational analysis of integrated genome-scale models of bacteria in the identification of new drug targets.

Many different strategies have been used to search for drug targets from system level models of bacterial metabolism. More often, essential genes are identified from single virtual knockouts where FBA is used to assess if this gene deletion is able to halt a selected function of the bacteria metabolism. Usually, such function is biomass production [74]. Other criteria can be combined to prioritize genes among candidate drug targets, such as existence of druggable pockets [75] or specificity to the bacteria as compared to the host proteins. An interesting example was recently reported where intracellular levels of all the proteins/enzymes coded by the predicted essential genes were assessed [76]. The rationale was that the lower the concentration of a target protein/enzyme for an inhibitor to bind, the lower would be this antibiotic's MIC (minimum inhibitory concentration).

Chaudhury [77] employed what they called a "staggered workflow" that combined drug target identification, target selection, in silico drug screening, and cell-based experimental validation for the rapid (32 weeks) identification of six compounds active on *Francisella tularensis* subspecies *tularensis Schu S4* strain, a highly infectious intracellular pathogen that is the causative agent of tularemia and is classified as a category A biological agent by the Centers for Disease Control and Prevention. These compounds present as putative target the protein pantetheine-phosphate adenylyltransferase, an enzyme involved in the biosynthesis of coenzyme A, encoded by gene *coaD*. The latter was identified from an initial set of 124 targets proposed to be essential for biomass accumulation upon knockout simulations with a genome-scale metabolic model for *F. tularensis*.

By predicting metabolic changes that result from genetic or environmental perturbations, integrated models of genome-wide gene regulatory and metabolic networks have several important applications, including diagnosing metabolic disorders and discovering novel drug targets. Nevertheless, this integration of transcriptional regulatory networks with the corresponding metabolic network represents a formidable challenge in obtaining accurate predictions. Methods developed so far consider that a perturbation to a TF results in alteration in the expression of its target genes. These are then mapped onto the metabolic network, and depending on the gene state, the fluxes through the reactions are constrained and the optimal growth rate are determined by using FBA (for a review see [78]). The simplest approach to this method is called rFBA (regulatory FBA) wherein genes and reaction fluxes can only have two states in the population: on or off (see section "Integrated Models of Bacteria"). In addition to the clear over simplification of this binary state model, the absence of an automated algorithm for determining the

Boolean rules for relating the regulator with its target represents a major limitation to its wider applicability.

To overcome this limitation, methods, such as GIMME [79, 80], E-Flux [81], Brandes [83], and iMAT [82], were developed, which extends FBA constraints to generate flux solutions that are consistent with a set of gene expression data. For instance, using E-Flux, Colijn [81] predicted the impact of drugs on MTB mycolic acid biosynthesis. Employing similar strategy, Chandrasekaran and Price [83] developed a method called probabilistic regulation of metabolism (PROM) which automatically quantify the regulatory interactions from high-throughput data (using conditional probabilities for modeling transcriptional regulation), thereby greatly increasing the capacity to generate genome-scale integrated models. The applicability of PROM to the identification of drug targets was demonstrated by the authors where of the 11 predicted essential genes for human pathogen *Mycobacterium tuberculosis* seven were known drug targets.

An improved version of the PROM method using an extended model of *M. tuberculosis* transcriptional regulation network and refined genome-scale metabolic model, so-called MTBPROM2.0, was published later [85] with improved TF essentiality prediction performance. Interestingly, the authors collected transcriptomic data for MTB exposure to known anti-TB agents and integrated these data into condition-specific metabolic models for the successful prediction of synergistic interactions between the drugs ethionamide and isoniazid through overexpression of the transcription factor *whiB4* that affects the growth of MTB. In order to answer questions about the accumulation or degradation of both intracellular and extracellular metabolites using the metabolic model of MTB, Garay [86] developed an extension of the E-Flux and PROM methods called E-Flux-MFC (E-Flux for maximum flux capacity).

8   Conclusion

Living organisms are extremely sophisticated, and understanding their behavior is a huge challenge. Even simpler organisms require complex models to reproduce their biology in silico. Therefore, for the computational modeling of biological systems, a large number of variables are required, as well as the adoption of suitable numerical methods; complementary computational tools, such as visualization tools; and advanced programming techniques. On the other hand, the advancement of computer technology, with the development of computational systems with larger capacity to process and store data, allows the execution of more complex systems in less time.

Computational models are useful in the analysis of complex systems and should be considered as approximations of reality. They are useful to reveal the main properties of the systems and highlight aspects considered relevant to certain phenomena. The major limitation to the construction of models is the difficulty in incorporating large amounts of data that describe the heterogeneity of biological systems.

The development of integrated computational models of living organism is still in its infancy. We can expect accelerated advancements in the coming years, with the creation of more powerful computational frameworks for model execution and the availability of libraries of models that can be reused in a *plug-and-play* approach. Up to now, integrated model development has focused single-cell organisms, essentially prokaryotes. Nevertheless, considering the advances described in this chapter, one can expect that computational models for multi-cell organisms and eukaryotic cells will be available in the coming years.

Acknowledgment This study was supported by fellowships from CAPES to FMF and from the Oswaldo Cruz Institute (https://pgbcs.ioc.fiocruz.br/) to TG. We also thank FAPERJ and CAPES for financial support.

References

1. Brasil. Ministério da Saúde. Agência Nacional de Vigilância Sanitária. Boletim de segurança do paciente e qualidade em serviços de saúde n° 14: Avaliação dos indicadores nacionais das Infecções Relacionadas à Assistência à Saúde (IRAS) e resistência microbiana do ano de 2015. Brasília (DF): Ministério da Saúde. (In portuguese) Available at: (https://www20.anvisa.gov.br/segurancadopaciente/index.php/publicacoes/item/boletim-de-seguranca-do-paciente-e-qualidade-em-servicos-de-saude-n-13-avaliacao-dos-indicadores-nacionais-das-infeccoes-relacionadas-a-assistencia-a-saude-iras-e-resistencia-microbiana-do-ano-de-2015) 2016.
2. World Health Organization. Global priority list of antibiotic-resistant bacteria to guide research, discovery, and development of new antibiotics. (http://www.who.int/medicines/publications/WHO-PPL-Short_Summary_25Feb-ET_NM_WHO.pdf?ua=1) 2017.
3. Covert M, Xiao N, Chen T, Karr J. Integrating metabolic, transcriptional regulatory and signal transduction models in *Escherichia coli*. Bioinformatics. 2008;24(18):2044–50.
4. Silveira M, Albano R, Asensi M, Assef A. The draft genome sequence of multidrug-resistant *Pseudomonas aeruginosa* strain CCBH4851, a nosocomial isolate belonging to clone SP (ST277) that is prevalent in Brazil. Mem Inst Oswaldo Cruz. 2014;109(8):1086–7.
5. Carrera J, Covert M. Why build whole-cell models? Trends Cell Biol. 2015;25(12):719–22.
6. Karr J, Sanghvi J, Macklin D, Gutschow M, Jacobs J, Bolival B, et al. A whole-cell computational model predicts phenotype from genotype. Cell. 2012;150(2):389–401.
7. Pier GB, Ramphal R. Pseudomonas aeruginosa. In: Mandell GL, Bennett JE, Dolin R, editors. Mandell, Douglas, and Bennett's principles and practice of infectious diseases. 7th ed. Philadelphia: Churchill Livingstone Elsevier; 2010. p. 2835–60.
8. Driscoll J, Brody S, Kollef M. The epidemiology, pathogenesis and treatment of *Pseudomonas aeruginosa* infections. Drugs. 2007;67(3):351–68.
9. Lee K, Yoon SS. *Pseudomonas aeruginosa* biofilm, a programmed bacterial life for fitness. J Microbiol Biotechnol. 2017;27(6):1053–64.
10. Balasubramanian D, Schneper L, Kumari H, Mathee K. A dynamic and intricate regulatory network determines *Pseudomonas aeruginosa* virulence. Nucleic Acids Res. 2012;41(1):1–20.
11. Engel J, Balachandran P. Role of *Pseudomonas aeruginosa* type III effectors in disease. Curr Opin Microbiol. 2009;12(1):61–6.
12. Tomita M, Hashimoto K, Takahashi K, Shimizu TS, Matsuzaki Y, Miyoshi F, Saio K, Tanida S, Yugi K, Venter J, Hutchison CA. E-CELL: software environment for whole-cell simulation. Bioinformatics. 1999;15(1):72–84.

13. Kerr K, Snelling A. *Pseudomonas aeruginosa*: a formidable and ever-present adversary. J Hosp Infect. 2009;73(4):338–44.
14. Kung V, Ozer E, Hauser A. The accessory genome of *Pseudomonas aeruginosa*. Microbiol Mol Biol Rev. 2010;74(4):621–41.
15. Vallet-Gely I, Boccard F. Chromosomal organization and segregation in *Pseudomonas aeruginosa*. PLoS Genet. 2013;9(5):e1003492.
16. Silveira M, Albano R, Asensi M, Carvalho-Assef A. Description of genomic islands associated to the multidrug-resistant *Pseudomonas aeruginosa* clone ST277. Infect Genet Evol. 2016;42:60–5.
17. Oliver A, Mulet X, López-Causapé C, Juan C. The increasing threat of *Pseudomonas aeruginosa* high-risk clones. Drug Resist Updat. 2015;21–22:41–59.
18. Cornaglia G, Giamarellou H, Rossolini G. Metallo-β-lactamases: a last frontier for β-lactams? Lancet Infect Dis. 2011;11(5):381–93.
19. Nascimento A, Ortiz M, Martins W, Morais G, Fehlberg L, Almeida L, et al. Intraclonal genome stability of the metallo-β-lactamase SPM-1-producing *Pseudomonas aeruginosa* ST277, an endemic clone disseminated in brazilian hospitals. Front Microbiol. 2016;7:1946.
20. Cavalcanti F, Almeida A, Vilela M, Morais M, Morais JM. Changing the epidemiology of carbapenem-resistant *Pseudomonas aeruginosa* in a Brazilian teaching hospital: the replacement of São Paulo metallo-β-lactamase-producing isolates. Mem Inst Oswaldo Cruz. 2012;107(3):420–3.
21. Gales A, Menezes L, Silbert S, Sader H. Dissemination in distinct Brazilian regions of an epidemic carbapenem-resistant *Pseudomonas aeruginosa* producing SPM metallo-β-lactamase. J Antimicrob Chemother. 2003;52(4):699–702.
22. Fonseca E, Freitas F, Vicente A. The Colistin-only sensitive Brazilian *Pseudomonas aeruginosa* clone SP (sequence type 277) is spread worldwide. Antimicrob Agents Chemother. 2010;54(6):2743.
23. Salabi A, Toleman M, Weeks J, Bruderer T, Frei R, Walsh T. First report of the metallo-β-lactamase SPM-1 in Europe. Antimicrob Agents Chemother. 2009;54(1):582.
24. Hopkins K, Findlay J, Mustafa N, Pike R, Parsons H, Wright L, et al. SPM-1 metallo-β-lactamase-producing *Pseudomonas aeruginosa* ST277 in the UK. J Med Microbiol. 2016;65(7):696–7.
25. Galán-Vásquez E, Luna B, Martínez-Antonio A. The regulatory network of *Pseudomonas aeruginosa*. Microb Inf Exp. 2011;1(1):3.
26. Babaei P, Ghasemi-Kahrizsangi T, Marashi S. Modeling the differences in biochemical capabilities of pseudomonas species by flux balance analysis: how good are genome-scale metabolic networks at predicting the differences? Sci World J. 2014;2014:1–11.
27. Brent M. Genome annotation past, present, and future: how to define an ORF at each locus. Genome Res. 2005;15(12):1777–86.
28. Richardson E, Watson M. The automatic annotation of bacterial genomes. Brief Bioinform. 2012;14(1):1–12.
29. Verli H. Bioinformática: da biologia à flexibilidade molecular. 1st ed. São Paulo: SBBq; 2014.
30. Campbell M, Yandell M. An introduction to genome annotation. Curr Protocol Bioinforma. 2015;52:4.1.1–4.1.17.
31. Delcher A. Improved microbial gene identification with GLIMMER. Nucleic Acids Res. 1999;27(23):4636–41.
32. Besemer J, Lomsadze A, Borodovsky M. GeneMarkS: a self-training method for prediction of gene starts in microbial genomes. Implications for finding sequence motifs in regulatory regions. Nucleic Acids Res. 2001;29(12):2607–18.
33. Tatusova T, DiCuccio M, Badretdin A, Chetvernin V, Nawrocki E, Zaslavsky L, et al. NCBI prokaryotic genome annotation pipeline. Nucleic Acids Res. 2016;44(14):6614–24.
34. Lagesen K, Hallin P, Rødland E, Stærfeldt H, Rognes T, Ussery D. RNAmmer: consistent and rapid annotation of ribosomal RNA genes. Nucleic Acids Res. 2007;35(9):3100–8.
35. Lowe T, Eddy S. tRNAscan-SE: a program for improved detection of transfer RNA genes in genomic sequence. Nucleic Acids Res. 1997;25(5):955–64.

36. Laslett D, Canback B. ARAGORN, a program to detect tRNA genes and tmRNA genes in nucleotide sequences. Nucleic Acids Res. 2004;32(1):11–6.
37. Kinouchi M, Kurokawa K. [Special issue: fact databases and freewares] tRNAfinder: a software system to find all tRNA genes in the DNA sequence based on the cloverleaf secondary structure. J Comput Aided Chem. 2006;7:116–24.
38. Overbeek R, Olson R, Pusch G, Olsen G, Davis J, Disz T, et al. The SEED and the rapid annotation of microbial genomes using subsystems technology (RAST). Nucleic Acids Res. 2013;42(D1):D206–14.
39. Seemann T. Prokka: rapid prokaryotic genome annotation. Bioinformatics. 2014;30(14): 2068–9.
40. Zimin A, Marçais G, Puiu D, Roberts M, Salzberg S, Yorke J. The MaSuRCA genome assembler. Bioinformatics. 2013;29(21):2669–77.
41. Otto T, Dillon G, Degrave W, Berriman M. RATT: rapid annotation transfer tool. Nucleic Acids Res. 2011;39(9):e57.
42. Thiele I, Palsson B. A protocol for generating a high-quality genome-scale metabolic reconstruction. Nat Protoc. 2010;5(1):93–121.
43. The Uniprot Consortium: the universal protein knowledgebase. Nucleic Acids Res. 2017;45(D1):D158–D169.
44. Barthelmes J, Ebeling C, Chang A, Schomburg I, Schomburg D. BRENDA, AMENDA and FRENDA: the enzyme information system in 2007. Nucleic Acids Res. 2007;35(Database):D511–4.
45. Kanehisa M, Goto S, Hattori M, Aoki-Kinoshita K, Itoh M, Kawashima S, et al. From genomics to chemical genomics: new developments in KEGG. Nucleic Acids Res. 2006;34(90001):D354–7.
46. Heavner B, Price N. Transparency in metabolic network reconstruction enables scalable biological discovery. Curr Opin Biotechnol. 2015;34:105–9.
47. Oberhardt M, Puchalka J, Fryer K, Martins dos Santos V, Papin J. Genome-scale metabolic network analysis of the opportunistic pathogen *Pseudomonas aeruginosa* PAO1. J Bacteriol. 2008;190(8):2790–803.
48. Vital-Lopez F, Reifman J, Wallqvist A. Biofilm formation mechanisms of *Pseudomonas aeruginosa* predicted via genome-scale kinetic models of bacterial metabolism. PLoS Comput Biol. 2015;11(10):e1004452.
49. Bartell J, Blazier A, Yen P, Thøgersen J, Jelsbak L, Goldberg J, et al. Reconstruction of the metabolic network of *Pseudomonas aeruginosa* to interrogate virulence factor synthesis. Nat Commun. 2017;8:14631.
50. Moreno-Hagelsieb G, Latimer K. Choosing BLAST options for better detection of orthologs as reciprocal best hits. Bioinformatics. 2008;24(3):319–24.
51. Novichkov P, Kazakov A, Ravcheev D, Leyn S, Kovaleva G, Sutormin R, et al. RegPrecise 3.0 – a resource for genome-scale exploration of transcriptional regulation in bacteria. BMC Genomics. 2013;14(1):745.
52. Bailey T, Boden M, Buske F, Frith M, Grant C, Clementi L, et al. MEME SUITE: tools for motif discovery and searching. Nucleic Acids Res. 2009;37.(Web Server:W202–8.
53. Hwang S, Kim C, Ji S, Go J, Kim H, Yang S, et al. Network-assisted investigation of virulence and antibiotic-resistance systems in *Pseudomonas aeruginosa*. Sci Rep. 2016;6(1):26223.
54. Jeong H, Mason S, Barabási A, Oltvai Z. Lethality and centrality in protein networks. Nature. 2001;411(6833):41–2.
55. Shannon P, Markiel A, Owen O, Nitin SB, Jonathan TW, Daniel R, et al. Cytoscape: a software environment for integrated models of biomolecular interaction networks. Genome Res. 2003;13(11):2498–504.
56. Trindade dos Santos M, Nascimento A, Medeiros Filho F, Silva F. Modeling gene transcriptional regulation. Theor Appl Asp Syst Biol. 2018;27:27–39.
57. Carrera J, Estrela R, Luo J, Rai N, Tsoukalas A, Tagkopoulos I. An integrative, multi-scale, genome-wide model reveals the phenotypic landscape of *Escherichia coli*. Mol Syst Biol. 2014;10(7):735.

58. Goldberg A, Chew Y, Karr J. Toward scalable whole-cell modeling of human cells. Proceedings of the 2016 annual ACM Conference on SIGSIM Principles of Advanced Discrete Simulation – SIGSIM-PADS '16. 2016.
59. Covert M, Palsson B. Transcriptional regulation in constraints-based metabolic models of *Escherichia coli*. J Biol Chem. 2002;277(31):28058–64.
60. Karr J, Sanghvi J, Macklin D, Arora A, Covert M. WholeCellKB: model organism databases for comprehensive whole-cell models. Nucleic Acids Res. 2013;41(D1):D787–92.
61. Karr J, Phillips N, Covert M. WholeCellSimDB: a hybrid relational/HDF database for whole-cell model predictions. Database. 2014;2014:bau095.
62. Lee R, Karr J, Covert M. WholeCellViz: data visualization for whole-cell models. BMC Bioinf. 2013;14(1):253.
63. Waltemath D, Karr J, Bergmann F, Chelliah V, Hucka M, Krantz M, et al. Toward community standards and software for whole-cell modeling. IEEE Trans Biomed Eng. 2016;63(10): 2007–14.
64. Ottino J. Engineering complex systems. Nature. 2004;427(6973):399.
65. Carothers C, Bauer D, Pearce S. ROSS: a high-performance, low-memory, modular time warp system. J Parallel Distrib Comput. 2002;62(11):1648–69.
66. Macklin D, Ruggero N, Covert M. The future of whole-cell modeling. Curr Opin Biotechnol. 2014;28:111–5.
67. Abreu R, Castro M, Silva F. Simulation step size analysis of a whole-cell computational model of bacteria. AIP Conf Proc. 2016;1790(1):100014.
68. Hansen J. GNU octave beginner's guide. Birmingham: Packt Publishing; 2011.
69. McPhillie M, Cain R, Narramore S, Fishwick C, Simmons K. Computational methods to identify new antibacterial targets. Chem Biol Drug Des. 2015;85(1):22–9.
70. Pujol A, Mosca R, Farrés J, Aloy P. Unveiling the role of network and systems biology in drug discovery. Trends Pharmacol Sci. 2010;31(3):115–23.
71. Schadt E, Friend S, Shaywitz D. A network view of disease and compound screening. Nat Rev Drug Discov. 2009;8(4):286–95.
72. Xie L, Li J, Xie L, Bourne P. Drug discovery using chemical systems biology: identification of the protein-ligand binding network to explain the side effects of CETP inhibitors. PLoS Comput Biol. 2009;5(5):e1000387.
73. Murabito E, Smallbone K, Swinton J, Westerhoff H, Steuer R. A probabilistic approach to identify putative drug targets in biochemical networks. J R Soc Interface. 2010;8(59):880–95.
74. Rienksma R, Suarez-Diez M, Spina L, Schaap P. Martins dos Santos V. Systems-level modeling of mycobacterial metabolism for the identification of new (multi-)drug targets. Semin Immunol. 2014;26(6):610–22.
75. Kozakov D, Hall D, Napoleon R, Yueh C, Whitty A, Vajda S. New frontiers in druggability. J Med Chem. 2015;58(23):9063–88.
76. Vashisht R, Bhat A, Kushwaha S, Bhardwaj A, Consortium O, Brahmachari S. Systems level mapping of metabolic complexity in *Mycobacterium tuberculosis* to identify high-value drug targets. J Transl Med. 2014;12(1):263–81.
77. Chaudhury S, Abdulhameed M, Singh N, Tawa G, D'haeseleer P, Zemla A, et al. Rapid countermeasure discovery against *Francisella tularensis* based on a metabolic network reconstruction. PLoS One. 2013;8(5):e63369.
78. Lewis N, Nagarajan H, Palsson B. Constraining the metabolic genotype–phenotype relationship using a phylogeny of in silico methods. Nat Rev Microbiol. 2012;10(4):291–305.
79. Becker S, Palsson B. Context-specific metabolic networks are consistent with experiments. PLoS Comput Biol. 2008;4(5):e1000082.
80. Shlomi T, Cabili M, Herrgård M, Palsson B, Ruppin E. Network-based prediction of human tissue-specific metabolism. Nat Biotechnol. 2008;26(9):1003–10.
81. Colijn C, Brandes A, Zucker J, Lun D, Weiner B, Farhat M, et al. Interpreting expression data with metabolic flux models: predicting *Mycobacterium tuberculosis* mycolic acid production. PLoS Comput Biol. 2009;5(8):e1000489.

82. Zur H, Ruppin E, Shlomi T. iMAT: an integrative metabolic analysis tool. Bioinformatics. 2010;26(24):3140–2.
83. Chandrasekaran S, Price N. Probabilistic integrative modeling of genome-scale metabolic and regulatory networks in *Escherichia coli* and *Mycobacterium tuberculosis*. Proc Natl Acad Sci. 2010;107(41):17845–50.
84. Brandes A, Lun D, Ip K, Zucker J, Colijn C, Weiner B, et al. Inferring carbon sources from gene expression profiles using metabolic flux models. PLoS One. 2012;7(5):e36947.
85. Ma S, Minch K, Rustad T, Hobbs S, Zhou S, Sherman D, et al. Integrated modeling of gene regulatory and metabolic networks in *Mycobacterium tuberculosis*. PLoS Comput Biol. 2015;11(11):e1004543.
86. Garay C, Dreyfuss J, Galagan J. Metabolic modeling predicts metabolite changes in *Mycobacterium tuberculosis*. BMC Syst Biol. 2015;9(1):57.
87. Toleman MA, Simm AM, Murphy TA, Gales AC, Biedenbach DJ, Jones RN, Walsh TR, Molecular characterization of SPM-1, a novel metallo-?-lactamase isolated in Latin America: report from the SENTRY antimicrobial surveillance programme. J Antimicrob Chemother. 2002;50(5):673–9.

System Biology to Access Target Relevance in the Research and Development of Molecular Inhibitors

Larissa Catharina, Marcio Argollo de Menezes, and Nicolas Carels

Abstract This review focuses on how system biology may assist techniques that are used in pharmacological research, such as high-throughput screening, high-throughput analytical characterization of biological samples, preclinical and clinical trials, as well as targets and drug validation in order to reach patients at the lowest possible cost in a translational perspective. In signaling networks, targets can be assessed through topological criteria such as their connectivity and/or centrality. In metabolic networks, the relevance of a target for drug development may rather be assessed through some sort of enzymatic specificity resulting from remote homology, analogy, or specificity in its strict sense. The concept of specificity is especially valuable in the context of a host-parasite relationship where targeting a protein specific of a parasite compared to its host is expected to minimize the noxious collateral effects of the inhibitor to the host. The relevance of putative molecular target must be proven through bench and animal validations prior to going through clinical trials. Flux balance analysis and other modeling methods of system biology enable to assess whether a molecular target can be considered as pathway's choke or not in a network context, which may facilitate the decision of developing drugs for it.

1 Introduction

The quality of pharmaceutical products started to be regulated with the inception, in the USA, of the Food and Drug Administration (FDA) in 1906. From a broad perspective, there had been a continuous growth in the number of annual new

L. Catharina · N. Carels (✉)
Laboratório de Modelagem de Sistemas Biológicos, Centro de Desenvolvimento Tecnológico em Saúde, Fundação Oswaldo Cruz, Rio de Janeiro, Brazil
e-mail: nicolas.carels@cdts.fiocruz.br

M. A. de Menezes
Instituto de Física, Universidade Federal Fluminense, Rio de Janeiro, Brazil
e-mail: marcio@mail.if.uff.br

© Springer International Publishing AG, part of Springer Nature 2018
F. A. B. da Silva et al. (eds.), *Theoretical and Applied Aspects of Systems Biology*,
Computational Biology 27, https://doi.org/10.1007/978-3-319-74974-7_12

221

therapeutic entities approved worldwide since 1940 [1]. However, on a shorter time scale, the pharmaceutical industry has experienced a dramatic decrease in productivity between the 1980s and 2010 that was principally due to the cost burden in investments of research and development (R&D) of new drugs [2]. At the apogee, these costs were estimated to peak at $1.7 billion each [3].

Far from the explosion of new drugs predicted to follow the human genome sequencing, the perceived failure of drug discovery, with an average of only two to three small-molecule drugs per year, has been attributed, in the 1990s, to the trend of pharmaceutical industry to direct its research more and more toward target-directed drug discovery. Here, by *target*, we understand polymers of amino (proteins) or nucleic (RNA) acids to which effector molecules (drugs or biopharmaceuticals: antibody, siRNA) bind to promote their biological activity. Indeed, this genotype approach (also known as *reverse pharmacology* or *reverse chemical biology*) has been very successful when applied to well-validated targets [4]. Such approach was tempting, since it allowed the rational drug development through structure-activity relationship (SAR) in the context of *one drug, for one target*, i.e., the *genes-to-drug* concept, through high-throughput biochemical assays. This strategy has the advantage of testing drug candidates directly on the target, but with the drawback of neglecting possible off-target activity. Off-target(s) indicate(s) unknown secondary target(s) that positively react(s) with a compound under testing and may produce(s) collateral effect(s). It may actually be hard to diagnose if the compound under consideration is producing effect by on-target or off-target action. Many drugs were actually discovered through their unwanted effects on off-targets rather than on the expected on-target, such as in the famous case of Viagra [5]. Another drawback of the target-oriented strategies is that complex diseases usually rely on more than one target; by consequence, the inactivation of one target by drug treatment is generally null or not enough to cure patients. As a result, the activity of a rough extract on a bioassay, such as a biochemical reaction or a cell culture, often vanishes during bioguiding, i.e., the process of guiding the extraction of a compound from an extract by following its activity on a bioassay at each purification step. In addition, the effect of selective interaction of a drug with a single target may be limited by network redundancies or paralogous proteins and may result in an increased risk of adaptive resistance by mutation [6].

A competitive strategy is to select compound activity based on cell phenotype (*forward pharmacology* or *classical pharmacology*) through high-throughput screening (HTS), which eliminates the need of the time-consuming process of suitable target identification. Through phenotype screening, one screens, in one shot, compounds that are active on a phenotype but that also demonstrate their ability to reach their target(s) within the cell. A phenotype screening can be *up* or *down* according whether it, respectively, promotes the increase or the decrease of the readout used for the associated bioassay [7]. An obvious drawback of this strategy is that the precise protein targets or mechanisms of action responsible for the observed phenotypes remain to be determined, a process that is called deconvolution [8]. Phenotypic screening is often optimized against mechanistic and pharmacodynamic biomarker modulation, while target-based screen optimizes the activity of a drug against a given target, even if a few additional off-targets may also be affected, but without properly investigating the broader cellular activity of the agent [9].

The principle drawback of screening chemically diverse libraries with no previous knowledge of the protein that are indeed specific is the ratio of the global in vivo screening capacity to the total number of potentially available compounds. In contrast, chemogenomic-based drug discovery against specific targets is designed to test a large number of compounds in molecular biochemical or binding assays [10]. Both biochemical and cell essays (whatever 2D, 3D, as well as even 3D cell printing or ex vivo tissues [11–13]) are simplified contexts, which fit HTS of million compounds by automated processes but neglect the complex tissue organization and homeostasis of vertebrates. To correct such simplification, animal experimentation is necessary; however, animals are not humans, and methods of target validation based on mouse genetics cannot fully predict human biology [14]. There are several basic differences that may affect the transfer of results from the animal model to humans:

(i) Many drugs are cross-reactive to nonhuman primates.
(ii) Treatment in animal model is frequently initiated before symptom onset, which is not the case of patients.
(iii) For ethical reasons, the number of animals used is kept at minimum.
(iv) Different potential side effects between animals and humans may affect the effective dosage.
(v) Lack of understanding of the biology underlying the disease.
(vi) Incomplete disease correlation between animals and humans.
(vii) Different drug metabolisms in animals and humans in case of drug needing biological activation [15].

As a consequence, a higher number of drugs interacting with less validated targets have entered clinical development during the last decade, which led to a decline in the success rate of drugs during clinical development. Discrepancies between inferences drawn from animal model experimentation and clinical trials led to failures (attrition) in phases II and III due to unfavorable efficacy, lack of commercial viability, and poor safety. Such compound termination late in the process of drug development has been understood as one of the main reasons of the cost burst around 2010 [1].

Another criticism that was made to pharmaceutical industry was to adhere to the *one target*, *one drug* paradigm, which has been hold responsible for productivity decline [16]. It has been argued that it should be replaced by a *multitarget*, *multidrug* model [17], which is by definition the area of polypharmacology, i.e., the design or use of pharmaceutical agents that act on multiple targets or disease pathways. Actually, it has been suggested that even the mild inhibition of multiple molecular nodes of a cellular network can be more efficient than the complete inhibition of a single target [18], which has prompted attempts to generate global blockades of biological processes and pathways by simultaneously targeting multiple nodes in the underlying network [19]. Actually, the impact of only one targeted protein on the whole pathway may be little, while the impact on that pathway will be much larger if it contains many targeted proteins and the expected impact on the whole organism could be huge [20].

In response to the unsustainable situation of the pharmaceutical industry as it peaked in 2010, the process of drug R&D has been shifted from how these activities were addressed as health-care priorities in the past to approaches that are dominated by their potential market value. Even if the situation reverted to the figure of the 1980s (as defined by the number of new chemicals licensed by the FDA), concerns still exist regarding future decision-making, which requires a new paradigm for the management of R&D activities to attend to global needs [21].

Another adaptation that occurred in response to the past experience has been to look for more stringent success criteria during the non-clinical stages in order to gain further confidence in clinical translatability. The most accepted criteria for target validation during drug discovery are based on three categories: (i) demonstration of the target protein expression in relevant cell types or in the target tissues from animal models or patients, (ii) demonstration that modulation of the target in cell systems results in the desired functional effect, and (iii) demonstration that the target has a causal role in producing the disease phenotype in animal models and/or patients [22].

In addition, to effectively fight against costly termination of drugs in the clinical phase, the pharmaceutical industry has been keen to invest in theoretical and computational modeling to promote the drug discovery process [23–27], which enabled a recursive process through hypothesis testing and bench experimentation [28, 29]. Models are fast to execute and able to reduce the use of animals and offer cheap predictive solutions for drug pharmacokinetics (PK) and pharmacodynamics (PD) as well as patient population responses.

Finally, the access to financial sources for research can be critical in some cases, such as neglected diseases, for instance, and *open science* and data sharing have received a growing interest as a mean of leveraging and combining the available resources to accelerate drug discovery efforts [30]. This community-based concept for a new drug discovery model led to the *London Declaration on Neglected Tropical Diseases* in 2012 for the control, elimination, or eradication of neglected tropical diseases (http://unitingtocombatntds.org).

2 Modeling Strategies

Most human diseases involve sophisticated mechanistic relationships between proteins necessitating thoroughly annotated drugs by experimental means and the incorporation of data modeling for drug combinations [31], drug and target networks [32], and polypharmacology [33]. These approaches are of particular importance in cancer and infectious disease, for which heterogeneity and evolution under the selective pressure of standard-of-care drugs result in the emergence of drug resistance.

The success of mechanism-based drug discovery relies on unambiguous evidences of the therapeutic action of drugs through clear biomolecular association, which is actually not always straightforward because of high affinity of a drug to an alternative target (off-target), for instance. This situation did evidence the need to map drugs on their protein target landscape, which led to identify protein families privileged for their historical contribution to drug discovery. In addition to privileged families, druggable protein did distribute into diverse, structurally unrelated protein families with small numbers of members mostly represented by diverse enzyme functions. Privileged families account for 44% of all human protein targets with G protein-coupled receptors (GPCRs) (12%), ion channels (19%), kinases (10%), and nuclear receptors (3%) and are responsible for the therapeutic effect of 70% of small-molecule drugs [9].

As outlined above, selecting the best model organism to study a particular disease or to validate a target involves that this model organism be able to develop the disease with sufficient similarities to the human pathology under consideration. One approach for this purpose is to take the set of gene products that are modulated by current drugs and to compile the list of their orthologs in the referred model organisms. These proteins can then be analyzed for their relative affinity to their associated drugs, and these drugs are compared for suitability to the therapeutic indication in order to infer which therapeutic areas are potentially best mimicked by which model organism. Overall, vertebrates (dog, pig, rat, mouse, and zebra fish) provide comparatively good coverage of human drug targets; as would be expected from the larger evolutionary distance and differences in anatomical systems, *Drosophila melanogaster* and *Caenorhabditis elegans* contain fewer orthologues for human disease targets [9]. For pathogen targets, it is often argued that the absence of the corresponding protein in the host organism is an important prerequisite for success and searching for proteins that are specific of the pathogen is often applied in bioinformatics filtering of potential targets; however, the task of identifying such suitable targets is not easy, since there are a number of proteins that are also present in humans [34].

Another challenge is how to assign targets to drugs reported to have broad mechanistic effects. The quest for these attractive multitarget medicines is progressively engaging the field of rational drug design [35] and should soon have a major role in molecular therapies. This process is assisted by several databases and algorithms that provide data on drug-target interactions with different scopes and foci. Among such tools, one may cite (i) DrugBank [36] is the most widely used and maps drugs to proteins, which have been reported to bind to them; (ii) SuperTarget [37] is a text-mining compilation of direct and indirect drug targets; (iii) the Potential Drug Target Database (http:// www.dddc.ac.cn/pdtd/) that gathers known and potential drug targets with structures from the Protein Data Bank (PDB) [38], which allows the potential binding site prediction with the structure-based druggability search engine provided by EMBL-EBI (https://www.ebi.ac.uk/chembl/drugebility); (iv) VisANT, a network platform integrating genes, drugs, diseases, and therapies [39]; (v) ChemProt, a disease chemical biology database [40]; (vi)

DINIES, a web interface for drug-target interaction prediction [41]; (vii) VNP, a database used for visualizing the disease-target-drug interaction network [42]; etc.

Large-scale integration of genomic, proteomic, signaling, and metabolomic data allowed the construction of complex networks of interacting entities (proteins, small molecules, DNA elements, disease features and symptoms, etc.) by modeling different types of relationships [43], involving (i) physical protein-protein interactions (PPIs), (ii) regulatory interactions, (iii) protein-ligand interactions, (iv) natural metabolites [44], or (v) drugs [45]. These networks are providing a new framework for understanding the molecular basis of physiological or pathophysiological cell states. Because of the emerging polypharmacology, it appeared that drugs, targets, and disease spaces can be correlated and that their interrelationships can be exploited for designing drugs or cocktails, which can effectively target one or more disease states [17].

High-throughput data generation has led to the development of algorithms able to process massive amounts of information. Uncountable methods are already available and in development to retrieve targets and their ligands. These methods can be classified in (i) ligand-based approach such as pharmacophore mapping, which uses quantitative structure-activity relationship (QSAR) to predict protein-ligand interaction by comparing a new ligand to the known ligands of a target protein [46, 47]. A pharmacophore can be identified by direct method (using receptor-ligand complexes) or by indirect method (using only a collection of ligands that are known to interact with a given receptor). The pharmacophore is then used as an abstraction of any ligand matching the target's docking side to screen large in silico compound datasets. The receptor-based approach for pharmacophore generation involves the mapping of physicochemical features at the active site and their spatial relationships; an active representation of this map is then used to construct the pharmacophore model. The method takes 3D target structure as an input and a set of ligands with known activity to generate an *interaction map* that is a complement of the target's docking site. This approach is based on the assumption that the selected pharmacophore is a representative for the observed activity [48]. (ii) Target-based approach involving docking simulation and relying on the 3D structure of proteins to predict protein-ligand interaction [49]. A more recent option is to screen drug target proteins based on primary sequence information using machine learning methods. This process inherently suffers from overfitting [50] because the training datasets have two classes. One class is called the positive dataset, which is supposed to be made of true positives, and the other is called the negative dataset, which is supposed to be made of true negatives. The accuracy and completeness of the predictions are limited by the inability to be sure that proteins in the negative dataset are indeed true negative proteins [51]. However, levels of accuracy and sensitivity as high as $\geq 90\%$ were recently reached by Li et al. (2017) [52] with such machine learning methods. (iii) Text mining that is based on keyword searching in literatures [53], but the synonymy in the name of the genes or compounds in the literatures is a major concern of this approach [54].

The integration of receptor- and ligand-based approaches through machine learning methods has let to *network pharmacology*, which is dedicated to the investigation of the complexity of polypharmacology [55]. Other networks with a higher level of integration are also emerging following the same trend with, for instance, network medicine [56], disease network [57], etc. Network pharmacology has also revealed significant correlations between drug structure similarity, target sequence similarity, and the drug-target interaction network topology, which may be used to predict unknown drug-target interaction networks from chemical structure and genomic sequence information simultaneously and on a large scale. This formalization of the drug-target interaction inference as a supervised learning problem for a bipartite graph enables the unification of the chemical and genomic entities in one space called the *pharmacological space* [58].

The combination of high-throughput experimental projects for analyzing the genome, transcriptome, and proteome has allowed a better understanding of the genomic spaces populated by targets of different protein classes. In parallel, the high-throughput screening of large-scale chemical compound libraries with various biological assays enabled the exploration of the chemical space for potential drugs [59–61]. The aim of chemical genomics is to relate the chemical space with the genomic one in order to identify potentially useful compounds. Actually, the number of compounds with information on their target protein is still limited, which implies that many potential interactions between chemical and genomic spaces remain to be discovered.

Effective in silico prediction methods are being developed to assist bench experimentation for the identification of compound-protein interactions or potential drug-target interactions that remains time-consuming, costly, and challenging to carry out [58]. Network mining is a component of quantitative systems pharmacology that aimed at modeling the efficacy prediction of drugs addressing known or novel targets on clinical end points and biomarkers. Ideally, quantitative systems pharmacology is combined with disease models in a way that certain specific hypothesis is generated that can subsequently be assessed in experimental animals and finally be fed back into the in silico model to potentially refine the hypothesis [15].

Since network descriptions often lack the molecular details necessary to understand how the different molecular processes function, a last development has been to connect the information of high-resolution 3D structures in order to provide chemical means to modulate such complex systems. Actually, it has been demonstrated that a 3D model template is available for almost all of the known interactions for which there is a structure for the two monomers [62]. Thus, Duran-Frigola et al. (2013) [63] combined domain-domain structural templates with a high-confidence human interactome, which allowed the rationalization of disease mutations, and provide hypotheses explaining their effects at molecular level. The Interactome3D (http://interactome3d.irbbarcelona.org) is a resource built by these authors that provides over 12,000 PPI in various model organisms with structural details at atomic resolution.

As just pointed out, it is important to understand how mutations translate into the 3D structure of gene products and their consequences for drug binding because of perturbation of the binding side leading to an alteration in the treatment response.

Resources that map mutations on drug targets [64] and drug-metabolizing enzymes [65], or in a proteome-wide scale, are now emerging to include this information in the rational design of personalized therapies. Genetic variation may also cause changes in the topology of cellular networks by changing protein expression or by affecting interactions in molecular networks.

The *druggability* of a protein target is its ability to be modulated by a high affinity small molecule. There are strong evolutionary arguments why proteins have evolved molecular recognition capabilities to avoid unwanted functional disruption in the vast sea of small-molecule metabolites in which they exist. Current estimates, from analysis of the pharmaceutical industry screening data, suggest that only approximately 15% of proteins expressed by an organism's genome have any inferred evidence of being potentially modulated by drug-like compounds. Additionally druggability is an attribute that is likely to be independent of lethality. Many genome-scale comprehensive knockout studies in model organisms have consistently identified around 19% of genes to be individually essential. Thus, targets that are both lethal and druggable represent an intersection of less than 3% of the proteins expressed in a genome, assuming lethality and druggability are not correlated factors [10].

Here, it is important to emphasize that signal networks differ from metabolic networks by the fact that the interacting protein interfaces in PPI are difficult targets as they are usually large and flat and often lack the cavities present at the surface of small-molecule protein receptors or enzymes [66]. However, significant progresses were made in the past years [67], but there is still a need for using the knowledge of successful cases to better rationalize the chemical space [68]. For instance, Koes and Camacho [69] recently developed a strategy that explores the properties of PPI interfaces to discover promising starting points for small-molecule design, and several algorithms were introduced to predict PPI hot spots [70].

3   Enzymes as Targets

Enzymes are biological catalysts able to work under mild conditions of temperature and pressure that accelerate chemical reactions making them compatible with life [71]. They hold a preeminent position among protein targets because of the essentiality of their activity in many disease processes and because the structural determinants of enzyme catalysis are suitable to inhibition by small drug-like molecules. Not surprisingly, enzyme inhibitors represent almost half the drugs in clinical use today [72]. Actually, inhibitors of enzyme reactions are among the most potent and effective drugs known when their mode of action is based on competition with the original enzyme substrate [73].

Target selectivity is generally considered an important attribute for the avoidance of off-target-based toxicities [74]. In preclinical compound evaluation, selectivity is most commonly measured as the ratio of drug affinity for binding to a collateral protein (off-target) to that for the target of interest (on-target) [75].

In 2002 newly launched drugs that reached almost half (47%) of the drugs approved by FDA were enzyme inhibitors [76]. Since that time, this proportion has diminished drastically with 70% of approved drugs in 2012 being small-molecule inhibitors from which only 3% were directed toward enzymes [77].

The attractiveness of enzymes as drug targets resulted not only from the essentiality of their catalytic activity but also from their suitability for inhibition by small molecular weight drug-like molecules (druggability). The fact that a protein contains a druggable binding pocket does not necessarily imply that it is a good target for drug discovery; there must be some expectation that the protein plays some pathogenic role in the disease under consideration so that its inhibition will lead to a disease modification. Further the binding pocket engaged by the drug must be critical to the biological activity of the molecular target, such that interactions between the drug and the target's binding pocket lead to an attenuation of biological activity. The active sites of enzymes are usually located in surface clefts and crevices. This substrate-target interaction effectively excludes bulk solvent (water), which would otherwise reduce the catalytic activity of the enzyme. In other words, the substrate molecule is desolvated upon binding. Solvation by water is replaced by specific interactions with the protein depending on the atom arrangement in the enzyme active site, which in some way complements the structure of the substrate molecule. The inhibitor binds to the enzyme with or without competition with the substrates, but in any case leading it to decrease its catalytic activity [72].

Many proteins are potential targets for drug interventions that control human diseases. The most recent number of drug targets was estimated to be in the hundreds, based on an analysis made before 2007 [78]. However, the number of druggable proteins is substantially greater according to the DrugBank database website (http://www.drugbank.ca/). The current version of this database (5.0) contains 8206 drug entries that are linked to 4333 nonredundant (nr) protein sequences (i.e., drug target/enzyme/transporter/carrier).

The lack of progress, the small number of drugs available in the antiparasitic class [9], and the relative toxicity of the existing ones are motivations to explore the opportunities raised by post-genomic strategies to boost pharmaceutical research in that area.

The major categories of drug targets include enzymes, GPCRs, nuclear hormone receptors, transporters, ion channels, and nucleic acids. Except for nucleic acids, all these targets elicit biological functions through ligand binding. Enzymes, however, are catalysts that make and break covalent chemical bonds, and nature has optimized enzymes for chemical transformations rather than ligand binding. This makes enzyme drug targets different and offers opportunities for drug design that take advantage of catalysis rather than binding [73]. Among target enzymes, kinases and proteases deserve attention from pharmaceutical industry at moment because of their involvement in (i) signaling cascade [79] and (ii) multiple biological processes in all living organisms [80], respectively.

Kinases, also known as phosphotransferases, are the most intensively studied protein drug target category in current pharmacological research, as evidenced by the large number of kinase-targeting agents enrolled in active clinical trials [81].

Kinase activity plays a key role in many cellular processes, such as cell cycle progression, apoptosis, differentiation, and signal transduction [82]. Eukaryotic protein kinases are related by a homologous catalytic domain of approximately 250–300 amino acids [83] and can be grouped into the serine/threonine and tyrosine kinases, which are responsible for phosphorylating the hydroxyl oxygen of their respective amino acids. Due to the pivotal role of kinases in the regulation of many cellular processes, aberrant kinase activity has been associated with a variety of diseases and the majority of human cancers [84]. Aberrant kinase activity is implicated in a variety of human diseases, in particular those involving inflammatory or proliferative responses, such as cancer, rheumatoid arthritis, cardiovascular and neurological disorders, asthma, and psoriasis [85].

Proteases are encoded by more than 550 human genes [86–88] that regulate growth factors, cytokines, chemokines, and cellular receptors, through activation and inactivation leading to downstream intracellular signaling and gene regulation [89]. They have an important role in many signaling pathways and represent potential drug targets for diseases ranging from cardiovascular disorders to cancer, as well as for combating many parasites and viruses [90].

Upregulation of proteolysis is commonly associated with different types of cancer and is linked to tumor metastasis, invasion, and growth [91]. Dysregulated proteolysis is also a feature of various inflammatory and other diseases.

4 System Biology to Predict Protein Targets in Infectious Diseases

As pointed out by Fauci and Morens [92] "great pandemics and local epidemics alike have influenced the course of wars, determined the fates of nations and empires, and affected the progress of civilization, making infections compelling actors in the drama of human history." Today, thanks to scientific achievements, pandemics are under control; however, in developed countries, pathogens are evolving toward the acquisition of antimicrobial resistance mainly because of the overuse of antibiotics in animals [92]. The increasing rate of emerging and reemerging pathogens is being observed as a consequence of the global circulation of populations from different ecoclimatic conditions as well as climatic changes [93]. As a matter of fact, emerging and neglected infectious diseases disproportionately afflict the poorest members of the global society. The World Health Organization (WHO) estimated that one in six of the world's population suffers from one or more neglected infectious diseases, such as onchocerciasis, trypanosomiasis, lymphatic filariasis, schistosomiasis, soil-transmitted helminthiasis, blinding trachoma, malaria, tuberculosis, and human immunodeficiency virus. Contagious diseases account for 50% of diseases in the developing countries, which represent 4.8 billion people and 80% of the world population [10]. Thus, systems of high productivity discovery that can be applied to a large number of pathogens are needed. The growing availability of parasite genome data provides the basis for the analysis of the pharmacological landscape of an infectious disease. However, infectious disease informatics is necessary for the rapid generation of plausible, novel medical hypotheses of testable pharmacological experiments [10].

The introduction of high-throughput sequencing or *next-generation sequencing* (NGS) has revolutionized the ability to detect novel infectious agents whose genomic sequences are completely unknown or are present in extremely low numbers [94]. NGS holds the promise of identifying all potential pathogens in a single assay by comparing reads of sufficiently large size to sequences of microbial genomes stored in well-annotated reference databases. Despite the deep sequence coverage of NGS, only a small fraction of short reads in clinical metagenomic samples may match pathogen sequences, and such sparse reads often do not overlap sufficiently to enable de novo assembly into larger contiguous sequences, also called *contigs* [95]. However, gene panels [96] now exist for pathogens and may be more appropriated to recover rare sequences by Ampliseq.

The most widely used approach of sample annotation in case of whole sample sequencing is to computationally subtract contigs with the host genome sequence (e.g., human), followed by their alignment to reference databases that contain sequences from candidate pathogens using Basic Local Alignment Search Tool (BLAST) [97], or some other algorithms. Another faster strategy that may be used is to detect the coding frame of contigs using codon bias, which is independent of the codon usage [98], and translate them into protein sequences prior to perform a BLASTp [99] or a SparkBLAST [100].

In addition, NGS and computational biology facilitate the identification of host factors that predispose or affect its response to a disease. Therefore, therapies can now be tailored according to the genetic makeup of a host and the characteristics of a microbe responsible for disease. For example, *genome-wide association studies* (GWAS) enabled scientists to identify human subpopulations that have genetic variants associated with different patterns of disease progression. However, a limitation of this technique is the huge sample size necessary for statistical validation of rare mutations [94].

Pathogens must overcome host defenses and then reproduce in order to propagate pathogen proteins that interact with host proteins to either suppress or hijack the normal host protein functions. Identification of these PPIs is not only critical for understanding the biology of infection but can also allow the identification of new targets in treatments against human pathogens [101].

Thus, comparative signaling network investigations have also been conducted in host-parasite system and seem to be more productive in potential protein targets than enzyme-based targets in terms of numbers. An example of this is the comparison of PPI in humans and *Leishmania major* that provided 140 targets that are specific to the protozoan parasite [102].

In order to maximize the benefit to patients of therapies against infectious agents, drugs should be as much as possible on-target for the parasite. Thus, pathogens' protein target should be (i) specific, that is, the selected target should be ideally absent from the human host because its presence could promote toxic effects from the treatment, and (ii) essential for the survival of the parasite [103, 104]. Thus, a drug that is able to inactivate a specific protein/enzyme of a pathogen is probably safe for its host. By contrast, a drug able to inactivate a protein/enzyme that is present in both the pathogen and the host may cause toxic effects on the host. This trivial observation leads to the concept of target specificity. One may distinguish several types of specificities in protein targets in a comparison between a parasite and its host:

(i) *Strict specificity* when the function associated with the protein is found in the parasite, but not in its host. As a consequence, an on-target drug for that target is expected to have only few or no side effect for the host.

(ii) *Functional specificity*, when paralogous proteins exist in both the parasite and its host but associated with different functions. Since both proteins are homologous, but with 3D structure optimized to interact with different substrates, a certain level of drug specificity is expected. However, the same is also true for expected collateral effects to the host since a drug could interact, even if with a different rate, with the other member of the homologous pair [34].

(iii) *Analogy*, when two proteins do have the same function but resulted from convergent evolution. Analogy is often associated with different 3D structures and may offer physicochemical specificity suitable for drug development [105].

These concepts were used to search for specific and analogous enzymes in several host-parasites systems [106] and automated in a pipeline called Analogous Enzyme Pipeline (AnEnPi) (http://anenpi.fiocruz.br/). Based on the interesting source of target represented by analogous enzymes, a pan-genomic investigation has been carried out in humans [107]. Similar pipeline was produced independently for human vs. *Escherichia coli* host-parasite relationship [108]. On a higher level of integration, several multi-omic approaches have integrated proteomics and metabolomics to obtain a system-level understanding of metabolic pathway regulation upon infection [109, 110]. In these studies, the integration of protein to metabolic pathways was used to identify specific proteins that may be targeted by pathogens to cause metabolic alterations. By analyzing network topology, one can identify functional relations between nodes in the network and key regulators of a system. In an early example of multi-omic network analysis during infection with hepatitis C virus (HCV), proteomic and lipidomic data were used to generate a network relating proteins and lipids through abundance correlations [111]. As new omics methods continue to be developed, their integration with other omics approaches will provide additional levels of information that will benefit pathogenic research.

5 Protein Network

Because the number of interactions of a single protein with its neighbors may vary from one or two to dozens, a common strategy of network representation is graphs where topological features become evident at a glance. These features may, for instance, be connection hubs or cliques, modules, and other regulatory motives [112].

PPIs have a key role in regulating many biological processes in metabolic, regulatory, and signaling pathways. The dysfunction of these pathways may lead to cellular diseases such as cancer and neurological disorders. Thus, network-based approaches have gained importance in drug R&D because they enable to investigate diseases as complex systems. Disease phenotypes reflect pathobiological processes that interact in a complex network [56]. The high interconnectivity in disease-associated networks suggests that it is better to target entire pathways rather than single proteins, which has promoted the raise polypharmacological approaches [113].

When inside cells, proteins typically do not function in their native state alone, but rather by interacting in concert with other proteins and metabolites, generating a high-dimensional network with a complicated structure, which justify system biology to identify essential protein targets for drug or vaccine development [102].

A PPI network involves nodes that correspond to proteins and edges, which denote interactions among proteins. High-throughput technologies allow rapid identification of PPIs and their networks [114].

PPI networks are generally considered to be scale-free, and such networks typically have a few hubs (nodes with high connection degree). The hubs play a central role in a network by connecting several nodes together; thus, the inactivation of their corresponding protein, in theory, may disrupt a number of essential pathways and disarticulated a large part of the signaling network. As a result, hubs may be considered for further evaluation for being potential drug targets [115].

Nodes with high betweenness centrality could be considered as initial candidates as drug targets. Betweenness centrality characterizes the degree of influence a protein has in communicating between protein pairs and is defined as the fraction of shortest paths going through a given node [116]. One can also employ *flux balance analysis* (FBA) of metabolic networks to find critical genes involved in the maintenance of a pathological metabolism or organism that could serve as drug targets [117].

Investigations carried out with experimental PPI networks of yeast and *C. elegans* have confirmed the effectiveness of topological metrics in predicting protein essentiality and demonstrated strong correlation between inferences and knockout or knockdown data [118, 119]. A single network covering all the genes of an organism may guide predictions at the level of individual cells or even tissues.

In yeast, Jeong et al. [118] showed the phenotypic consequence of a single-gene deletion that resulted in extensive topological alterations in the complex hierarchical web of molecular interactions in a PPI network [118].

Protein connectivity in the signaling network of *C. elegans* was demonstrated to be a parameter of essentiality that is evolutionarily conserved, and it has allowed the prediction of which genes are important for the majority of systematically tested phenotypes in this worm [119].

6 Metabolic Networks

Advances in post-genomic era opened the door for understanding physiology from the outcome of biologically plausible constraints imposed on metabolic reaction fluxes at the molecular level. In particular, genome-scale reconstruction of metabolic networks provided valuable tools for assessing organism-wide changes resulting from point modifications of metabolic reaction fluxes both at gene and posttranscriptional levels [120].

Reconstruction of genome-scale metabolic networks containing all of the known metabolic reactions in an organism and the genes encoding each enzyme is now possible [121, 122]. In particular, constraint-based methods like FBA enable the prediction of phenotypic traits like optimal growth rate from the maximization of an objective function written in terms of reaction fluxes and biologically plausible constraints [123]. The most general constraint underlying FBA is mass balance or stationarity of metabolite concentrations inside the cell, which is observed when a bacterial population doubles at regular intervals [124] and the concentration of each intracellular metabolite remains constant in time through the balance between production and consumption rates.

Cell duplication can be abstracted as the accumulation of biomass precursors in relative amounts determined by experimental assays [125]. A pseudoreaction with substrates as biomass precursors can be added to the reconstruction to enable cell growth modeling; each metabolite is given a stoichiometric index corresponding to its relative cellular amount with a negative or positive value if the metabolite is the substrate or product of a reaction, respectively, and scored as zero if not present. The flux of this reaction, f_b, can be interpreted as the growth rate, and arguments based on evolution and adaptation suggest that, under favorable conditions, cells should maximize growth rates [126]. Thus, reaction rates in this organism should satisfy the FBA conditions determined by formula 1:

$$\text{MAX } f_b \tag{1}$$

Given formula 2,

$$\sum_{j=1}^{N} S_{ij} f_j = 0 \text{ for all metabolites } i, \tag{2}$$

where f_j is the flux of reaction j and S_{ij} the stoichiometry of the metabolite i in the reaction j. All reactions have upper bounds with some of which that can be experimentally determined or computationally predicted. In practice, metabolic fluxes are limited on their positive side, i.e., metabolite production, and reversible reactions are split in two parts according to the rate of conversion from substrate to product and vice versa.

FBA is a powerful tool for the prediction of essential reactions or essential genes from an organism. By *essential*, we mean here that the inactivation of the referred reaction or gene will break biomass production and growth rate of the modeled organism. Actually, FBA states that a gene is essential if it is constraining fluxes of associated reaction(s) to zero, which leads to a zero biomass production, i.e., MAX $f_b = 0$ (see Eq. 1), in the above definition [127–129]. This statement has the corollary that critical genes can be search in a given organism in comparison to another, which is what one would expect of a molecular target in a host-parasite relationship. However, in order to be able to consider the target as potentially suitable (lead candidate), it must bear some grade of specificity of the parasite compared to its host (see discussion above). This second requirement is needed to reduce as much as possible on-target toxicity on the host. Of course, the referred specificity does not eliminate toxicity sources from off-target noxious effect on the host, but this is an issue that must be evaluated by rational drug design in relation to the parasite target. New strategies for protein targeting are now available and are commonly grouped in the category of *biopharmaceuticals* (RNAi, peptides, antibodies, etc.). However, the suitability of these molecules must be analyzed case by case because they may eventually fail to reach their intracellular target by being unable to cross the outer layer of the targeted cell. Among the promising techniques, one may cite gene silencing or posttranscriptional gene regulation by RNA interference [130], where interception and degradation of mRNA are induced by synthetic double-stranded RNA molecules (dsRNA) homologous to the target gene. Thus, in order to be considered an efficient drug, the engineered dsRNA molecules must (i) target essential genes from the pathogen, (ii) share little or no homology with the host's genes, and (iii) be able to reach its molecular target.

With the large number of entirely sequenced organisms and genome-scale reconstruction of metabolic networks, it is now possible to put host-pathogen interactions in a metabolic network perspective to predict possible organism-wide metabolic changes resulting from the underlying molecular interactions [131]. We propose the combination of FBA methodology to find essential genes with similarity tests to determine which genes are specific to a pathogen. Genes which are good targets for RNA interference or any other means of molecular target inactivation (small-molecule inhibitors, peptides, haptamers, antibodies) are those that are (i) critical for biomass production, in the FBA sense, and (ii) specific to the pathogen's genome as diagnosed through sequence homology comparison measured by e-value and/or score.

7 Conclusion

Information systems that could automatically search for associations between proteins, compounds, and diseases are not any more an unreachable dream. First, data on disease parameters such as (i) protein targets that are effectively druggable, (ii) group of genes or of gene products that are key in the management of a disease, (iii) profiling on individual metabolic response to drug, (iv) off-target action of drugs and their noxious collateral effect, and (v) mutation frequencies in key genes of humans and parasites are accumulating. Second, system biology is extending its analytical and processing toolbox through the development of (i) high-throughput methodologies of molecular characterization, such as cheap DNA or RNA sequencing, proteome mass spectrometry, metabolome mass spectrometry, and magnetic nuclear resonance (MNR), and of (ii) automated treatments through the algorithms of bioinformatics and computational biology in the context of omics and artificial intelligence, such as machine learning, probabilistic models (Bayesian and Markovian), Boolean models, kinetic modeling through nonlinear differential equation, etc. The integration of these technologies forms the core of the system biology solution to the problem of human disease and illustrates the path that humanity is taking toward a molecular approach of precision medicine with maximum reduction of collateral effects to patients.

Essentiality and synthetic lethality (when a combination of deficiencies in the expression of two or more genes leads to cell death) can be addressed by polypharmacology to beat biochemical network wiring and dynamic rewiring (promoting escaping or compensating pathways) in pathogens as well as in cell diseases such as cancer. Druggability and selectivity are physicochemical concepts, while efficacy depends on the holistic functioning of the host-pathogen system, i.e., it extends across essential genes, synthetically lethal gene combinations, virulence factors, and other factors. To achieve the desired lethal or static effect, different degrees of inhibition may be required, which implies the necessity of tuning the type of therapeutic agents used as well as their combination and dosage.

If system biology is promising, it is also true that the labor to reach the dream outlined above is huge and has ultimately to go through clinical phase approval to be transformed in translational action. Clinical trial step is an integral part of the feedback loop in which system biology (hypothesis generation) and bench experimentation (preclinical validation) are engaged. The reproducibility in that experimentation cycle is an essential element that goes through what is called *good laboratory practice* and formed by rules of conduct to respect to safely reach a given purpose. Rules form themselves through learning curves, and we give, here, the recommendations for NGS-based experimentation [132], as an example.

Acknowledgment This study was supported by a fellowship from *Coordenação de Aperfeiçoamento de Pessoal de Nível Superior* (http://www.capes.gov.br/) to LCC.

References

1. Knight-Schrijver VR, Chelliah V, Cucurull-Sanchez L, Le Novère N. The promises of quantitative systems pharmacology modelling for drug development. Comput Struct Biotechnol J. 2016;14:363–70.
2. Arrowsmith J. A decade of change. Nat Rev Drug Discov. 2012;11(1):17–8.
3. Paul SM, Mytelka DS, Dunwiddie CT, Persinger CC, Munos BH, Lindborg SR, et al. How to improve R&D productivity: the pharmaceutical industry's grand challenge. Nat Rev Drug Discov. 2010;9(3):203–14.

4. Butcher EC. Can cell systems biology rescue drug discovery? Nat Rev Drug Discov. 2005;4(6):461–7.
5. Osterloh IH. The discovery and development of Viagra® (sildenafil citrate). In: Dunzendorfer U, editor. Sildenafil. Milestones in drug therapy MDT. Basel: Birkhäuser; 2004. https://doi.org/10.1007/978-3-0348-7945-3_1.
6. Priest BT, Erdemli G. Phenotypic screening in the 21st century. Front Pharmacol. 2014;5:264.
7. Kaelin WG Jr. Common pitfalls in preclinical cancer target validation. Nat Rev Cancer. 2017;17:425–40.
8. Lee J, Bogyo M. Target deconvolution techniques in modern phenotypic profiling. Curr Opin Chem Biol. 2013;17(1):118–26.
9. Santos R, Ursu O, Gaulton A, Bento AP, Donadi RS, Bologa CG, et al. A comprehensive map of molecular drug targets. Nat Rev Drug Discov. 2017;16(1):19–34.
10. Hopkins AL, Richard Bickerton G, Carruthers IM, Boyer SK, Rubin H, Overington JP. Rapid analysis of pharmacology for infectious diseases. Curr Top Med Chem. 2011;11(10): 1292–300.
11. Soldatow VY, LeCluyse EL, Griffith LG, Rusyn I. In vitro models for liver toxicity testing. Toxicol Res. 2013;2(1):23–39.
12. Luni C, Serena E, Elvassore N. Human-on-chip for therapy development and fundamental science. Curr Opin Biotechnol. 2014;25:45–50.
13. Peng W, Unutmaz D, Ozbolat IT. Bioprinting towards physiologically relevant tissue models for pharmaceutics. Trends Biotechnol. 2016;34(9):722–32.
14. Andrade EL, Bento AF, Cavalli J, Oliveira SK, Freitas CS, Marcon R, et al. Non-clinical studies required for new drug development-part I: early in silico and in vitro studies, new target discovery and validation, proof of principles and robustness of animal studies. Braz J Med Biol Res. 2016;49(11):e5644.
15. Denayer T, Stöhr T, Van Roy M. Animal models in translational medicine: validation and prediction. New Horizons Transl Med. 2014;2(1):5–11.
16. Wermuth CG. Multitargeted drugs: the end of the "one-target-one-disease"philosophy? Drug Discov Today. 2004;9(19):826–7.
17. Masoudi-Nejad A, Mousavian Z, Bozorgmehr JH. Drug-target and disease networks: polypharmacology in the post-genomic era. In silico Pharmacol. 2013;1(1):17.
18. Ágoston V, Csermely P, Pongor S. Multiple weak hits confuse complex systems: a transcriptional regulatory network as an example. Phys Rev E. 2005;71(5):51909.
19. Winter GE, Rix U, Carlson SM, Gleixner KV, Grebien F, Gridling M, et al. Systems-pharmacology dissection of a drug synergy in imatinib-resistant CML. Nat Chem Biol. 2012;8(11):905–12.
20. Chen S, Jiang H, Cao Y, Wang Y, Hu Z, Zhu Z, et al. Drug target identification using network analysis: taking active components in Sini decoction as an example. Sci Rep. 2016;6. https://doi.org/10.1038/srep24245.
21. Kinch MS, Haynesworth A, Kinch SL, Hoyer D. An overview of FDA-approved new molecular entities: 1827–2013. Drug Discov Today. 2014;19(8):1033–9.
22. Winkler H. Target validation requirements in the pharmaceutical industry. Targets. 2003;2(3):69–71.
23. Milligan PA, Brown MJ, Marchant B, Martin SW, Graaf PH, Benson N, et al. Model-based drug development: a rational approach to efficiently accelerate drug development. Clin Pharmacol Ther. 2013;93(6):502–14.
24. Visser SAG, Manolis E, Danhof M, Kerbusch T. Modeling and simulation at the interface of nonclinical and early clinical drug development. CPT Pharmacometrics Syst Pharmacol. 2013;2(2):1–3.
25. Visser SAG, Aurell M, Jones RDO, Schuck VJA, Egnell A-C, Peters SA, et al. Model-based drug discovery: implementation and impact. Drug Discov Today. 2013;18(15):764–75.
26. Cook D, Brown D, Alexander R, March R, Morgan P, Satterthwaite G, et al. Lessons learned from the fate of AstraZeneca's drug pipeline: a five-dimensional framework. Nat Rev Drug Discov. 2014;13(6):419–31.

27. Visser SAG, Alwis DP, Kerbusch T, Stone JA, Allerheiligen SRB. Implementation of quantitative and systems pharmacology in large pharma. CPT Pharmacometrics Syst Pharmacol. 2014;3(10):1–10.
28. Kumar N, Hendriks BS, Janes KA, de Graaf D, Lauffenburger DA. Applying computational modeling to drug discovery and development. Drug Discov Today. 2006;11(17):806–11.
29. Brodland GW. How computational models can help unlock biological systems. Semin Cell Dev Biol. 2015;47–48:62–73.
30. Bombelles T, Coaker H. Neglected tropical disease research: rethinking the drug discovery model. Future Med Chem. 2015;7(6):693–700.
31. Al-Lazikani B, Workman P. Unpicking the combination lock for mutant BRAF and RAS melanomas. Cancer Discov. 2013;3(1):14–9.
32. Workman P, Clarke PA, Al-Lazikani B. Blocking the survival of the nastiest by HSP90 inhibition. Oncotarget. 2016;7(4):3658.
33. Paolini GV, Shapland RHB, van Hoorn WP, Mason JS, Hopkins AL. Global mapping of pharmacological space. Nat Biotechnol. 2006;24(7):805–15.
34. Catharina L, Lima CR, Franca A, Guimarães ACR, Alves-Ferreira M, Tuffery P, Derreumaux P, et al. A computational methodology to overcome the challenges associated with the search for specific enzyme targets to develop drugs against. Bioinform Biol Insights. 2017;11. https://doi.org/10.1177/1177932217712471.
35. Wei D, Jiang X, Zhou L, Chen J, Chen Z, He C, et al. Discovery of multitarget inhibitors by combining molecular docking with common pharmacophore matching. J Med Chem. 2008;51(24):7882–8.
36. Wishart DS, Knox C, Guo AC, Shrivastava S, Hassanali M, Stothard P, et al. DrugBank: a comprehensive resource for in silico drug discovery and exploration. Nucleic Acids Res. 2006;34(suppl 1):D668–72.
37. Günther S, Kuhn M, Dunkel M, Campillos M, Senger C, Petsalaki E, et al. SuperTarget and matador: resources for exploring drug-target relationships. Nucleic Acids Res. 2007;36(suppl_1):D919–22.
38. Gao Z, Li H, Zhang H, Liu X, Kang L, Luo X, et al. PDTD: a web-accessible protein database for drug target identification. BMC Bioinformatics. 2008;9(1):104.
39. Hu Z, Chang Y-C, Wang Y, Huang C-L, Liu Y, Tian F, et al. VisANT 4.0: integrative network platform to connect genes, drugs, diseases and therapies. Nucleic Acids Res. 2013;41(W1):W225–31.
40. Kim Kjærulff S, Wich L, Kringelum J, Jacobsen UP, Kouskoumvekaki I, Audouze K, et al. ChemProt-2.0: visual navigation in a disease chemical biology database. Nucleic Acids Res. 2012;41(D1):D464–9.
41. Yamanishi Y, Kotera M, Moriya Y, Sawada R, Kanehisa M, Goto S. DINIES: drug–target interaction network inference engine based on supervised analysis. Nucleic Acids Res. 2014;42:W39–45.
42. Hu Q, Deng Z, Tu W, Yang X, Meng Z, Deng Z, et al. VNP: interactive visual network pharmacology of diseases, targets, and drugs. CPT Pharmacometrics Syst Pharmacol. 2014;3(3):1–8.
43. Barabasi A-L, Oltvai ZN. Network biology: understanding the cell's functional organization. Nat Rev Genet. 2004;5(2):101–13.
44. Yamada T, Bork P. Evolution of biomolecular networks—lessons from metabolic and protein interactions. Nat Rev Mol Cell Biol. 2009;10(11):791–803.
45. Yıldırım MA, Goh K-I, Cusick ME, Barabási A-L, Vidal M. Drug—target network. Nat Biotechnol. 2007;25(10):1119–26.
46. Butina D, Segall MD, Frankcombe K. Predicting ADME properties in silico: methods and models. Drug Discov Today. 2002;7(11):S83–8.
47. Byvatov E, Fechner U, Sadowski J, Schneider G. Comparison of support vector machine and artificial neural network systems for drug/nondrug classification. J Chem Inf Comput Sci. 2003;43(6):1882–9.

48. Khedkar SA, Malde AK, Coutinho EC, Srivastava S. Pharmacophore modeling in drug discovery and development: an overview. Med Chem. 2007;3(2):187–97.
49. Morris GM, Huey R, Lindstrom W, Sanner MF, Belew RK, Goodsell DS, et al. AutoDock4 and AutoDockTools4: automated docking with selective receptor flexibility. J Comput Chem. 2009;30(16):2785–91.
50. Knox C, Law V, Jewison T, Liu P, Ly S, Frolkis A, et al. DrugBank 3.0: a comprehensive resource for "omics" research on drugs. Nucleic Acids Res. 2010;39:D1035–41.
51. Wang JT, Liu W, Tang H, Xie H. Screening drug target proteins based on sequence information. J Biomed Inform. 2014;49:269–74.
52. Li Z, Han P, You Z-H, Li X, Zhang Y, Yu H, et al. In silico prediction of drug-target interaction networks based on drug chemical structure and protein sequences. Sci Rep. 2017;7(1):11174.
53. Lam MPY, Venkatraman V, Xing Y, Lau E, Cao Q, Ng DCM, et al. Data-driven approach to determine popular proteins for targeted proteomics translation of six organ systems. J Proteome Res. 2016;15(11):4126–34.
54. Zhu S, Okuno Y, Tsujimoto G, Mamitsuka H. A probabilistic model for mining implicit "chemical compound–gene"relations from literature. Bioinformatics. 2005;21(suppl 2):ii245–51.
55. Wang Z, Li J, Dang R, Liang L, Lin J. PhIN: a protein pharmacology interaction network database. CPT Pharmacometrics Syst Pharmacol. 2015;4(3):160–6.
56. Barabási A-L, Gulbahce N, Loscalzo J. Network medicine: a network-based approach to human disease. Nat Rev Genet. 2011;12(1):56–68.
57. Goh K-I, Cusick ME, Valle D, Childs B, Vidal M, Barabási A-L. The human disease network. Proc Natl Acad Sci. 2007;104(21):8685–90.
58. Yamanishi Y, Araki M, Gutteridge A, Honda W, Kanehisa M. Prediction of drug–target interaction networks from the integration of chemical and genomic spaces. Bioinformatics. 2008;24(13):i232–40.
59. Stockwell BR. Chemical genetics: ligand-based discovery of gene function. Nat Rev Genet. 2000;1(2):116–25.
60. Dobson CM. Chemical space and biology. Nature. 2004;432(7019):824–8.
61. Kanehisa M, Goto S, Hattori M, Aoki-Kinoshita KF, Itoh M, Kawashima S, et al. From genomics to chemical genomics: new developments in KEGG. Nucleic Acids Res. 2006;34:D354–7.
62. Kundrotas PJ, Zhu Z, Janin J, Vakser IA. Templates are available to model nearly all complexes of structurally characterized proteins. Proc Natl Acad Sci. 2012;109(24):9438–41.
63. Duran-Frigola M, Mosca R, Aloy P. Structural systems pharmacology: the role of 3D structures in next-generation drug development. Chem Biol. 2013;20(5):674–84.
64. Yang JO, Oh S, Ko G, Park S-J, Kim W-Y, Lee B, et al. VnD: a structure-centric database of disease-related SNPs and drugs. Nucleic Acids Res. 2010;39(suppl 1):D939–44.
65. Preissner S, Kroll K, Dunkel M, Senger C, Goldsobel G, Kuzman D, et al. SuperCYP: a comprehensive database on cytochrome P450 enzymes including a tool for analysis of CYP-drug interactions. Nucleic Acids Res. 2009;38(suppl 1):D237–43.
66. Fuller JC, Burgoyne NJ, Jackson RM. Predicting druggable binding sites at the protein–protein interface. Drug Discov Today. 2009;14(3):155–61.
67. Schlecht U, Miranda M, Suresh S, Davis RW, Onge RPS. Multiplex assay for condition-dependent changes in protein–protein interactions. Proc Natl Acad Sci U S A. 2012;109(23):9213–8.
68. Sperandio O, Reynès CH, Camproux A-C, Villoutreix BO. Rationalizing the chemical space of protein–protein interaction inhibitors. Drug Discov Today. 2010;15(5):220–9.
69. Koes DR, Camacho CJ. PocketQuery: protein–protein interaction inhibitor starting points from protein–protein interaction structure. Nucleic Acids Res. 2012;40:W387–92.
70. Taboureau O, Baell JB, Fernández-Recio J, Villoutreix BO. Established and emerging trends in computational drug discovery in the structural genomics era. Chem Biol. 2012;19(1):29–41.

71. Świderek K, Tuñón I, Moliner V, Bertran J. Computational strategies for the design of new enzymatic functions. Arch Biochem Biophys. 2015;582:68–79.
72. Copeland RA. Evaluation of enzyme inhibitors in drug discovery: a guide for medicinal chemists and pharmacologists. Hoboken: Wiley; 2013.
73. Robertson JG. Enzymes as a special class of therapeutic target: clinical drugs and modes of action. Curr Opin Struct Biol. 2007;17(6):674–9.
74. Copeland RA. The dynamics of drug-target interactions: drug-target residence time and its impact on efficacy and safety. Expert Opin Drug Discov. 2010;5(4):305–10.
75. Copeland RA. Evaluation of enzyme inhibitors in drug discovery: a guide for medicinal chemists and pharmacologists. New York: Wiley-Interscience; 2005. p. 178–213.
76. Hopkins AL, Groom CR. The druggable genome. Nat Rev Drug Discov. 2002;1(9):727–30.
77. Thomas D. A big year for novel drugs approvals [Internet]. 2013. Available from: http://www.biotech-now.org/business-and-investments/inside-bio-ia/2013/01/a-big-year-for-novel-drugs-approvals.
78. Bull SC, Doig AJ. Properties of protein drug target classes. PLoS One. 2015;10(3):e0117955.
79. Burkhard K, Shapiro P. Use of inhibitors in the study of MAP kinases. MAP Kinase Signal Protoc. 2010;661:107–22.
80. López-Otín C, Bond JS. Proteases: multifunctional enzymes in life and disease. J Biol Chem. 2008;283(45):30433–7.
81. Rask-Andersen M, Zhang J, Fabbro D, Schiöth HB. Advances in kinase targeting: current clinical use and clinical trials. Trends Pharmacol Sci. 2014;35(11):604–20.
82. Manning G, Whyte DB, Martinez R, Hunter T, Sudarsanam S. The protein kinase complement of the human genome. Science. 2002;298(5600):1912–34.
83. Hanks SK, Hunter T. Protein kinases 6. The eukaryotic protein kinase superfamily: kinase (catalytic) domain structure and classification. FASEB J. 1995;9(8):576–96.
84. Engh RA, Bossemeyer D. Structural aspects of protein kinase control—role of conformational flexibility. Pharmacol Ther. 2002;93(2):99–111.
85. Melnikova I, Golden J. Targeting protein kinases. Nat Rev Drug Discov. 2004;3(12):993–4.
86. Puente XS, Sánchez LM, Overall CM, López-Otín C. Human and mouse proteases: a comparative genomic approach. Nat Rev Genet. 2003;4(7):544–58.
87. Puente XS, López-Otín C. A genomic analysis of rat proteases and protease inhibitors. Genome Res. 2004;14(4):609–22.
88. Puente XS, Sanchez LM, Gutierrez-Fernandez A, Velasco G, Lopez-Otin C. A genomic view of the complexity of mammalian proteolytic systems. Biochem Soc Trans. 2005;33(Pt 2):331–4.
89. Rodríguez D, Morrison CJ, Overall CM. Matrix metalloproteinases: what do they not do? New substrates and biological roles identified by murine models and proteomics. Biochim Biophys Acta. 2010;1803(1):39–54.
90. Drag M, Salvesen GS. Emerging principles in protease-based drug discovery. Nat Rev Drug Discov. 2010;9(9):690–701.
91. Duffy MJ, McGowan PM, Gallagher WM. Cancer invasion and metastasis: changing views. J Pathol. 2008;214(3):283–93.
92. Fauci AS, Morens DM. The perpetual challenge of infectious diseases. N Engl J Med. 2012;366(5):454–61.
93. Fontana JM, Alexander E, Salvatore M. Translational research in infectious disease: current paradigms and challenges ahead. Transl Res. 2012;159(6):430–53.
94. Wooller SK, Benstead-Hume G, Chen X, Ali Y, Pearl FMG. Bioinformatics in translational drug discovery. Biosci Rep. 2017;37(4):BSR20160180.
95. Naccache SN, Federman S, Veeraraghavan N, Zaharia M, Lee D, Samayoa E, et al. A cloud-compatible bioinformatics pipeline for ultrarapid pathogen identification from next-generation sequencing of clinical samples. Genome Res. 2014;24(7):1180–92.
96. Frey KG, Bishop-Lilly KA. Next-generation sequencing for pathogen detection and identification. Methods Microbiol. 2015;42:525–54.

97. Altschul SF, Gish W, Miller W, Myers EW, Lipman DJ. Basic local alignment search tool. J Mol Biol. 1990;215(3):403–10.
98. Carels N, Frias D. A statistical method without training step for the classification of coding frame in transcriptome sequences. Bioinform Biol Insights. 2013;7:35.
99. Carels N, Gumiel M, da Mota FF, de Carvalho Moreira CJ, Azambuja P. A metagenomic analysis of bacterial microbiota in the digestive tract of triatomines. Bioinform Biol Insights. 2017;11. https://doi.org/10.1177/1177932217733422.
100. de Castro MR, dos Santos TC, Dávila AMR, Senger H, da Silva FAB. SparkBLAST: scalable BLAST processing using in-memory operations. BMC Bioinformatics. 2017;18(1):318.
101. Beltran PMJ, Federspiel JD, Sheng X, Cristea IM. Proteomics and integrative omic approaches for understanding host–pathogen interactions and infectious diseases. Mol Syst Biol. 2017;13(3):922.
102. Flórez AF, Park D, Bhak J, Kim B-C, Kuchinsky A, Morris JH, et al. Protein network prediction and topological analysis in *Leishmania major* as a tool for drug target selection. BMC Bioinf. 2010;11(1):484.
103. Cascante M, Boros LG, Comin-Anduix B, de Atauri P, Centelles JJ, Lee PW-N. Metabolic control analysis in drug discovery and disease. Nat Biotechnol. 2002;20(3):243–9.
104. Haanstra JR, Gerding A, Dolga AM, Sorgdrager FJH, Buist-Homan M, Du Toit F, et al. Targeting pathogen metabolism without collateral damage to the host. Sci Rep. 2017;7:40406.
105. Capriles PVSZ, Baptista LPR, Guedes IA, Guimarães ACR, Custódio FL, Alves-Ferreira M, et al. Structural modeling and docking studies of ribose 5-phosphate isomerase from *Leishmania major* and *Homo sapiens*: a comparative analysis for leishmaniasis treatment. J Mol Graph Model. 2015;55:134–47.
106. Otto TD, Guimarães ACR, Degrave WM, de Miranda AB. AnEnPi: identification and annotation of analogous enzymes. BMC Bioinf. 2008;9:544.
107. Piergiorge RM, de Miranda AB, Guimarães AC, Catanho M. Functional analogy in human metabolism: enzymes with different biological roles or functional redundancy? Genome Biol Evol. 2017;9(6):1624–36.
108. Mondal SI, Ferdous S, Jewel NA, Akter A, Mahmud Z, Islam MM, et al. Identification of potential drug targets by subtractive genome analysis of *Escherichia coli* O157: H7: an in silico approach. Adv Appl Bioinf Chem AABC. 2015;8:49.
109. Su L, Zhou R, Liu C, Wen B, Xiao K, Kong W, et al. Urinary proteomics analysis for sepsis biomarkers with iTRAQ labeling and two-dimensional liquid chromatography–tandem mass spectrometry. J Trauma Acute Care Surg. 2013;74(3):940–5.
110. Villar M, Ayllón N, Alberdi P, Moreno A, Moreno M, Tobes R, et al. Integrated metabolomics, transcriptomics and proteomics identifies metabolic pathways affected by *Anaplasma phagocytophilum* infection in tick cells. Mol Cell Proteomics. 2015;14(12):3154–72.
111. Diamond DL, Syder AJ, Jacobs JM, Sorensen CM, Walters K-A, Proll SC, et al. Temporal proteome and lipidome profiles reveal hepatitis C virus-associated reprogramming of hepatocellular metabolism and bioenergetics. PLoS Pathog. 2010;6(1):e1000719.
112. Salazar GA, Meintjes A, Mazandu GK, Rapanoël HA, Akinola RO, Mulder NJ. A web-based protein interaction network visualizer. BMC Bioinformatics. 2014;15(1):129.
113. Pujol A, Mosca R, Farrés J, Aloy P. Unveiling the role of network and systems biology in drug discovery. Trends Pharmacol Sci. 2010;31(3):115–23.
114. Hormozdiari F, Salari R, Bafna V, Sahinalp SC. Protein-protein interaction network evaluation for identifying potential drug targets. J Comput Biol. 2010;17(5):669–84.
115. Holme P, Kim BJ, Yoon CN, Han SK. Attack vulnerability of complex networks. Phys Rev E. 2002;65(5):56109.
116. Estrada E. Protein bipartivity and essentiality in the yeast protein−protein interaction network. J Proteome Res. 2006;5(9):2177–84.
117. Kell DB. Systems biology, metabolic modelling and metabolomics in drug discovery and development. Drug Discov Today. 2006;11(23):1085–92.
118. Jeong H, Mason SP, Barabási A-L, Oltvai ZN. Lethality and centrality in protein networks. Nature. 2001;411(6833):41–2.

119. Lee I, Lehner B, Crombie C, Wong W, Fraser AG, Marcotte EM. A single gene network accurately predicts phenotypic effects of gene perturbation in Caenorhabditis elegans. Nat Genet. 2008;40(2):181–8.
120. Kaltdorf M, Srivastava M, Gupta SK, Liang C, Binder J, Dietl A-M, et al. Systematic identification of anti-fungal drug targets by a metabolic network approach. Front Mol Biosci. 2016;3:22.
121. Joyce AR, Palsson BØ. The model organism as a system: integrating 'omics' data sets. Nat Rev Mol Cell Biol. 2006;7(3):198–210.
122. Oberhardt MA, Palsson BØ, Papin JA. Applications of genome-scale metabolic reconstructions. Mol Syst Biol. 2009;5(1):320.
123. Orth JD, Thiele I, Palsson BØ. What is flux balance analysis? Nat Biotechnol. 2010;28(3):245–8.
124. Neidhardt FC. Bacterial growth: constant obsession withdN/dt. J Bacteriol. 1999;181(24):7405–8.
125. Feist AM, Palsson BO. The biomass objective function. Curr Opin Microbiol. 2010;13(3):344–9.
126. Ibarra RU, Edwards JS, Palsson BO. *Escherichia coli* K-12 undergoes adaptive evolution to achieve in silico predicted optimal growth. Nature. 2002;420(6912):186–9.
127. Joyce AR, Palsson BØ. Predicting gene essentiality using genome-scale in silico models. Microb Gene Essentiality Protoc Bioinf. 2008;416:433–57.
128. Sylke M. Comprehensive analysis of parasite biology: from metabolism to drug discovery. Vol. 7. Weinheim: Wiley; 2016. 576 p.
129. Tobalina L, Pey J, Rezola A, Planes FJ. Assessment of FBA based gene essentiality analysis in cancer with a fast context-specific network reconstruction method. PLoS One. 2016;11(5):e0154583.
130. Koch A, Biedenkopf D, Furch A, Weber L, Rossbach O, Abdellatef E, et al. An RNAi-based control of fusarium graminearum infections through spraying of long dsRNAs involves a plant passage and is controlled by the fungal silencing machinery. PLoS Pathog. 2016;12(10):e1005901.
131. Peyraud R, Dubiella U, Barbacci A, Genin S, Raffaele S, Roby D. Advances on plant–pathogen interactions from molecular toward systems biology perspectives. Plant J. 2017;90(4):720–37.
132. Aziz N, Zhao Q, Bry L, Driscoll DK, Funke B, Gibson JS, et al. College of American Pathologists' laboratory standards for next-generation sequencing clinical tests. Arch Pathol Lab Med. 2014;139(4):481–93.

Mathematical-Computational Modeling in Behavior's Study of Repetitive Discharge Neuronal Circuits

Celia Martins Cortez, Maria Clicia Stelling de Castro, Vanessa de Freitas Rodrigues, Camila Andrade Kalil, and Dilson Silva

Abstract Mathematical-computational modeling is a tool that has been widely used in the field of Neuroscience. Despite considerable advances of Physiological Sciences, the neuronal mechanisms involved in the abilities of central nervous system remain obscure, but they can be revealed through modeling. Significant amount of experimental data already available has facilitated the development of models that combine experimentation with theory. They allow to evaluate hypotheses and to seek understanding of neuronal circuit functioning capable of explaining neurophysiological deficits. To model the behavior of repetitive discharge of neuronal circuits, we have used differential equations, graph theory, and other mathematical methods. Through computational simulations, using programs developed in C and C ++ language and neurophysiological data obtained in the literature, we can test the model's behavior in face of numerical variations of their parameters, trying to observe their characteristics.

1 Introduction

It has been known since the nineteenth century that neurons in the nervous system group together and connect to each other through synapses. They form complex circuits and neuronal networks, which are extremely complex structures and have an enormous capacity for processing and storing information [1, 2]. Each neuronal grouping presents its own organization, which allows the processing of signals in an own and unique way, thus allowing associations among different groups perform the multiplicity of nervous system functions [3, 4].

Almost all nerve synapses are chemical synapses. A nerve path transmits signals in a single sense, from the neuron called presynaptic to the postsynaptic neuron. The

C. M. Cortez · M. C. S. de Castro (✉) · V. d. F. Rodrigues · C. A. Kalil · D. Silva
Universidade do Estado do Rio de Janeiro, UERJ – Rua São Francisco Xavier, Rio de Janeiro, Brazil
e-mail: clicia@ime.uerj.br; dilsons@uerj.br

© Springer International Publishing AG, part of Springer Nature 2018
F. A. B. da Silva et al. (eds.), *Theoretical and Applied Aspects of Systems Biology*,
Computational Biology 27, https://doi.org/10.1007/978-3-319-74974-7_13

signals depend on chemical processes involving a neurotransmitter. This substance is produced in presynaptic neuron. Once it is released into synaptic cleft (or gap junction, space between two neurons), it interacts with a specific receptor, which is a struct located on postsynaptic neuron membrane. A synapse may be considered excitatory or inhibitory, according to the effect on postsynaptic neuron membrane. Excitatory synapses work to promote transmission of information in a nerve pathway. Inhibitory synapses generate resistance to transmission of information [5].

Anatomical structure of brain circuits and networks, including interneuronal connectivity and interregional connectivity, and the mapping of detailed connectivity patterns are objectives of Neuroscience [6–12]. With a mathematical model capable of representing the phenomenon or system, various simulations can be conducted to observe their behavior, testing hypotheses and verifying their responses to individual variations of their parameters [13–16]. The ultimate goal of most modeling efforts is to obtain a fully predictable description for modeling system [17].

The neurons network and circuits analysis offers new ways to quantitatively characterize anatomical and physiological patterns [7, 10–12, 14, 15, 18–21]. In neuronal networks there are grouped circuits forming clusters, maintaining short distance integrative characteristics. At present, detailed maps of links within human and other animal brains are being generated with new technologies available. Mathematical functions have been used to understand the general organizational characteristics of these structures [22, 23].

Over the years, artificial neural network models have sought to bring the processing of computers closer to that of the brain. Much knowledge has been gained by modeling neurophysiological circuits using experimental data in simulations to observe the behavior and limitations of models [14, 15, 18–21, 24]. Based on neural network characteristics, it is believed that new generations of computational systems will emerge, much more efficient and intelligent than the current systems. Undoubtedly, since the computer evolution, a great desire of man has been the creation of a machine that can operate free of human control, that is, an autonomous machine that operates according to its own learning and has the ability to interact with environments [25–27].

We have developed studies to deepen the knowledge about operation of repetitive discharge neuronal circuits. Initially we implemented a signal transmission model in a neuron small network composed of a reverberant circuit. In addition, we develop mathematical-computational modeling based on differential equations. Later, we apply the theory of graphs in neuronal circuits. Programs in C language were developed, and simulations were done using neurophysiological data obtained in the literature.

2 Neuronal Circuits and Reverberation

A neuron may have several connections to other neurons. Note in Fig. 1 that the soma surface and dendrites of postsynaptic neuron A are covered by many synapses, which may be both excitatory and inhibitory. These synapses are contact points

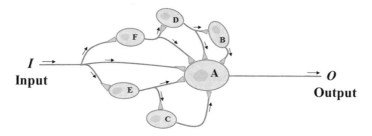

Fig. 1 Parallel neuronal circuit showing multiple synapses in a single output neuron

between neuron *A* and various neurons that surround it. In this way, all neurons in figure *converge* information to neuron *A*, and the synapses can be activated simultaneously.

In Fig. 1, *I* is the circuit entry pathway and *O* (axon of neuron *A*) is the exit path. Each nerve fiber can branch out for hundred to thousand times, generating many endings that are distributed over large area inside and outside the circuit, widely distributing information it carries. Generally, within the neuronal grouping, the dendrites also branch out and spread to thousands of micrometers.

2.1 Synaptic Transmission

In the central nervous system, a synapse consists of a contact zone between two neurons, functioning as a nervous *chip*, as it is capable of not only transmitting messages between two neurons but also blocking or modifying them entirely. Thus, in synapse, a truly information processing is performed, which can end up by modulating input information [4].

The information transmission at synapse or synaptic transmission depends on complex mechanisms involving five basic elements. These elements are presynaptic membrane, neurotransmitter, synaptic enzymes, postsynaptic membrane, and synaptic receptor.

Synaptic transmission process begins with the presynaptic membrane depolarization, by the arrival of action potential (AP) in presynaptic membrane (presynaptic neuron axon end). Action potential causes the opening of synaptic vesicles, and neurotransmitter flows into synaptic cleft (space between pre- and postsynaptic membranes). Upon reaching postsynaptic membrane, the neurotransmitter interacts with receptor and generates a change in the permeability of this membrane, through which a specific ion, which gives rise to postsynaptic potential (PSP), flows. At excitatory synapse, the neurotransmitter-receptor interaction generates a polarity inversion (or depolarization) of postsynaptic membrane, establishing an excitatory postsynaptic potential (EPSP). In inhibitory synapses, the interaction causes membrane hyperpolarization, generating an inhibitory postsynaptic potential (IPSP).

Soon after its formation, the postsynaptic potential (EPSP and IPSP) begins to vary exponentially with time, according to a time constant. The value of this constant depends on neurotransmitter type, synaptic receptor, and properties of postsynaptic membrane [5].

Synaptic enzymes have the function of limiting action time of each presynaptic discharge, by neurotransmitter inactivation. Thus, not all neurotransmitter molecules released into slit will have action on the receptor [2].

An excitatory synapse characteristic is that a single AP is not capable of causing the postsynaptic neuron firing, that is, transmitting information later, since the postsynaptic potential is a localized event. EPSP and IPSP occurring simultaneously can be added temporally and spatially. If this sum reaches the firing level or threshold, one (or more) AP(s) propagates the axon of postsynaptic neuron. The firing frequency depends on this summation (5). The postsynaptic potentials' algebraic sum of various active synapses is that it elevates neuronal membrane potential to a threshold.

Synaptic transmission has very specific characteristics that must be considered in its modeling and can be considered as problem contour conditions [14, 19]. In addition to signal transmission in a single direction, each synapse in a neuronal pathway represents a delay in information transmission. This time varies from 0.25 to 1 ms in the central nervous system (CNS). Depending on presynaptic discharge frequency, the synapse may be *fatigued* by depletion of neurotransmitter stores and decreased receptor sensitivity by very prolonged exposure (seconds to few minutes) to its neurotransmitter [5].

Post-tetanic facilitation (after high-frequency firings) is another important feature. This causes the postsynaptic neuron firing to be *facilitated* after a short-lived repetition of stimuli, since it leaves the membrane potential close to firing threshold, while post-tetanic or long-term potentiation generates a pronounced and prolonged potentiation of excitatory postsynaptic potential after tetanic stimulation, of a few seconds, increasing the likelihood of a postsynaptic neuron firing with subsequent stimulation [4].

2.2   Repetitive Discharge Circuits

The repetitive discharge circuits are mounted in such a way that output neuron fires repeatedly for each input signal. Two basic examples of circuits are parallel circuit and reverberant circuit.

Reverberant circuit is one of the most important circuits in the entire nervous system. Its assembly allows an internal positive feedback, so that an input signal feedbacks the circuit itself for a certain time, generating a reverberation effect [3, 28]. There are many reverberant circuit variations [5, 29]. A simple example is shown in Fig. 2 which involves only two neurons. The output neuron (*B*) sends a collateral branch back to input neuron (*A*), reinforcing its stimulation [30].

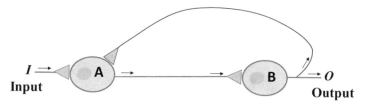

Fig. 2 Single reverberant circuit

One type of parallel circuit can be seen in Fig. 1. In this figure an input signal stimulates a sequence of neurons that connect with a single output cell. Thus, the output neuron receives several pulses, separated by a short interval, that result from a single input signal. Circuit output due to a single input signal ceases only after stimulation of the last two neurons [5].

3 Mathematical Models for Reverberating Neuronal Circuit

We can observe in Fig. 2 one circuit where the nerve impulse entering through I reaches neuron A and is processed. The generated signal follows to B. In turn, the signal generated in B, after processing, goes to O and simultaneously returns to A. In the following, we present two types of mathematical modeling for this type of circuit.

3.1 Using Differential Equations to Model the Reverberating Neuronal Circuit

In this modeling, the synaptic transmission phases, with molecular and electrical effects of synaptic phenomenon, are represented by differential equations [19, 21, 24].

It is known that neurotransmitter continuous discharges by presynaptic membrane progressively decrease the number of available postsynaptic receptors, increasing the effect on the postsynaptic potential over time, with a temporal summation occurring. On the other hand, the neurotransmitter-receptor interaction has limited duration, and receptors are soon vacated and occupied again. Another factor that limits the effect of each neurotransmitter pool is the enzyme action in synaptic cleft, which act with specificity by deactivating the excess neurotransmitter molecules [5].

Analyzing the phenomenon in an excitatory postsynaptic membrane, it can be considered that the variation of postsynaptic effect E (electric effect), due to neurotransmitter-receptor interaction, is proportional to membrane electric potential, having as proportionality constant the time constant τ_E. Thus, we can write that

$$dE/dT = -E/\tau_E,$$

being the solution of this equation:

$$E = E_0 \exp \left[\frac{-(t - t_0)}{\tau_E} \right], \tag{1}$$

where E_0 is the characteristic value of EPSP. Considering that in each t_i, the propagation of an AP at presynaptic terminal may or not occur (with a consequent neurotransmitter pool discharge), depending on signal frequency propagating at that terminal, we can write

$$E\ (t_n) = \sum_{i=1}^{n} E_0 \exp \left[\frac{-(t_i - t_0)}{\tau_E} \right], \tag{2}$$

where t_i is the arrival time of the *ith* AP and n is the total number of APs since t_0. Let E_j the EPSP value of synapse j, for the set of k synapses, being the weight of each synapse (W_{Ej}); we can write that synaptic effect is [24]

$$E\ (t_n) = \sum_{j}^{k} W_{Ej} E_{0j} \sum_{i=1}^{n} \exp \left[\frac{-(t_i - t_0)}{\tau_E} \right], \text{ for } t_n \geq t_0 \tag{3}$$

In analogy with this result, we can write for m action potentials arriving at q inhibitory synapses of synaptic weights W_{Ij} and time constants τ_I that

$$I\ (t_n) = \sum_{j}^{q} W_{Ij} I_{0j} \sum_{i=1}^{m} \exp \left[\frac{-(t_i - t_0)}{\tau_I} \right], \text{ for } t_n \geq t_0 \tag{4}$$

Equations 3 and 4 allow us to calculate excitatory postsynaptic effects caused by n APs arriving at k excitatory synapses and m APs arriving at q inhibitory synapses simultaneously in a postsynaptic neuron. We know that the postsynaptic potential variation and AP fire depend on the membrane intrinsic characteristics, i.e., resting potential (P_R), reversion potential (P_0), resting threshold potential (T_R), and AP potential post-firing threshold (T_p). Thus, we need at least two equations to model the postsynaptic neuron behavior:

(i) An equation to represent the membrane potential, $V(t)$, which is the net conductance variation of cytoplasmic membrane, involving excitatory and inhibitory synaptic effects along its surface, initial potential condition, and membrane characteristics P_R and P_0

(ii) Another to represent the resistance to the membrane potential development in direction of a threshold value, $T(t)$, thus involving and considering that membrane conditions before and after the AP trigger are different, already that AP is accompanied by a period of refractoriness, which lasts for a few milliseconds

Thus, it is understood that AP will form when $V(t) > T(t)$.

3.1.1   Membrane Potential: $V(t)$

Disregarding synaptic effects caused by neurotransmitter-receptor interaction, one can write that the membrane potential at any instant t is the sum of resting potential (P_P) and potential during refractory period (P_R):

$$V(t) = P_R + P_P. \tag{5}$$

As the P_P change in the direction of rest is proportional to the own P_P,

$$dP_P/dt = -P_R/\tau_P,$$

whose solution, considering that the value of P_P can vary from P_R to P_0, is

$$P_p = (P_0 - P_R) \exp[-(t - t_0)/\tau_P].$$

Substituting in Eq. 5 and considering the synaptic effects (Eqs. 3 and 4), we obtain

$$V(t) = P_R + (P_0 - P_R) \exp\left[-(t_i - t_0)/\tau_p\right] + E(t) + I(t). \tag{6}$$

where τ_P is the time constant of hyperpolarization period after AP. In this equation,

$$(P_0 - P) \exp\left[-(t_i - t_0)/\tau_P\right]$$

is the term referring to the variation of the membrane potential during hyperpolarization time. Thus, membrane potential $V(t)$ is the sum of four important terms due to (1) resting potential P_R; (2) difference between reversal potentials and resting potential, considering the time variation term, which involves time constant of hyperpolarization period after PA; (3) excitatory effect; and (4) inhibitory effect.

3.1.2   Threshold Equation: $T(t)$

Figure 3a illustrates the neuron membrane behavior potential in its various phases, showing the resting potential (P_R), the action potential (AP), and the refractory periods that influence the firing threshold of the neuron, absolute refractory period

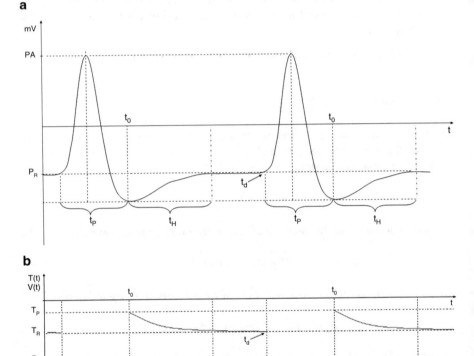

Fig. 3 (**a**) Neuron membrane potential model and its diverse phases. Resting potential (P_R), action potential (*AP*), absolute refractory period phase (t_P), relative refractory period phase (t_H), or hyperpolarization post-AP or threshold refund. Total refractory period: $t_R = t_P + t_H$. (**b**) Firing threshold variation, $T(t)$, according to represented model in Eq. 7. T_P= post-AP threshold potential, t_H = relative refractory period, T_H= relative threshold potential, T_R= resting threshold potential

(t_P) and relative refractory period (t_H), or post-AP hyperpolarization phase, until a threshold potential is restored.

In period t_P, it is not possible to deflagrate an AP, regardless of presynaptic discharge frequency, since t_P coincides with repolarization phase. It can be considered that the post-AP threshold potential (T_P) tends to infinity in this period. In period t_H, the relative threshold potential (T_H) varies with time, decreasing until reaching the value of resting threshold potential (T_R). So we can write that

$$R_P = t_P + t_H \text{ and } T(t) = T_R + T_H$$

Taking into account that the fall of T_H is proportional to T_H, we have to

$$dT_H/dt = -T_H/\tau_H.$$

Considering that T_H varies T_p from T_R,

$$T_H = (T_P + T_R) \exp\left[\frac{-(t - t_0)}{\tau_H}\right].$$

Substituting this expression into $T(t) = T_R + T_H$, we find the term referring to evolution of post-potential threshold of action up to the resting threshold:

$$T(t) = T_R + (T_P - T_R) \exp\left[- (t_i - t_0) / \tau_H\right], \tag{7}$$

where τ_H is the decay time constant of relative refractory period. Thus, the potential $T(t)$ depends on two terms: (1) a term of firing threshold at rest and (2) a term involving the difference between the firing threshold at rest and the firing threshold post-AP. The latter varies with time, involving a decay time constant of relative refractory period.

Figure 3b shows the behavior of membrane threshold potential $T(t)$, according to the model given by Eq. 8. Comparing Figs. 3 and 4, we can observe the exponential increase of P_o toward P_R, while the T_P threshold decreases toward T_R, from time t_0, instant post-AP moment.

3.1.3 Modeling the Reverberation

Considering that D is the distance between two neurons in Fig. 2, the propagation time (Δt) of the AP from one neuron to another can be given by [24], where v is the propagation velocity along it and t_0 is the synaptic delay ($t_0 = 6$ ms). In feedback loop synapses, two different situations are considered: (i) fixed synaptic weight and (ii) synaptic weight varying with time. In situation 2, when the postsynaptic effect (Eqs. 3 and 4) reached values E_{0J} /2 or I_{0J}/2, synaptic weights began to vary in accordance to the following equations:

$$W_{Ej} = \exp\left(\frac{-k}{t - \tau_w}\right) \tag{8}$$

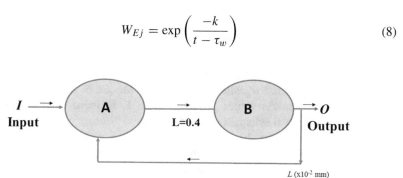

Fig. 4 Reverberation circuit representation as a graph structure. L is the distance between neurons A and B

where τ_H is a decay constant of synaptic weight. Thus, the synaptic weight decreased inversely with both time and frequency of recurrent presynaptic discharges, adjusting the postsynaptic neuron firing.

Computational model based on this mathematical model is assembled so that, using data from the literature, EPSP, IPSP, $V(t)$, and $T(t)$ can be calculated, considering that when $V(t) > T(t)$, the neuron B fires an AP that simultaneously exits the circuit and feeds neuron A, generating repetitive discharges.

3.2 Modeling the Reverberating Neural Circuit by Graph Theory

Graph theory is an increasingly popular computational framework for analyzing network data. Graphical analysis can be applied to neuronal circuits at different spatial levels [22, 23]. According to graph theory, structural brain networks can be described as graphs that are composed of nodes (vertices) denoting neural elements (neurons or brain regions) that are connected by edges that represent physical connections (axonal projections in synapses) [1].

Note that the circuit of Fig. 2 can be seen as a graph (Fig. 4), where axons form the edges and the contact points in the bodies of neurons are nodes. Through the nodes, synaptic transmission occurs. Each node can be associated with at least two edges, one input and one output. Edge node contacts can be positive or negative. Positive contacts represent excitatory synapses, where the signals add up. Negative contacts represent inhibitory synapses, where the operation is subtraction between input signals [31].

In this case, the computational model must contain numerical values and conditions that define the circuit: distances between neurons, signal characteristics, and nodes, among others.

4 Evaluating and Discussing Models

Figures 5 and 6 show the results obtained from modeling of reverberant neuronal circuit using differential equations to represent phases of synaptic transmission based on the parametric values given in Table 1. In these experiments, it was considered that the two synapses were excitatory.

In Fig. 5 we can see the behavior of membrane potential $V(t)$ (Eq. 6) and firing threshold $T(t)$ (Eq. 7) as a time function. After the neuronal firing, $V(t)$ and $T(t)$ gradually returned to their respective resting values. After 10–15 ms, both the resting and threshold potentials have been restored, reproducing the refractory period normally found in membranes of neurons.

Figure 6 shows the case where an excitatory stimulus $E(t)$ occurs outside the neuron membrane resting state. Therefore, the membrane potential and firing

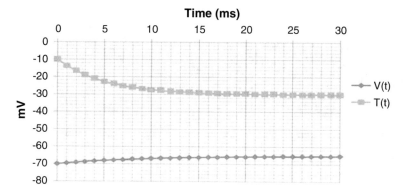

Fig. 5 Membrane potential variation $V(t)$ (Eq. 6) and the membrane firing threshold $T(t)$ (Eq. 7) in time [24]

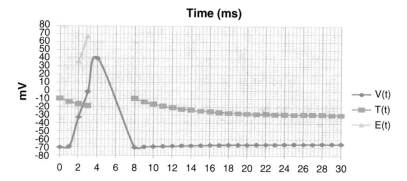

Fig. 6 Membrane potential variation and AP generation, according to Eqs. 6 and 7

Table 1 Parametric values used in our experiments

Parameters (symbols)	Value	References
Resting potential (P_R)	-65 mV	[34]
Reversal potential (P_0)	-70 mV	[35]
Time constant of excitatory effect (τ_E)	7.3 ms	[36]
Time constant of inhibitory effect (τ_I)	13 ms	[37]
Decay time constant of refractory period (τ_H)	1.2 mso	[38]
Resting threshold potential (T_R)	-30 mV	[38]
Post-AP threshold potential (T_P)	-10 mV	[38]

Source: Adapted from [14]

threshold are not at their resting values. In this situation, this first stimulus was not enough to generate an AP. However, the arrival of a second later stimulus depolarized the membrane sufficiently, so that potential exceeded the membrane firing threshold, thus generating the AP in postsynaptic neuron.

Time (ms)

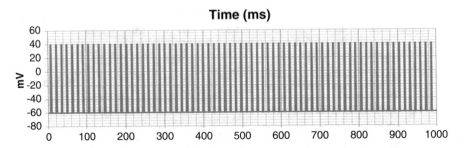

Fig. 7 Firing frequency spectrum of output neuron during 1 s, considering the values of resting and reversal potentials and time constants showed in Table 1 [24]

Time (ms)

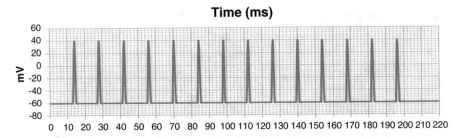

Fig. 8 Neuron firing frequency spectrum when the weight of synaptic connections (w_{Ej}) was varied according to Eq. 8 and using the values of resting and reversal potentials and time constants showed in Table 1 [24]

The firing frequency spectrum of output neuron for an input pulse is shown in Fig. 7, considering that synaptic connections had the same weight and values relative to reversal and rest potentials and time constants. In addition, the value of synaptic weights was the same for all synapses. Analyzing the spiking pattern of spectra shown in this figure, we can see that a single supraliminal input was able to initiate a reverberating process, whose output frequency increased with each cycle. In that condition, one action potential generated in neuron *A* propagated to neuron *B*, which could, in turn, send a feedback to neuron *A*. This one then generated secondary, tertiary signals, and so on, forming recurrent cycles.

The reverberating circuit theory considers that, when a high-frequency stimulus is applied directly to the surface of cerebral cortex, the excited area continues sending rhythmic signals for a short period of time, even after this stimulus has been suspended. In a first analysis, this theory may be associated to the concept of immediate memory, since the fatigue effect in reverberating circuits could explain the weakening of memory and the temporal data replacement with entry of new signals into the circuit [28, 32, 33].

In Fig. 8 we can see the output frequency spectrum of output neuron, when all synapses were excitatory, and the synaptic weights (w_{Ej}) of feedback loops varied in Eq. 8. For a single input, the reverberation period was 196 ms, displaying a temporally limited reverberation process with fixed frequency.

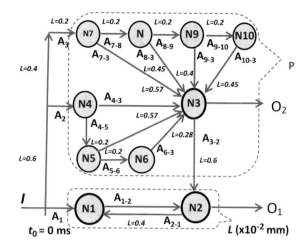

Fig. 9 Parallel and reverberant neuronal network of repetitive discharges

We know that the reverberating time of an input signal and its power to modify an information have a strong dependency, not only on the synaptic weights but also on the number of neurons composing the circuit, because of delay time of each synapse. However, these results show that, for variable w model, the output pattern in synaptic loops could be controlled and maintained at a given level, evidencing synaptic weight's capability to control the postsynaptic effect of recurrent signal, even for an exclusively excitatory circuit. It is important to note that the values of reverberating period mentioned above (Fig. 8) are near to those found by [28]. These authors have shown that reverberating activity in lateral amygdala (a basal nucleus within temporal lobe) could be prolonged for more than 40 ms in vitro and the apparent polysynaptic activity of the awake rat may reach 240 ms.

We also evaluated a parallel and reverberant neuronal network using graph theory [31]. The small neuronal network of repetitive discharges shown in Fig. 9 is composed of two circuits through which a nerve impulse entering by E can propagate. R is a simple reverberant circuit, and the signal passing by $A1$ is processed in $N1$. The signal generated in $N1$ follows to $N2$, and then the signal generated in this one goes to $S1$ and simultaneously returns to $N1$. From this return a reverberant process is formed. The synapses between these two neurons are also excitatory. P is a parallel circuit that also generates repetitive discharges. The signal passing $A2$ and $A3$ reaches $N4$ and $N7$, and the processing in neurons $N4$, $N5$, ..., $N10$ generates signals converging on a single output neuron, the $N3$. All synapses in this circuit are excitatory.

Figure 10a and b shows the results obtained from modeling reverberant neuronal circuit using graph theory [31]. The spectrum is generated at the circuit output of Fig. 9 in response after the first 5 pulses entering by I, for a total of 100 pulses (1 pulse/5 ms), for two different values of refractory period (RP), 5 ms (Fig. 10a) and 2 ms (Fig. 10b), considering a 2 ms synaptic delay (SR) at all nodes. The output

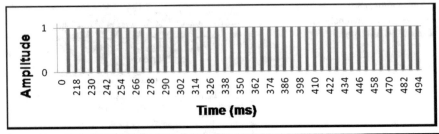

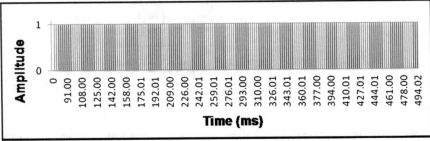

Fig. 10 Spectrum generated at the circuit output (Fig. 9) for the model using graph theory. For 1 input peaks every 5 ms, where $RS = 2$ ms and $RP = 5$ (a) and 2 ms (b)

frequencies for these RP values were 97 Hz and 354 Hz. For $RP = 2$ ms, the first peak appeared at 212 ms after the first input pulse and at 77 ms, for $RP = 5$ ms.

In our model we considered that the circuit occupies an area of cortex equivalent to 10^{-2} mm^2 and distances traveled by the signal (L) of an element for other element are equal to $L = 0.4 \times 10^{-2}$ mm. In addition, signal characteristics, its propagation, and processing in the circuit were considered as follows:

(a) Fixed-amplitude pulse train (or APs) was represented as a table of 0 and 1, being 0 without signal and 1 with sign.
(b) Rhythm and frequency could vary (depending on the distribution of 0 and 1).
(c) Duration of each AP was 1 ms.
(d) Propagation velocity at each constant edge.
(e) Propagation time from one node to the next depending on fiber's distance and characteristic speed (edges).
(f) Propagation velocity adopted was 20 m/s.

Between arrival time of a signal and output of processed signal, there was a 2 ms interval, which is synaptic delay (SR).

The signals arriving at each edge of a neuron are summed algebraically until the sum reaches threshold value (20 positive pulses), then an AP is deflagrated by neuron (or node), following exit edge. It was also considered that a new signal arriving entered the algebraic sum only if the time interval between its arrival and the last sum did not exceed the time of 40 ms. When the interval between two signals exceeded the value and their algebraic sum had not reached firing threshold, the sum started again, and everything that had been added until then was discarded.

With the RP reduction in nodes, we observe in Fig. 8 the important change in the spectrum shape, for the same stimulus level, 1 pulse every 5 ms. The frequency value for $RP = 2$ ms (Fig. 10a), 354 Hz, was approximately 3.6 times greater (97 Hz) than the value observed for $RP = 5$ ms (Fig. 10b), that is, an increase of 2.5 times in RP value caused a reduction of more than 260% in output frequency.

This important change observed in frequency at output 1 is due, in part, to the change in time required for appearance of first pulse, which initiates reverberation process. Increasing RP of 2.5 times, it was verified the necessity of a time almost 2.7 times smaller (77 ms, for $RP = 5$ ms, and 212 ms, for $RP = 2$ ms) for firing of first pulse.

5   Conclusions

Both models were able to simulate the operation of a reverberation circuit and were designed to simulate the dynamics of a neuron network that includes a reverberant circuit, in order to deepen the knowledge about the operation of a similar biological circuit.

The first model, using differential equations to represent the synaptic transmission, was able to represent the functioning of real synapses; the computational model were designed to simulate the occupancy-vacancy dynamics of binding sites in postsynaptic receptors and the post-action potential refractory period.

However, the results related to the discharge of the reverberation circuit using graph theory were different from results presented with the first model [24], which simulated the operation of such a circuit using a mathematical model in which the potential post-membranous growth increased exponentially with presynaptic discharges. In this, for a single input pulse, the reverberatory circuit entered a looping that lasted for at least 1 s. This phenomenon was not observed in the present model, in which we experienced resting potentials (RPs) within the same order of magnitude.

References

1. Bullmore E, Sporns O. Complex brain networks: graph theoretical analysis of structural and functional systems. Nat Rev Neurosci. 2009;10(4):312. https://doi.org/10.1038/nrn2575.
2. Cooper GM, Hausman RE. The cell: a molecular approach. Q Rev Biol. 2007;82(1):44. Fourth Edition. The University of Chicago Press. ISSN 0-87893-219-4. https://doi.org/10.1086/513338.
3. Marrink SJ, Risselada HJ, Yefimov S, Tieleman DP, de Vries AH. The MARTINI force field: coarse grained model for biomolecular simulations. J Phys Chem B. 2007;111(27):7812–24. American Chemical Society. ISSN 1520-6106. https://doi.org/10.1021/jp071097f.
4. Levitan IB, Kaczmarek LK. The neuron: cell and molecular biology. Oxford: Oxford University Press; 2002.
5. Cortez CM, Silva D. Fisiologia Aplicada a Psicologia. Guanabara Koogan. 1ª. Rio de Janeiro: Edição; 2008.

6. Dai Z, Yan C, Li K, Wang Z, Wang J, Cao M, Lin Q, Shu N, Xia M, Bi Y, He Y. Identifying and mapping connectivity patterns of brain network hubs in Alzheimer's disease. Cereb Cortex. 2014;25(10):3723–42. Oxford University Press

7. Sorger PK, Lopez C, Flusberg D, Eydgahi H, Bachman J, Sims J, Chen W, Spencer S, Gaudet S. Measuring and modeling life-death decisions in single cells. FASEB J. 2012; (26), no. 1 Supplement 228.1. Federation of American Societies for Experimental Biology.

8. Tao H, Guo S, Ge T, Kendrick KM, Xue Z, Liu Z, Feng J. Depression uncouples brain hate circuit. Mol Psychiatry. 2011;18:101–11. https://doi.org/10.1038/mp.2011.127.

9. Richardson RJ, Blundon JA, Bayazitov IT, Zakharenko SS. Connectivity patterns revealed by mapping of active inputs on dendrites of thalamorecipient neurons in the auditory cortex. J Neurosci. 2009;29(20):6406–17. https://doi.org/10.1523/JNEUROSCI.0258-09.

10. Sporns O, Tononi G, Edelman GM. Theoretical neuroanatomy and the connectivity of the cerebral cortex. Behav Brain Res. 2002;135:69–74. https://doi.org/10.1016/S0166-4328 (02)00157-2.

11. Sporns O, Chialvo DR, Kaiser M, Hilgetag CC. Organization, development and function of complex brain networks. Trends Cogn Sci. 2004;8:418–25. https://doi.org/10.1016/j.tics.2004. 07.008.

12. Sporns O, Honey CJ, Kötter R. Identification and classification of hubs in brain networks. PLoS One. 2007;2:e1049.

13. Pinto TM, Wedemann RS, Cortez CM. Modeling the electric potential across neuronal membranes: the effect of fixed charges on spinal ganglion neurons and neoroblastoma cells. PLoS One. 2014;9(5):e96194.

14. Cardoso FRG, Cruz FAO, Silva D, Cortez CM. Computational modeling of synchronization process of the circadian timing system of mamals. Biol Cybern. 2009;100:385–93.

15. Cardoso FRG, Cruz FAO, Silva D, Cortez CM. A simple model for the circadian timing system of Mammals. Braz J Med Biol Res. 2009;42:122–7.

16. Velarde MG, Nekorkin VI, Kazantsev VB, Makarenko VI, Llinás R. Modeling inferior olive neuron dynamics. Neural Netw. 2002;15:5–10. https://doi.org/10.1016/S0893-6080(01) 00130-7.

17. Ingalls BP. Mathematical modelling in systems biology: an introduction. 2015. Available http:/ /www.math.uwaterloo.ca/~bingalls/MMSB/MMSB_w_solutions.pdf. Accessed Oct 2017.

18. Dalcin BLG, Cruz FAO, Cortez CM, Passos EL. Applying backpropagation neural network in the control of medullary reflex pattern. International conference of computational methods in science and engineering (ICCMSE 2015). Athens. 2015;1702:130006–130006-4.

19. Dalcin BLG, Cruz FAO, Cortez-Maghelly C, Passos EL. Computer modeling of a spinal reflex circuit. Braz J Phys. 2005;4(35):987–94.

20. Cruz FAO, Cortez CM. Computer simulation of a central pattern generator via Kuramoto model. Physica A. 2005;353:258–70.

21. Cruz FAO, Silva D, Cortez CM. Simulation of a spinal reflex circuit model controlled by a central pattern generator. Far East J Appl Math. 2008;33:307–36.

22. Hadley JA, Kraguljac NV, White DM, Ver Hoef L, Tabora J, Lahti AC. Change in brain network topology as a function of treatment response in schizophrenia: a longitudinal resting-state fMRI study using graph theory. NPJ Schizophr. 2016;2:16014. https://doi.org/10.1038/npjschz.2016.14.

23. Stobb M, Peterson JM, Mazzag B, Gahtan E. Graph theoretical model of a sensorimotor connectome in zebrafish. PLoS One. 2012;7(5):e37292.

24. Rodrigues VF, Castro MCS, Wedemann RS, Cortez CM. A model for reverberating circuits with controlled feedback. AIP Conference Proceedings 2015; 1702:130005. Published by the AIP Publishing. https://doi.org/10.1063/1.4938912.

25. Hecht-Nielsen R. Neurocomputing. Reading: Addison-Wesley Publish Co; 1990. https://doi.org/10.1038/359463a0.

26. Brooks RA. Elephants don't play chess. Robot Auton Syst. 1990;6:3–15.

27. Nocks L. The robot: the life story of a technology. Westport: Greenwood Publishing Group; 2007.

28. Johnson LR, Ledoux JE, Doyère V. Hebbian reverberations in emotional memory micro circuits. Front Neurosci. 2009;3:198–205. https://doi.org/10.3389/neuro.01.027.
29. Di Lazzaro V, Ziemann U. The contribution of transcranial magnetic stimulation in the functional evaluation of microcircuits in human motor cortex. Front Neural Circ. 2013;7:18. https://doi.org/10.3389/fncir.2013.00018.
30. Guyton AC, Hall JE. Textbook of medical physiology. 11th ed. Amsterdam: Elsevier Saunders; 2006.
31. Kalil CA, Castro MCS, Cortez CM. Computer model of reverberant and parallel circuit coupling. International Conference of Computational Methods in Science and Engineering (ICCMSE 2017). AIP Conference Proceedings. 2017. Published by the AIP Publishing.
32. Mongillo G, Barak O, Tsodyks M. Synaptic theory of working memory. Science. 2008;319:1543–6.
33. Takeuchi T, Duszkiewicz AJ, Morris RGM. The synaptic plasticity and memory hypothesis: encoding, storage and persistence. Philos Trans R Soc Lond Ser B Biol Sci. 2014;369:20130288.
34. Walsh IB, Van Den Berg RJ, Marani E, Rietveld WJ. Spontaneous and stimulated firing in cultured rat suprachiasmatic neurons. Brain Res. 1992;588:120–31.
35. Ef B, Barret JN, Crill E. Voltage-sensitive outward currents in cat motoneurones. J Physiol. 1980;304:251–76.
36. Kim YI, Dudek FE. Membrane properties of rat suprachiasmatic nucleus neurons receiving optic nerve input. J Physiol. 1993;464:229–43.
37. Kim YI, Dudek FE. Intracellular electrophysiological study of suprachiasmatic nucleus neurons in rodents: inhibitory synaptic mechanisms. J Physiol. 1992;458:247–60.
38. Carp JS. Physiological properties of primate lumbar motoneurons. J Neurophysiol. 1992;68:1121–32.

Printed in the United States
By Bookmasters